Variational Convergence for Functions and Operators

Variational Convergence for Functions and Operators

H Attouch
University of Perpignan

Pitman Advanced Publishing Program
BOSTON · LONDON · MELBOURNE

PITMAN PUBLISHING LIMITED
128 Long Acre, London WC2E 9AN

PITMAN PUBLISHING INC
1020 Plain Street, Marshfield, Massachusetts 02050

Associated Companies
Pitman Publishing Pty Ltd, Melbourne
Pitman Publishing New Zealand Ltd, Wellington
Copp Clark Pitman, Toronto

First published 1984

AMS Subject Classifications: 35B30, 35B35, 35B40, 47H05, 49A22, 49A50, 49D, 49G

ISSN 0743-0353

Library of Congress Cataloging in Publication Data

Attouch, H.
Variational convergence for functions and operators.

(Applicable Mathematics Series)
Bibliography: p.
1. Calculus of variations. 2. Convergence.
3. Sequences (mathematics). 4. Functions. 5. Operator theory. I. Attouch, H. II. Title. III. Series.
QA315.A87 1984 515′.64 84-20602
ISBN 0-273-08583-2

British Library Cataloguing in Publication Data

Attouch, H.
Variational convergence for functions and operators.—(Applicable mathematics, ISSN 0743–0353)
1. Sequences (Mathematics) 2. Convergence
I. Title
515′.24 QA292

ISBN 0-273-08583-2

Reproduced and printed by photolithography
in Great Britain by Biddles Ltd, Guildford

Contents

Preface

During the past twenty years (1964-84) new concepts of convergence for sequences of functions and operators have been appearing in mathematical analysis. These concepts are specially designed to approach the limit of sequences of variational problems and are called "variational convergences". With each type of variational problem (minimization, maximization, min-max, the saddle-value problem,...) is associated to a particular concept of convergence.

In this book we focus our attention on *minimization problems* and develop a convergence theory for sequences of functions, called *epi-convergence*, which may be regarded as the "weakest" notion which allows to approach the limit in the corresponding minimization problems. This concept of convergence thus has natural applications in all branches of optimization theory - from stochastic optimization, optimal control, numerical analysis and approximation to calculus of variations and perturbation problems in physics.

The book is divided into three parts. In Chapter 1, epi-convergence is introduced in a general topological setting as the natural concept of convergence that allows to approach the limit of sequences of minimization problems. Relying only on its definition, we show how the technique of epi-convergence may be used to solve various limit problems in analysis. Examples have been chosen which, because of their physical interest and the difficulties they present (in these examples, intuition by itself is not of much help in identifying the right limit problem), have contributed to the development of well-adapted mathematical tools, to which we refer when we speak of variational convergences.

Composite materials (fibred, stratified, porous,...) play an important role in many branches of physical engineering. A good approximation of the macroscopic behaviour of such materials may be obtained by letting the parameter ε, which describes the fineness of structure, approach zero in the equations describing phenomena such as heat conduction and elasticity. This limit analysis process is called *homogenization*. Many examples considered here come from the works of Bensoussan, Lions and Papanicolaou [1], Sanchez-Palencia [9],

Cioranescu and Murat [1], Tartar [2], and Marchenko and Hruslov [1]. Epi-convergence provides a precise and flexible tool for such problems. It enhances their topological aspects and can easily be combined with other tools such as convex analysis and measure theory. For some nonlinear problems, such as homogenization of elastoplastic torsion (cf. Attouch [6], Carbone and Salerno [1]) and homogenization of fissured elastic materials (cf. Attouch and Murat [2]) it is the only proof we have, at the present time, of this limit analysis process.

A second type of example comes from perturbation theory: *"singular" perturbation* problems arise naturally when a physical parameter (such as conduction, viscosity, mean free path of a particle, etc), or an economic one (such as a cost) becomes very small or very large with respect to the others. Typical applications are reinforcement problems and shells in mechanical engineering.

Our purpose in Chapter 2 is to give a complete exposition of the topological properties of epi-convergence. A large part of this chapter owes much to De Giorgi and Wets. We pay particular attention to the Moreau-Yosida approximation by inf-convolution which is developed for general real-valued functions defined on a metric space:

$$\forall \lambda > 0 \quad \forall u \in X \quad F_\lambda(u) = \inf_{v \in X} \{F(v) + \frac{1}{2\lambda} d^2(u,v)\}.$$

Epi-limits of sequences of functions can be re-expressed in terms of pointwise limit of their Moreau-Yosida approximates. A topology is naturally attached to the pointwise convergence of these approximates. In the locally compact case (or in an equivalent way when considering uniformly "inf-compact" functions), one can prove that this topology induces epi-convergence. It should be noted that, in general, epi-convergence is not attached to a topology. It is only in certain particular cases (like those described above which, indeed, cover a large number of applications) that there exists a topology $\mathfrak{T}$ for which

$$F = \tau\text{-}\lim_e F^n \Longleftrightarrow F^n \overset{\mathfrak{T}}{\to} F.$$

In a number of applications such as stochastic optimization a precise approach is to consider functionals as elements of such a topological (compact metric) space (cf. Salinetti and Wets [4], Dal Maso and Modica [3]).

The third and final chapter is devoted to the study of epi-convergence of sequences of convex functions. For simplicity, we restrict our attention to the case of X, a reflexive Banach space. In this case, two topologies play an important role: the strong and the weak topologies.

In this infinite-dimensional framework the continuity property of the Young-Fenchel transformation can be formulated as follows (refer to Theorems 3.7 and 3.9 for precise assumptions):

$$F = w\text{-}\lim_e F^n \Leftrightarrow F^* = s\text{-}\lim_e F^{n*}.$$

Therefore weak and strong topologies are exchanged when considering epi-convergence of convex functions and of their conjugates. These considerations lead to the introduction of the so-called Mosco-convergence, which is epi-convergence for both strong and weak topologies and which, from the above considerations, has the following basic property:

$$F^n \to F \text{ in Mosco sense} \Leftrightarrow F^{n*} \to F^* \text{ in Mosco sense.}$$

This property explains the importance of this concept in the study of stability properties, approximation, etc, in convex optimization. Historically, it appeared (when considering infinite-dimensional spaces) earlier than the more general concept of epi-convergence with respect to a given topology. Let us give its formulation: a sequence $F^n : X \to]-\infty, +\infty]$ Mosco-converges to F if

$$\begin{cases} \text{for every } x \in X, \text{ there exists } (x_n)_{n \in \mathbb{N}} \text{ strongly converging to } x \text{ in } X \\ \text{such that } F^n(x_n) \to F(x); \\ \\ \text{for every weakly converging sequence } x_n \to x,\ F(x) \leq \liminf F^n(x_n). \end{cases}$$

Mosco-convergence is indeed equivalent to the pointwise convergence of the Moreau-Yosida approximates:

$$F^n \to F \text{ in Mosco sense} \Leftrightarrow \text{for every } \lambda > 0, \text{ for every } x \in X\ F^n_\lambda(x) \to F_\lambda(x)$$

where

$$F^n_\lambda(x) = \min_{u \in X} \{F^n(u) + \frac{1}{2\lambda} \|x-u\|^2_X\}.$$

Therefore there exists a topology on the class of closed convex functions called the topology of Mosco-convergence, inducing this convergence. When X is separable, this topological space is a Polish space that is metrizable, separable and complete for a metric inducing the topology.

In convex analysis, in addition to the conjugation operation, another concept plays a fundamental role: this is the subdifferential operation

$$F \to \partial F$$

where $\partial F : X \to X^*$, the subdifferential of F, is given by

$$\partial F = \{(u,f) \in X \times X^* / F(v) \geq F(u) + \langle f, v-u \rangle \text{ for every } v \in X\}.$$

A natural question is: given a sequence of closed convex functions $F^n : X \to]-\infty, +\infty]$ which is epi-convergent, what is the corresponding notion of convergence for the sequences of operators $\{\partial F^n : X \to X^*; n = 1,2,...\}$? Indeed, historically it was the converse of this question that was asked: subdifferentials of convex functions form an important subclass (nonlinear version of self-adjoint operators) of maximal monotone operators. For such operators a good convergence concept is *graph-convergence* which is equivalent to the pointwise convergence of the *resolvents*. This concept, introduced by Kato [1] for linear monotone operators, has been extended by Browder [2], Brezis [1], [2] to nonlinear maximal monotone operators and by Benilan [1] to accretive operators. It makes it possible to attack convergence of semi-groups, approximation and perturbation of evolution equations governed by such operators.

The equivalence between graph-convergence of subdifferential operators $\{\partial F^n; n = 1,2,3...\}$ and Mosco-convergence of functions $\{F^n; n = 1,2,...\}$ was proved by the author around 1976 (Attouch [2], [4], cf. also Matzeu [1], Zolezzi [4], Sonntag [1]). This links the two theories, convergence of functions and convergence of operators, which in the convex case turn out to be equivalent. Moreover one obtains convergence of elements attached to such operators, such as spectrum (in the linear case), semigroups.

In the last few years many extensions and promising new fields of application of variational convergence have appeared in the literature:

- Convergence of saddle-value and min-sup problems: Attouch and Wets [3], [4], Cavazzuti [1], with a view to applications to critical point problems in economics and mechanics.

- The study of limit analysis problems for systems and higher-order problems in mechanics: Brillard [2] for Stokes equations and Darcy's law in porous media, Aze [1] for elasticity and the dual formulation of the homogenization formula expressed in terms of constraint tensor (cf. also Suquet [2]), Picard [1] for the biLaplacian, Attouch and Murat [2] for homogenization of fissured elastic materials, etc.
- The study of variational convergence in non-reflexive Banach spaces: Picard [1] for minimal surface problems with varying unilateral or bilateral constraints, extensions to capillarity, plasticity etc.
- The study of stochastic optimization problems, for example in statistical decision theory (Salinetti and Wets [4]), in stochastic homogenization (Dal Maso and Modica [3]) etc.
- The study of convergence problems for evolution equations, control problems, rate of convergence (Attouch and Wets [5]) etc.

Acknowledgements

I should like to express my sincere thanks to H. Brezis and R. Wets, who encouraged me to write this book, and to J.L. Lions, who initially suggested to me this direction of research.

In the preparation of this book I have received valuable advice from many friends, especially from E. De Giorgi and U. Mosco and from the active Italian school of mathematics - M. Biroli, L. Boccardo, G. Buttazzo, L. Carbone, E. Cavazzuti, G. Dal Maso, P. Marcellini, L. Modica, A. Negro, C. Sbordone, T. Zolezzi - to whom I owe a large part of my knowledge of variational convergence.

For providing both a warm and a challenging scientific atmosphere I have to thank all my colleagues and friends in Paris - D. Caillerie, D. Cioranescu, A. Damlamian, F. Murat, C. Picard, E. Sanchez Palencia, L. Tartar - and also Ph. Bénilan (Besançon), A. Fougères (Perpignan), J.-L. Joly (Bordeaux) and Y. Sonntag (Marseilles) for their contribution to the development of the theory of convergence. Finally I should like to thank the publishers, Pitman, for improving my English and for preparing the final typescript, D. Aze for helping me in the final lecture and Annie for her patience and for the many lost holidays and movie shows!

H. A.

Introduction

The concept of variational convergence seems to appear for the first time in the work of Wijsman [2] (1964-66) in statistical decision theory, where it arose in the study of the continuity properties of the application

$$C \mapsto s(C)$$

which associates with a closed convex subset of $\mathbb{R}^m$ its support function

$$s(C,x^*) = \sup_{x \in C} \langle x^*,x \rangle .$$

When considering convex subsets C contained in a fixed ball of $\mathbb{R}^m$, Hausdorff metric provides a quite natural measure for perturbations of sets, and

$$C^n \xrightarrow[n \to +\infty]{} C \text{ for Hausdorff metric} \Leftrightarrow \forall x^* \in \mathbb{R}^m \; s(C^n,x^*) \to s(C^n,x^*).$$

However, when considering possible unbounded subsets of $\mathbb{R}^m$, Hausdorff metric is no longer an adequate concept and one has to introduce the more general notion of set convergence, also called Kuratowski convergence (equality between lim inf and lim sup). The question that naturally arose was that of finding the corresponding concept of convergence for the associated support functions. Here it is not the pointwise convergence of the support functions (as in the case of the Hausdorff metric) that gives the correct answer, but rather the epi-convergence (called by Wijsman "infimal convergence").

Epi-convergence thus stems naturally from set convergence theory and has been introduced to study the continuity properties of duality operations. Indeed, epi-convergence of a sequence of functions $F = \lim_e F^n$ is equivalent to the set convergence of their epigraphs, epi $F^n \to$ epi F, where epi F = $\{(x,\lambda) \in X \times \mathbb{R} / \lambda \geq F(x)\}$. This justifies the terminology!

Moreover, the continuous dependence of the support function $s(C,\cdot)$ on C turns out to be a particular case of the following fundamental result: the Young-Fenchel transformation is continuous with respect to epi-convergence

$$F = \lim_e F^* \Leftrightarrow F^* = \lim_e F^{n*}$$

where F^n, F are closed convex functions and $F^*(x^*) = \sup_{x \in \mathbb{R}^m} \{\langle x^*, x\rangle - F(x)\}$.

At this stage the theory has been developed in a satisfactory way but only for the finite-dimensional case. This restriction blurs some important topological features of the theory that cannot be ignored in infinite-dimensional spaces. Active mathematical orientation research, mainly arising from applications to optimization and decision theory (Wets [1], Salinetti and Wets [1]..., Back [1], Vervaat [1], MacLinden [1]) bears a natural relation to the work of Wijsman.

The next step in the development of the theory came from quite a different direction and can be traced to the work of researchers such as Stampacchia and Lions on variational inequalities. In order to study the convergence of solutions of approximations of variational inequalities (such as Galerkin approximation), Mosco [2] (1967-73) and Joly [1] (1970-76) extended the earlier results to infinite-dimensional spaces. The theory was still limited to the case of convex functions and to topologies such as weak and strong topologies on (reflexive) Banach spaces.

The concept in a general topological setting has finally been delineated by De Giorgi [1] (1973-83) and the Italian mathematical school (Spagnolo [1], Carbone and Sbordone [2], Buttazzo [1], Dal Maso [1], Modica [1], Boccardo and Marcellini [1], etc). They were mostly concerned with the study of lower semicontinuity and perturbation problems in calculus of variations. To that end, a convergence theory for functions was developed, called Γ-convergence, (and a corresponding theory for operators, called G-convergence). This contains as a particular case epi-convergence, which can be regarded as Γ-convergence specially adapted to minimization problems. The corresponding concept for maximization problems is hypo-convergence, which can easily be derived from epi-convergence by changing functions F into their opposites in the definitions and statements. Recently the Γ-convergence theory for saddle-value problems, called epi-hypo-convergence (which includes the two above concepts), has been developed by Attouch and Wets [3] and Cavazzuti [1].

As with all Γ-convergence concepts, the definition of epi-convergence only requires a topological structure. Given (X,τ), a topological space (which for simplicity we assume here to be metrizable), and F^n, $F: X \to \bar{\mathbb{R}}$, a sequence of real (extended) valued functions, the sequence $\{F^n; n \to +\infty\}$ is said to be τ-epi-convergent to F at $x \in X$ if the two following conditions hold:

(i) there exists a convergent sequence $x_n \xrightarrow[(n\to+\infty)]{} x$ in (X,τ) such that

$F(x) \geq \limsup_{n\to+\infty} F^n(x_n)$;

(ii) for every convergent sequence $x_n \xrightarrow[(n\to+\infty)]{} x$ in (X,τ), $F(x) \leq \liminf_{n\to+\infty} F^n(x_n)$.

We then write $F(x) = (\tau\text{-}\lim_e F^n)(x)$. When this property holds for every $x \in X$, the sequence $\{F^n; n = 1,2,...\}$ is said to be τ-epi-convergent to F and $F = \tau - \lim_e F^n$.

Let us first notice that when such convergence holds, the limit function F is given by the formula

$$F(x) = \min \ \{\lim_{n\to+\infty} F^n(x_n);\ x_n \xrightarrow[n\to+\infty]{\tau} x\}.$$

When taking $F^n \equiv F^o$, a stationary sequence, F is equal to the τ-lower semi-continuous regularization of F^o. Thus, epi-convergence includes as a particular case the Γ-closure operation. This is the origin of the above terminology.

The fundamental variational property of epi-convergence can now be formulated: let us take $\{F^n, F:X \to \bar{\mathbb{R}}; n = 1,2,...\}$, a sequence of real (extended) functions which satisfies the condition that there exists a topology τ on X and a τ-relatively compact subset K of X such that, for every n = 1,2,...,

$$\inf_{x\in X} F^n(x) = \inf_{x\in K} F^n(x).$$

Then, $F = \tau - \lim_e F^n$ implies the convergence of the corresponding minimization problems (as $n \to +\infty$):

$$\inf_{x\in X} F^n(x) \xrightarrow[(n\to+\infty)]{} \inf_{x\in X} F(x)$$

and every τ-cluster point x of a minimizing sequence ($x_n \in \text{Argmin } F^n; n=1,2,...$) minimizes F. (When there is uniqueness, there is convergence of the whole sequence.) In general, epi-convergence is not implied by and does not imply pointwise convergence. They are two separate concepts. There is, in fact, one important case in which the two concepts coincide: this is when the sequences of functions are monotonically increasing (or decreasing). This explains the success of all monotone schemes in approximation theory.

1 Epigraph-convergence

In this chapter, epi-convergence is introduced in a variational way, as the "weakest" notion of convergence for sequences of functions which allows to approach the limit on corresponding minimization problems. On the way, various limit analysis problems are analyzed: some, such as homogenization of materials with many small holes and cracks and transmission through thin isolating layer, come from physics, and some from optimization and control theory. For all these examples epi-convergence provides a flexible tool and a deep insight.

The terminology is then justified: epi-convergence of a sequence of functions is equivalent to set-convergence of the corresponding epigraphs.

The concept is then placed in the general framework of Γ-convergence theory, as defined by De Giorgi, which covers a larger class of variational problems like saddle value problems, min-max, etc.

Properties of epi-convergence are discussed in the next chapter.

1.1 MODEL EXAMPLES

In this first section, we present various examples, selected because of their physical interest and striking character, which illustrate some concepts of *variational convergence*. For each of these examples, we announce a convergence theorem in terms of convergence of the solutions, and introduce the corresponding "variational convergence" property for functionals and operators governing these equations. Proofs of these theorems and rigorous mathematical formulations (in terms of epi-convergence of the corresponding functionals) are given in Section 1.3.

1.1.1 Homogenization of elliptic equations

Composite materials like concrete, materials with many small holes or fissures, fibred or stratified materials, etc., play an important role in modern engineering. Typically, in such materials, the *physical parameters* (such as conductivity, elasticity coefficients ...) are *discontinuous* and oscillate between the different values characterizing each of the components.

When these components are intimately mixed, these parameters oscillate very rapidly and the microscopic structure becomes complicated.

On the other hand, the material, from a macroscopic point of view, becomes quite simple, since it tends to behave like an ideal homogeneous material, called a *homogenized material* (in accordance with the well-known physical law). It is the purpose of homogenization theory to describe these limit processes, when ε, the parameter which describes the fineness of the microscopic structure, tends to zero.

For simplicity, let us start by describing the *homogenization of the stationary heat equation*(*) in a material whose structure is supposed ε-periodic in all directions, with ε a small parameter. In order to describe the conductivity coefficients, let us introduce Y, a basic cell in $\mathbb{R}^N$ (take $Y = [0,1]^N$ for simplicity) and $\{a_{ij}: \mathbb{R}^N \to \mathbb{R}^+;\ 1 \leq i, j \leq N\}$ satisfying (1.1) and (1.2):

$$\text{For every } 1 \leq i, j \leq N,\ a_{ij} \text{ belongs to } L^\infty(\mathbb{R}^N) \text{ and is Y-periodic} \tag{1.1}$$

$$\text{There exists some } \lambda_o > 0, \text{ such that } \sum_{i,j=1}^{N} a_{ij}(x)\xi_i\xi_j \geq \lambda_o|\xi|^2 \quad \begin{array}{l}\text{for every } \xi \in \mathbb{R}^N \\ \text{for a.e. } x \in \mathbb{R}^N\end{array} \tag{1.2}$$

Then, $\{a_{ij}(\frac{\cdot}{\varepsilon});\ 1 \leq i, j \leq N\}$ are εY-periodic functions, describing the conductivity coefficients of the material.

Let us assume the material occupies a domain Ω (bounded open subset of $\mathbb{R}^N$) whose boundary $\partial\Omega$ is maintained at a prescribed temperature $u = 0$. Then, for every external heat supply f, the temperature u_ε satisfies the following equation:

$$\begin{cases} -\displaystyle\sum_{i,j=1}^{N} \frac{\partial}{\partial x_i}\left(a_{ij}\left(\frac{x}{\varepsilon}\right)\frac{\partial u_\varepsilon}{\partial x_j}\right) = f \text{ on } \Omega \\ u_\varepsilon = 0 \text{ on } \partial\Omega. \end{cases} \tag{1.3}$$

(*) Equations are the same as for the electrostatic potential equation.

Thanks to coercivity assumption (1.2), this is a well-posed problem and for every $f \in L^2(\Omega)$, there exists a unique solution $u_\varepsilon \in H_0^1(\Omega)$ of (1.3).

In order to visualize this situation, Fig. 1.1 shows such a material with two isotropic components of respective conductivity coefficients α and β.

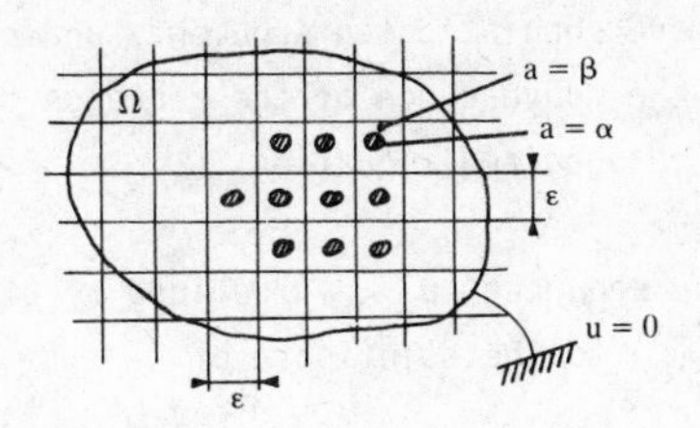

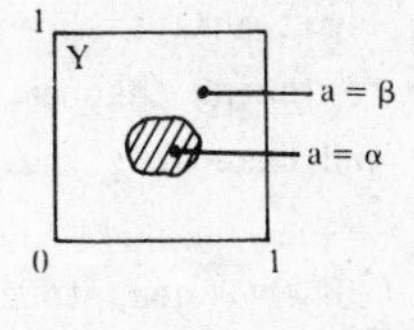

Figure 1.1

The real physical problem corresponds to $\varepsilon = \varepsilon_0$ *fixed.* But, as we have already explained, when ε is small with respect to the size of Ω, $|\Omega|$, the microscopic structure is rather complicated and hence u_ε is difficult to compute: the conductivity coefficients $a_{ij}(\frac{\cdot}{\varepsilon})$ have discontinuities through the interfaces separating the components in each small cell of size ε. So, in order to take account of the corresponding transmission conditions, one has to do a number of discretizations in each of these microscopic cells. When ε is small with respect to $|\Omega|$, this number exceeds the capabilities of computers. So, we are led to consider ε as a parameter and expect that when ε goes to zero, the solutions u_ε of equations (1.3) converge to some u and that u, in turn, will provide an approximation of u_ε, for ε small (in 1.6.3 we explain how, by introduction of explicit correctors $\{r_\varepsilon; \varepsilon \to 0\}$ and considering $u + r_\varepsilon$, one obtains a better approximation of u_ε).

In order to obtain the limit equation, a first naive approach would consist of replacing $a_{ij}^\varepsilon = a_{ij}(\frac{\cdot}{\varepsilon})$ in (1.3) by its weak $\sigma(L^\infty(\Omega), L^1(\Omega))$ limit, as $\varepsilon \to 0$, that is to say $\mathfrak{m}_Y(a_{ij})$, where $\mathfrak{m}_Y$ denotes the mean value over Y of a_{ij}. But this is not the correct answer: the coefficients a_{ij}^ε being only bounded in $L^\infty(\Omega)$, one can only expect the sequence $\{u_\varepsilon\}$ to be bounded in $H_0^1(\Omega)$. Thus, the derivatives of u_ε converge only weakly in $L^2(\Omega)$ and we have to go to the limit on the product of two weakly convergent sequences in $L^2(\Omega)$:

$$a_{ij}(\frac{x}{\varepsilon}) \longrightarrow \mathfrak{m}_Y(a_{ij}) \text{ weakly in } L^2(\Omega), \text{ as } \varepsilon \to 0.$$

$$\frac{\partial u_\varepsilon}{\partial x_j} \longrightarrow \frac{\partial u}{\partial x_j} \quad \text{weakly in } L^2(\Omega), \text{ as } \varepsilon \to 0.$$

In general, such a limit does not exist (even in the distribution sense). Indeed, the limit analysis requires more sophisticated arguments and concepts. The following theorem states the convergence of the sequence $\{u_\varepsilon; \varepsilon \to 0\}$ and describes the limit (homogenized) equation.

<u>THEOREM 1.1</u> When ε goes to zero, the sequence $\{u_\varepsilon\}$, u_ε defined by (1.3), converges for the weak topology of $H_0^1(\Omega)$ to the solution u of

$$\begin{cases} -\sum_{i,j=1}^{N} a_{ij}^{hom} \frac{\partial^2 u}{\partial x_i \partial x_y} = f \text{ on } \Omega, \\ u = 0 \text{ on } \partial\Omega \end{cases} \tag{1.4}$$

where $a_{ij}^{hom} = \mathfrak{m}_Y(a_{ij}) - \mathfrak{m}_Y(\sum_{k=1}^{N} a_{ik} \frac{\partial \chi^j}{\partial y_k})$ and χ^j is the solution of $(j=1,2,\ldots,N)$

$$\begin{cases} A\chi^j = Ay_j \\ \chi^j \text{ Y-periodic} \end{cases} \quad \text{with } Av = -\sum_{i,j} \frac{\partial}{\partial y_i}(a_{ij}(y)\frac{\partial v}{\partial y_i}).$$

Moreover, there is *convergence of the energies:* In the symmetric case $(a_{ij} = a_{ji})$

$$\int_\Omega \sum_{i,j=1}^{N} a_{ij}(\frac{x}{\varepsilon}) \frac{\partial u_\varepsilon}{\partial x_i} \frac{\partial u_\varepsilon}{\partial x_y} dx \to \int_\Omega \sum_{i,j} a_{ij}^{hom} \frac{\partial u}{\partial x_i} \frac{\partial u}{\partial x_y} dx, \text{ as } \varepsilon \text{ goes to zero.}$$

The coefficients a_{ij}^{hom} are constant and are called the *homogenized coefficients* or *effective coefficients*. Note that a_{ij}^{hom} differs from the mean value of a_{ij} via a correcting term whose computation involves *the resolution of variational problems* (for $N > 1$, it cannot be reduced to a finite number of algebraic and mean value operations). Formulation of these coefficients does not depend on f or Ω. They depend in a significant way on the properties of the material. The limit equation (1.4) is called the *homogenized equation* and gives an approximation of the macroscopic behaviour of the material: the heterogenous material has been replaced by a homogeneous one with quite similar macroscopic behaviour.

Case N = 1. When the dimension of the space $\mathbb{R}^N$ is equal to one, the situation and the formulae turn to be far simpler and a quite elementary direct approach can be used: take $\Omega = (0,1)$, then (1.3) specializes into

$$\begin{cases} -\frac{d}{dx}\left(a_\varepsilon(x)\frac{du_\varepsilon}{dx}\right) = f \text{ on } (0,1) \\ u_\varepsilon(0) = u_\varepsilon(1) = 0 \end{cases} \tag{1.5}$$

where we set $a_\varepsilon(x) = a(\frac{x}{\varepsilon})$. Via elementary estimates, one can extract subsequences (still denoted ε) such that

$$\begin{aligned} &\frac{du_\varepsilon}{dx} \longrightarrow \frac{du}{dx} \text{ in weak-}L^2(0,1), \text{ as } \varepsilon \to 0 \\ &\xi_\varepsilon = a_\varepsilon(x)\frac{du_\varepsilon}{dx} \longrightarrow \xi \text{ in weak-}L^2(0,1), \text{ as } \varepsilon \to 0. \end{aligned} \tag{1.6}$$

Going to the limit on (1.5) in the distribution sense, we obtain

$$-\frac{d\xi}{dx} = f. \tag{1.7}$$

The problem is to determine the relation between ξ and u. The trick is to write the relation between ξ_ε and u_ε in the following form:

$$\frac{du_\varepsilon}{dx} = \frac{1}{a_\varepsilon(x)}\cdot\xi_\varepsilon . \tag{1.8}$$

Noticing that the sequence $\{\xi_\varepsilon;\ \varepsilon \to 0\}$ is bounded in $L^2(\Omega)$, along with its derivative (from equation (1.5)), it is bounded in $H_o^1(\Omega)$ and hence relatively compact in $L^2(\Omega)$. Thus

$$\xi_\varepsilon \to \xi \text{ strongly in } L^2(\Omega), \text{ as } \varepsilon \to 0. \tag{1.9}$$

On the other hand, from Y-periodicity of a, and hence of $\frac{1}{a}$,

$$\frac{1}{a_\varepsilon} \longrightarrow \mathfrak{m}_Y(\tfrac{1}{a}) \text{ weakly in } L^2(\Omega), \text{ as } \varepsilon \to 0. \tag{1.10}$$

Combining these two last results (1.9) and (1.10), we can go to the limit in the distribution sense on (1.8) and obtain

$$\frac{du}{dx} = \mathfrak{m}_Y(\tfrac{1}{a})\xi.$$

Returning to (1.7), we finally obtain the homogenized equation satisfied by u:

$$- \frac{1}{\mathfrak{m}_Y(\frac{1}{a})} \frac{d^2u}{dx^2} = f. \tag{1.11}$$

One can easily verify that the above formula (1.11) can be obtained as a particular case of the general one given in Theorem 1.1: Let us compute

$$a^{hom} = \mathfrak{m}_Y(a(1 - \frac{d\chi}{dy})) \tag{1.12}$$

where χ is a 1-periodic function satisfying

$$\frac{d}{dy} (a(y)(\frac{d\chi}{dy} - 1)) = 0 \text{ on } (0,1).$$

It follows that

$$\frac{d\chi}{dy} - 1 = \frac{c}{a} \tag{1.13}$$

where C is a constant equal, from the 1-periodicity of χ, to $-1/\mathfrak{m}_Y(\frac{1}{a})$. Thus, from (1.13),

$$a(\frac{d\chi}{dy} - 1) = C = - \frac{1}{\mathfrak{m}_Y(\frac{1}{a})}$$

and formula (1.12) provides

$$a^{hom} = \frac{1}{\mathfrak{m}_Y(\frac{1}{a})}$$

which is in accordance with the above direct computation (1.11).

<u>Comments</u> (a) The limit coefficient $\mathfrak{m}_Y(a)$, obtained by pointwise limit operation on the operators $\{A^\varepsilon; \varepsilon \to 0\}$ which govern equations (1.5)

$$A^\varepsilon v = - \frac{d}{dx} (a(\frac{x}{\varepsilon}) \frac{dv}{dx}),$$

and the correct homogenized coefficient $1/\mathfrak{m}_Y(\frac{1}{a})$ are clearly different. But they always satisfy the inequality

$$a^{hom} = \frac{1}{\mathfrak{m}_Y(\frac{1}{a})} \leqslant \mathfrak{m}_Y(a)$$

which is indeed a consequence of Jensen's inequality and of convexity of the function $x \mapsto \frac{1}{x}$ from $]0,+\infty[$ into $]0,+\infty[$.

So, in dimension one, the effective conductivity coefficient is strictly less than the mean value of a.

As we shall see, this type of inequality can be extended to dimension N, when expressed in terms of inequality between quadratic forms:

$$\sum_{i,j} a_{ij}^{hom} \xi_i \xi_j < \sum_{i,j} m_Y(a_{ij})\xi_i\xi_j \text{ for every } \xi \in \mathbb{R}^N. \quad (1.14)$$

(b) One cannot expect for such a homogenization procedure to get strong convergence in $L^2(\Omega)$ of the derivatives of u_ε. Let us assume for a moment that

$$\frac{\partial u_\varepsilon}{\partial x_i} \to \frac{\partial u}{\partial x_i} \text{ strongly in } L^2(\Omega), \text{ as } \varepsilon \text{ goes to zero.}$$

Then, one would be able to go to the limit directly on (1.3) and obtain

$$-\sum_{i,j} m_Y(a_{ij}) \frac{\partial^2 u}{\partial x_i \partial x_j} = f \text{ on } \Omega.$$

But we know that this is not the limit equation. So, we get a contradiction. Indeed, from a numerical point of view, weak $H^1(\Omega)$ convergence is not completely satisfactory: one has to compute, in addition, corrections terms r_ε such that

$$u_\varepsilon \sim u + r_\varepsilon \text{ with } \lim_{\varepsilon\to 0} \|u_\varepsilon - u - r_\varepsilon\|_{H^1(\Omega)} = 0$$

(refer to Section 1.6.3)!

Let us now extract the concepts of *variational convergences* for functionals and operators underlying this situation:

Convergence of Operators is the more "natural" one (historically, it has appeared first).

Let us introduce $\{\mathbb{A}^\varepsilon : \varepsilon \to 0\}$, $\mathbb{A}^{hom}$ the operators governing respectively the approximating and limit equations (1.3), (1.4):

$$\mathbb{A}^\varepsilon : H_0^1(\Omega) \to H^{-1}(\Omega) \qquad \mathbb{A}^\varepsilon(u) = -\mathrm{div}(A^\varepsilon(\mathrm{grad}\, u))$$

$$\mathbb{A}^{hom} : H_0^1(\Omega) \to H^{-1}(\Omega) \qquad \mathbb{A}^{hom}(u) = -\mathrm{div}(A^{hom}(\mathrm{grad}\, u))$$

where matrices $\{a_{ij}(\frac{\cdot}{\varepsilon}): 1 < i, j < N\}$ and $\{a_{ij}^{hom}: 1 < i, j < N\}$ are denoted resp. by A^{ε} and A^{hom}. From Theorem 1.1,

$$\forall f \in H^{-1}(\Omega) \quad (\mathbb{A}^{\varepsilon})^{-1}(f) \xrightarrow[\varepsilon \to 0]{} (\mathbb{A}^{hom})^{-1}(f) \text{ in weak-}H_o^1(\Omega). \qquad (1.15)$$

When (1.15) is satisfied, the sequence of operators $\{\mathbb{A}^{\varepsilon}; \varepsilon \to 0\}$ is said to be G-*convergent* to $\mathbb{A}^{hom}$:

$$\mathbb{A}^{hom} = \text{G-}\lim_{\varepsilon \to 0} \mathbb{A}^{\varepsilon}.$$

This concept of G-convergence for sequences of elliptic operators was introduced by De-Giorgi & Spagnolo [1], Spagnolo [1], [2], [4]. In Chapter 3, we shall see it is a special case of *graph-convergence* (which justifies the G code letter) for sequences of maximal monotone operators. An equivalent terminology is *resolvent convergence* (Kato [1], Brezis [1], [2], Benilan [1], Attouch [1], [4]). A close concept to G-convergence, called H-*convergence* (H for homogenization) has been introduced by Murat & Tartar, cf. Murat [7].

Convergence of functions When matrices $A^{\varepsilon} = (a_{ij}^{\varepsilon})$ are symmetric $a_{ij}^{\varepsilon} = a_{ji}^{\varepsilon}$, problems (1.3) can be formulated as minimization ones:

$$\min_{u \in H_o^1(\Omega)} \{F^{\varepsilon}(u) - \int_{\Omega} fu\}$$

where

$$F(u) = \int_{\Omega} j(\frac{x}{\varepsilon}, \text{grad } u(x))dx$$

and

$$j(y,z) = \frac{1}{2} \sum_{i,j} a_{ij}(y) z_i z_j.$$

Denoting

$$F^{hom}(u) = \int_{\Omega} j^{hom}(\text{grad } u)dx$$

with

$$j^{hom}(z) = \frac{1}{2} \sum_{i,j} a_{i,j}^{hom} \; z_i z_j$$

we have

$$\forall f \in H^{-1}(\Omega) \min_{u \in H_0^1(\Omega)} \{F^\varepsilon(u) - \langle f,u\rangle\} \xrightarrow[\varepsilon\to 0]{} \min_{u \in H_0^1(\Omega)} \{F^{hom}(u) - \langle f,u\rangle\}$$

and convergence of the corresponding minimizing sequences.

It is natural to consider F^{hom} *as the "variational" limit of the sequence* $\{F^\varepsilon ; \varepsilon \to 0\}$.

This is a new concept of convergence. It is different from pointwise convergence, since

$$\forall u \in H_0^1(\Omega) \quad F^\varepsilon(u) \underset{\varepsilon\to 0}{\to} \int_\Omega \sum_{i,j} \mathfrak{m}_Y(a_{ij}) \frac{\partial u}{\partial x_i} \frac{\partial u}{\partial x_j} dx$$

which, for a general u, is strictly greater than $F^{hom}(u)$ (cf. (1.14)).

This convergence can be interpreted (refer to Section 1.4) as the set-convergence of the epigraphs of F^ε to the epigraph of F^{hom}. For this reason, it is called *epi-convergence*. With this terminology, the above result is formulated

$$F^{hom} = \lim_e F^\varepsilon \text{ as } \varepsilon \to 0 \text{ (cf. Definition 1.9)} \tag{1.16}$$

Epi-convergence is a particular case of the concept of Γ-*convergence* introduced by De Giorgi (cf. Section 1.5).

In Chapter 3 is established the equivalence between such variational convergence for sequences of convex functions and the graph convergence of their subdifferentials.

The epi-convergence approach indeed yields, in homogenization problems, a formulation of the limit function F^{hom}, expressed in terms of j^{hom}:

$$\text{for every } z \in \mathbb{R}^N,\ j^{hom}(z) = \min \{\int_Y j(y,\text{grad}w(y)+z)dy / w \text{ Y-periodic}\}. \tag{1.17}$$

This formulation is very flexible, the above formula (1.17) can be naturally extended as we shall see to many different situations: non-linear materials, materials with holes, fissures, elasto-plastic materials, etc. □

1.1.2 Domains with many small "holes"

In $\mathbb{R}^N$ ($N > 1$), let us consider a domain Ω (open bounded set) with "many small" holes. "Holes" are denoted by $T = \bigcup_{i\in I} T^i$ while the complement, the "material", is denoted by $\Omega^m = \Omega \setminus T$. So, Ω^m has a fragmented, disconnected boundary $\partial T \cup \partial\Omega$. Let us study the stationary heat equation in Ω^m, for various types of boundary conditions, on ∂T: Neumann (N), Dirichlet (D) ...:

$$(N)\begin{cases} -\Delta u = f \text{ on } \Omega^m \\ \dfrac{\partial u}{\partial n} = 0 \text{ on } \partial T \\ \text{Boundary condition on } \partial\Omega \end{cases}$$

$$(D)\begin{cases} -\Delta u = f \text{ on } \Omega^m \\ u = 0 \text{ on } \partial T \\ \text{Boundary condition on } \partial\Omega. \end{cases}$$

For any f in $L^2(\Omega^m)$ (for example) the above problems (N) and (D) are well posed: there exists a unique solution in $H^1(\Omega^m)$. But, when the number of the holes becomes important (which is the case in a number of physical situations: crushed ice, porous media,...) it is very difficult to compute such a solution numerically. So, for the same reasons as in the preceding example, in order to analyze the macroscopic behaviour of such materials, a limit analysis is required.

Let us introduce $\varepsilon > 0$, a small parameter describing the microscopic structure of the material, and assume, for simplicity, that in each small cell we have the same configuration (periodicity assumption) (Fig. 1.2).

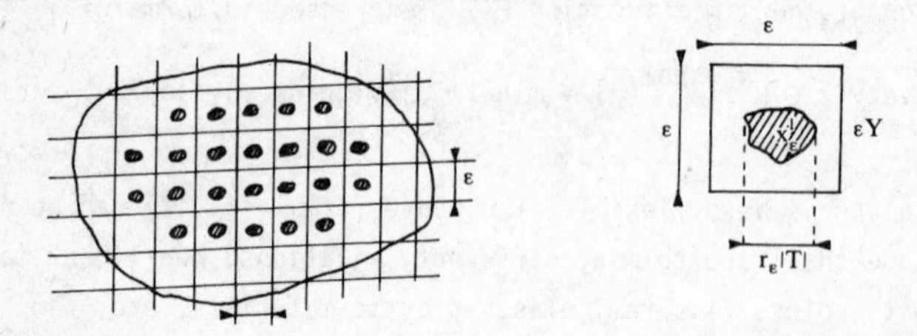

Figure 1.2

Let us assume that $T_\varepsilon = \bigcup_{i\in I} T^i_\varepsilon$ and $T^i_\varepsilon = r_\varepsilon . T + x^i_\varepsilon$ where T is a fixed "hole" strongly included in $Y =]-\frac{1}{2}, +\frac{1}{2}[^N$ (i.e. $c\ell(T) \subset Y$), where r_ε is a strictly positive parameter and x^i_ε is the centre of the cell of index $i \in I$. For each $\varepsilon > 0$, let us denote u_ε (resp. v_ε) the solution of (N_ε) (resp. (D_ε))

$$(N_\varepsilon)\begin{cases} -\Delta u_\varepsilon = f \text{ on } \Omega_\varepsilon \\ \dfrac{\partial u_\varepsilon}{\partial n} = 0 \text{ on } \partial T_\varepsilon + \text{boundary condition on } \partial\Omega \end{cases}$$

$$(D_\varepsilon)\begin{cases} -\Delta v_\varepsilon = f \text{ on } \Omega_\varepsilon \\ v_\varepsilon = 0 \text{ on } \partial T_\varepsilon + \text{boundary condition on } \partial\Omega. \end{cases}$$

where $\Omega_\varepsilon = \Omega_\varepsilon^m = \Omega \setminus T_\varepsilon$.

The density of holes and, hence, *the behaviour of* u_ε *and* v_ε *as* ε *goes to zero depend on the parameter* r_ε $(0 < r_\varepsilon < \varepsilon)$.

Completely different situations are produced according to the type of boundary condition (Neumann or Dirichlet) *imposed on the "moving boundary"* ∂T_ε. (The condition imposed on the fixed boundary $\partial\Omega$ does not play a role in the limit analysis argument. It is only used in order to assert existence and get estimations on (u_ε) and (v_ε). For simplicity take, for example, $u_\varepsilon = v_\varepsilon = 0$ on $\partial\Omega$).

A. Neumann boundary conditions

THEOREM 1.2 The solutions $\{u_\varepsilon;\ \varepsilon \to 0\}$ of problems (N_ε) satisfy

$$\sup_{\varepsilon>0} \|u_\varepsilon\|_{H^1(\Omega_\varepsilon)} < +\infty.$$

They admit uniformly bounded extensions

$$\{\mathbb{P}_\varepsilon(u_\varepsilon)\} \text{ in } H^1(\Omega) : \sup_{\varepsilon>0} \|\mathbb{P}_\varepsilon(u_\varepsilon)\|_{H^1(\Omega)} < +\infty.$$

(a) When $r_\varepsilon << \varepsilon$, the $\mathbb{P}_\varepsilon(u_\varepsilon) \to u$ strongly in $H^1(\Omega)$, as ε goes to zero, where u is the solution of

$$\begin{cases} -\Delta u = f \text{ on } \Omega \\ u = 0 \text{ on } \partial\Omega. \end{cases}$$

Moreover, there is *convergence of the energies*, that is

$$\int_{\Omega_\varepsilon} |\text{grad } u_\varepsilon|^2\, dx \to \int_\Omega |\text{grad } u|^2\, dx, \text{ as } \varepsilon \text{ goes to zero.}$$

(b) When $r_\varepsilon \sim \varepsilon$ (if $r_\varepsilon \sim k\varepsilon$ one can reduce to this situation changing T into kT), then $\mathbb{P}_\varepsilon(u_\varepsilon) \longrightarrow u$ weakly in $H^1(\Omega)$, as ε goes to zero, where u is the solution of the limit problem:

$$\begin{cases} A^{hom} u = \theta . f \text{ on } \Omega \\ u = 0 \text{ on } \partial\Omega. \end{cases}$$

The operator A^{hom} is the subdifferential of the convex functional F^{hom}

$$F^{hom}(u) = \int_\Omega j^{hom}(\text{grad } u)dx$$

with

$$j^{hom}(z) = \min \{\int_{Y \setminus T} |\text{grad } w(y) + z|^2 dy / w \text{ Y-periodic}\}, \tag{1.18}$$

and θ is equal to the Lebesgue measure of T.

Moreover, there is *convergence of the energies*, that is,

$$\int_{\Omega_\varepsilon} |\text{grad } u_\varepsilon|^2 dx \to \int_\Omega j^{hom}(\text{grad } u)dx \text{ as } \varepsilon \text{ goes to zero.}$$

Comments: The above situation describes the case of small holes distributed in volume in Ω, with Neumann boundary condition on their moving boundary. The limit analysis depends on the total volume of these holes, the case where the total volume tends to zero (resp. a strictly positive value) corresponding to $r_\varepsilon \ll \varepsilon$, (resp. $r_\varepsilon \sim \varepsilon$). In the first situation (case (a)) they completely disappear in the limit problem. The second situation, (case (b)) can be interpreted as a *homogenization process*. This is made clear when considering the corresponding *variational formulation*: For every $\varepsilon > 0$ $\mathbb{P}_\varepsilon(u_\varepsilon)$ actually minimizes over $H_0^1(\Omega)$, $u \to F^\varepsilon(u) - 2\int_\Omega a(\frac{x}{\varepsilon})uf\, dx$ where $F^\varepsilon(u) = \int_\Omega a(\frac{x}{\varepsilon})|\text{grad } u|^2 dx$, and a is the Y-periodic function equal to zero on T and one on $Y \setminus T$.

That is exactly the situation described in the previous example with the difference that a is no longer strictly positive (a is equal to zero on T, which physically means that T *is an isolating inclusion*). It is not surprising that the same conclusion holds: formula (1.17) specializes into

$$F^{hom} = \lim_e F^\varepsilon \text{ as } \varepsilon \text{ goes to zero (cf. Theorem 1.23)} \tag{1.19}$$

with

$$F^{hom}(u) = \int_\Omega j^{hom}(\text{grad } u)dx$$

and

$$j^{hom}(z) = \min \{\int_Y a(y)|\text{grad } w(y) + z|^2 \, dy/w \text{ Y-periodic}\},$$

which is precisely the conclusion (1.18) of the above theorem in case b). One can notice that j^{hom} is still a strictly positive definite quadratic form

$$j^{hom}(z) = \sum_{i,j} a_{ij}^{hom} z_i z_j.$$

B. Dirichlet boundary conditions. Cloud of ice

As already mentioned, the situation is completely different. The limit analysis depends now on the dimension of the space $\mathbb{R}^N$. Take, for example, $N = 3$.

THEOREM 1.3 For every $\varepsilon > 0$, let v_ε be the solution of (D_ε); $v_\varepsilon \in H_o^1(\Omega_\varepsilon)$ admits a natural extension (by zero) to $H_o^1(\Omega)$, that is still denoted v_ε. Then, the sequence $\{v_\varepsilon; \varepsilon \to 0\}$ remains bounded in $H_o^1(\Omega)$.

Depending on the behaviour, as $\varepsilon \to 0$, of $r_\varepsilon/\varepsilon^3$, three different situations can occur:

(a) When $r_\varepsilon \ll \varepsilon^3$, $v_\varepsilon \to v$ strongly in $H_o^1(\Omega)$, as $\varepsilon \to 0$, where v is the solution of

$$\begin{cases} -\Delta v = f \text{ on } \Omega \\ \quad v = 0 \text{ on } \partial\Omega. \end{cases}$$

(b) When $r_\varepsilon \sim \varepsilon^3$, $v_\varepsilon \rightharpoonup v$ weakly in $H_o^1(\Omega)$, where v is the solution of

$$\begin{cases} -\Delta v + C.v = f \text{ on } \Omega \\ \quad v = 0 \text{ on } \partial\Omega \end{cases}$$

and C is the positive constant equal to the capacity of T in $\mathbb{R}^3$:

$$C = \inf \{\int_{\mathbb{R}^3} |\text{grad } w|^2 dx;\ w \in H^1(\mathbb{R}^3),\ \tilde{w} \geq 1 \text{ quasi-everywhere on } T\}$$

Moreover, there is convergence of the energies:

$$\int_\Omega |\text{grad } v_\varepsilon|^2 dx \to \int_\Omega |\text{grad } v|^2 dx + c\int_\Omega v^2 dx \text{ as } \varepsilon \text{ goes to zero.}$$

(c) <u>When $r_\varepsilon >> \varepsilon^3$</u>, then v_ε converges strongly to zero in $H_0^1(\Omega)$.

<u>Comments:</u> Let us give a physical interpretation of the above results in the *"crushed ice problem"* (cf. Rauch & Taylor [1]): let us imagine that the $\{T_\varepsilon^i;\ i \in I_\varepsilon\}$ are little (spherical) coolers maintained at zero temperature. (Small pieces of ice for example). Because of the ε-periodicity of the problem in $\mathbb{R}^3$, their number is of order $\frac{|\Omega|}{\varepsilon^3}$.

One might expect that, as in the Neumann problem, it is the total volume of the coolers $\frac{4}{3}\pi r_\varepsilon^3 . \frac{1}{\varepsilon^3}$, i.e. $\frac{r_\varepsilon}{\varepsilon}$ which, is the significant parameter, or perhaps their total surface $4\pi r_\varepsilon^2 \times 1/\varepsilon^3$, i.e. $r_\varepsilon/\varepsilon^{3/2}$. This is not the case: the right "critical" parameter is $r_\varepsilon/\varepsilon^3$!

(a) When $r_\varepsilon << \varepsilon^3$, global effect of the coolers goes to zero with ε and, in the limit problem, it completely disappears.

(c) When $r_\varepsilon >> \varepsilon^3$, the *cloud of small ice coolers* behaves like a solid block of ice filling all of Ω: so, $v_\varepsilon \to 0$.

(b) When $r_\varepsilon \sim \varepsilon^3$, we are in an intermediate situation, with the appearance of an extra term "C.v" in the limit equation. The interpretation of this term and the computation of the constant C have puzzled mathematicians for many years: for this reason, this extra term was initially called the "strange term".

The *variational approach*, again, gives a quite natural interpretation of these results.

For every $\varepsilon > 0$, v_ε minimizes $v \to F^\varepsilon(v) - \int_\Omega fv$ over $H_0^1(\Omega)$, where

$$F^\varepsilon(v) = \Phi(v) + I_{K^\varepsilon}(v)$$

$$\Phi(v) = \int_\Omega |\text{grad } v|^2 dx$$

and $$I_{K^\varepsilon}(v) = \begin{cases} 0 & \text{if } v \in K^\varepsilon \\ +\infty & \text{otherwise} \end{cases}$$

is *the indicator functional of the "convex set of constraints"* K^ε

$$K^{\varepsilon} = \{v \in H_0^1(\Omega)/v = 0 \text{ on } T_{\varepsilon}\}.$$

Noticing that $I_{K^{\varepsilon}}$ can be written in an integral form

$$I_{K^{\varepsilon}}(v) = \int_{\Omega} a_{\varepsilon}(x)v^2(x)dx,$$

where

$$a_{\varepsilon}(x) = \begin{cases} +\infty \text{ if } x \in T_{\varepsilon} \\ 0 \text{ if } x \in \Omega \setminus T_{\varepsilon}, \end{cases}$$

the conclusions of Theorem 1.3 can be reformulated, in a unified way, in terms of epi-convergence (cf. Theorem 1.27):

$$\text{for every } v \in H_0^1(\Omega)$$
$$\Phi(v) + C\int_{\Omega} v^2(x)dx = \lim_e\{\Phi(v) + \int_{\Omega} a_{\varepsilon}(x)v^2(x)dx\} \text{ as } \varepsilon \to 0. \tag{1.20}$$

The cases $C = 0$, $C = +\infty$, C strictly positive finite, correspond respectively to $r_{\varepsilon} \ll \varepsilon^3$, $r_{\varepsilon} \gg \varepsilon^3$, $r_{\varepsilon} \sim \varepsilon^3$. With this approach, approximating and limit functionals turn into similar types of functionals. In the "critical case" $r_{\varepsilon} \sim \varepsilon^3$, the limit constraint functional $v \to C\int_{\Omega} v^2(x)\,dx$ may take values different from zero or $+\infty$. For this reason, this phenomenon can be interpreted as a *"relaxation" or "fuzzy" form of the constraint* $u = 0$. This type of phenomenon can be exhibited for all the equations of continuum mechanics and, like homogenization, naturally occurs in resolution of some control problems.

Notice that the critical size r_{ε} is equal to $\varepsilon^{N/N-2}$ when $N > 2$ and $\exp(-1/\varepsilon^2)$ when $N = 2$. (This phenomenon does not occur when $N = 1$). □

<u>REMARK 1.4</u> Following Brillard [1], one can treat in a unified way the limit analysis of Dirichlet and Neumann boundary conditions by considering the double-indexed sequence of functionals $\{F^{\varepsilon,\lambda}\}_{\substack{\varepsilon>0\\ \lambda>0}}$

$$F^{\varepsilon,\lambda}(u) = \int_{\Omega_{\varepsilon}} |\text{grad } u|^2 dx + \lambda\int_{\partial T_{\varepsilon}} u^2(\sigma)d\mu_{\varepsilon}(\sigma)$$

where μ_{ε} denotes the (N-1) Hausdorff measure supported by ∂T_{ε}. The cases

$\lambda = 0$, $\lambda = +\infty$, λ finite, correspond respectively to Neumann, Dirichlet, *mixed boundary conditions* on the fragmented boundary ∂T_ε. Considering the "absorption coefficient" λ and the "size of the holes parameter" r as functions of ε, one can determine critical values separating the different types of limit behaviour.

REMARK 1.5 As already mentioned. Theorem 1.3 can be interpreted as a convergence result for sequences of functionals $\{F_\varepsilon; \varepsilon \to 0\}$

$$F^\varepsilon = \Phi + I_{K^\varepsilon}$$

where $\Phi(v) = \int |\text{grad } v|^2 dx$ is a fixed energy functional, and $\{K^\varepsilon; \varepsilon \to 0\}$ are varying *bilateral constraints*. The bilateral problem is, in fact, intimately connected with the *unilateral (also called "obstacle")* one, which is described below: Take

$$C_\varepsilon = \{u \in H_0^1(\Omega)/u \geq 0 \text{ on } T_\varepsilon\}, \text{ and } G^\varepsilon = \Phi + I_{C^\varepsilon}.$$

Then

$$\lim_e G^\varepsilon = G \text{ exists as } \varepsilon \text{ goes to zero,}$$

with

$$G(v) = \int_\Omega |\text{grad } v|^2 dx + C \int_\Omega (v(x)^-)^2 dx.$$

The critical size r_ε and the corresponding values of C are the same as in the bilateral case.

In Section 2.4 we shall study in a systematic way the limit analysis of variational inequalities with varying obstacles g_ε, i.e. $C_\varepsilon = \{u \in H^1(\Omega)/u \geq g_\varepsilon\}$.

1.1.3 Transmission problems through thin layers

In physics (heat equation, electrostatics, elasticity, ...), as already noticed in previous examples, it frequently occurs that the problem is governed by elliptic equations or elliptic variational inequalities with discontinuous coefficients: the coefficients have jumps along certain hypersurfaces.

Typically, if Σ is the interface between Ω_1 and Ω_2, solution u of the problem satisfies limit conditions on Σ of the following type:

$$\begin{cases} u|_1 = u|_2 \\ \lambda_1 \frac{\partial u}{\partial n}\Big|_1 = \lambda_2 \frac{\partial u}{\partial n}\Big|_2 \end{cases}$$

where n is the normal at Σ and λ_i are physical parameters (conductivity, elasticity coefficient) relative to each Ω_i.

This. in fact, is one of the possible form of transmission one can obtain by limit analysis of a thin "isolating" layer:

A. Thin isolating layer(*)

Let $\Sigma_{\varepsilon,\lambda}$ be a thin isolating layer of thickness ε, which shrinks to Σ as $\varepsilon \to 0$, while simultaneously its conductivity coefficient λ goes to zero.

Following J.-L. Lions [2], Sanchez-Palencia [1], [4], Brezis et al. [1], let us describe the limit behaviour of such material as (ε,λ) goes to $(0,0)$.

Let us consider Σ a smooth manifold in $\mathbb{R}^3$ and Σ_ε be an ε-neighbourhood of Σ in $\mathbb{R}^3$ (that is equal to the union of all points whose distance from Σ is less than $\frac{\varepsilon}{2}$). (Fig. 1.3). We assume the conductivity coefficient $a = a_{\varepsilon,\lambda}$

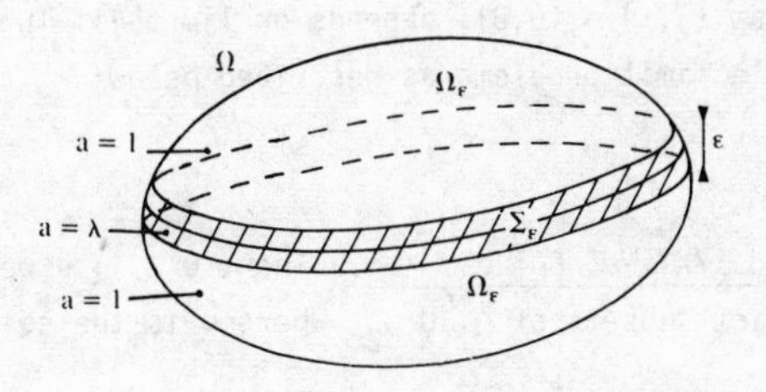

Figure 1.3

to be equal to

$$a_{\varepsilon,\lambda} = \left\langle \begin{array}{l} \lambda_1 = \lambda_2 = 1 \text{ on } \Omega_\varepsilon \\ \lambda > 0 \text{ on } \Sigma_\varepsilon. \end{array} \right.$$

(*) The opposite case of a highly conductive and thin layer is considered in Section 2.4.

Then, the solutions $\{u_{\varepsilon,\lambda};\ \varepsilon > 0,\ \lambda > 0\}$ of the corresponding elliptic problems verify:

$$\begin{cases} -\Delta u_{\varepsilon,\lambda} = f \text{ on } \Omega_\varepsilon = \Omega\setminus\Sigma_\varepsilon \\ -\lambda\Delta u_{\varepsilon,\lambda} = f \text{ on } \Sigma_\varepsilon \\ u_{\varepsilon,\lambda}\Big|_{\Omega_\varepsilon} = u_{\varepsilon,\lambda}\Big|_{\Sigma_\varepsilon} \text{ on } \partial\Omega_\varepsilon = \partial\Sigma_\varepsilon \\ \lambda \dfrac{\partial u_{\varepsilon,\lambda}}{\partial n}\Big|_{\partial\Sigma_\varepsilon} = \dfrac{\partial u_{\varepsilon,\lambda}}{\partial n}\Big|_{\partial\Omega_\varepsilon} \text{ on } \partial\Omega_\varepsilon = \partial\Sigma_\varepsilon \\ u_{\varepsilon,\lambda} = 0 \text{ on } \partial\Omega. \end{cases}$$

$u_{\varepsilon,\lambda}$ is the solution of the following minimization problem:

$$\min_{u\in H_0^1(\Omega)} \left\{ \frac{1}{2}\int_\Omega a_{\varepsilon,\lambda}(x)|\text{grad } u|^2 dx - \int_\Omega fu\, dx \right\}$$

where

$$a_{\varepsilon,\lambda}(x) = \begin{cases} 1 \text{ if } x \in \Omega_\varepsilon \\ \lambda \text{ if } x \in \Sigma_\varepsilon. \end{cases}$$

The limit behaviour, as $(\varepsilon,\lambda) \to (0,0)$, depends on $\lim (\frac{\varepsilon}{\lambda})$! There are actually three types of possible limit problems as described below:

THEOREM 1.6

(a) $(\varepsilon,\lambda) \to (0,0)$ and $\lambda/\varepsilon \to \alpha,\ 0 < \alpha < +\infty$. Then, $u_{\varepsilon,\lambda} \to u$ weakly in $L^2(\Omega)$ and uniformly on compact subsets of $\Omega_1 \cup \Omega_2$, where u is the solution of:

$$\begin{cases} -\Delta u = f \text{ on } \Omega_1 \cup \Omega_2 \\ \dfrac{\partial u}{\partial n}\Big|_1 = \dfrac{\partial u}{\partial n}\Big|_2 = \alpha[u] \text{ on } \Sigma \\ u|_{\partial\Omega} = 0 \end{cases}$$

where u *is possibly discontinuous along* Σ and $[u] = u|_{\partial\Omega_1} - u|_{\partial\Omega_2}$.

(b) $(\varepsilon,\lambda) \to (0,0)$ and $\lambda/\varepsilon \to 0$, the same conclusion holds: just take $\alpha = 0$ in the above formula, i.e.

$$\begin{cases} -\Delta u = f \text{ on } \Omega_1 \cup \Omega_2 \\ \left.\frac{\partial u}{\partial n}\right|_1 = \left.\frac{\partial u}{\partial n}\right|_2 = 0 \text{ on } \Sigma \\ u|_{\partial\Omega} = 0. \end{cases}$$

(c) $(\varepsilon,\lambda) \to (0,0)$ and $\lambda/\varepsilon \to +\infty$: then, the isolating inclusion vanishes at the limit and $u_{\varepsilon,\lambda} \to u$, the solution of

$$\begin{cases} -\Delta u = f \text{ on } \Omega \\ u = 0 \text{ on } \partial\Omega \end{cases}$$

(this corresponds to taking $\alpha = +\infty$ in (a)).

Moreover, in all these cases, there is *convergence of the energies*, i.e.

$$\int_\Omega a_{\varepsilon,\lambda}|\text{grad } u_{\varepsilon,\lambda}|^2 dx \to \int_{\Omega_1\cup\Omega_2} |\text{grad } u|^2 dx + \alpha\int_\Sigma [u]^2 d\sigma$$

with α respectively finite, zero.

Comments: This result has a clear physical interpretation: when ε goes to zero, depending on the isolating property of the thin layer that is, $\lambda \ll \varepsilon$ or $\lambda \gg \varepsilon$, Ω_1 and Ω_2 are completely disconnected or connected. The critical size and the corresponding limit equation can be, again, easily understood when *writing the problem in a variational form*:

For every $\varepsilon > 0$, $\lambda > 0$, $u_{\varepsilon,\lambda}$ actually minimizes $F^{\varepsilon,\lambda}(\cdot) - \int fu$ where

$$F^{\varepsilon,\lambda}(u) = \frac{1}{2}\int_{\Omega_\varepsilon} |\text{grad } u|^2 dx + \frac{1}{2}\int_{\Sigma_\varepsilon} \lambda|\text{grad } u|^2 dx.$$

When ε is small, this last integral can be approximated by

$dx \simeq \varepsilon\, d\sigma$ (where $d\sigma$ is the Hausdorff measure supported by Σ)

$$|\text{grad } u| \simeq \frac{1}{\varepsilon}\,(u|_{\partial\Omega_1} - u|_{\partial\Omega_2}) = \frac{[u]}{\varepsilon}.$$

Hence

$$F^{\varepsilon,\lambda}(u) \simeq \frac{1}{2}\int_\Omega |\text{grad } u|^2 dx + \frac{1}{2}\frac{\lambda}{\varepsilon}\int_\Sigma [u]^2 d\sigma.$$

We shall prove indeed that the sequence of functionals $\{F^{\varepsilon,\lambda};\ \varepsilon \to 0,\ \lambda \to 0\}$ converges in a variational sense epi-converges) to the limit functional F.

$$F(u) = \frac{1}{2}\int_{\Omega_1 \cup \Omega_2} |\mathrm{grad}\ u|^2 dx + \frac{\alpha}{2}\int_{\Sigma} [u]^2 d\sigma \text{ (with } \alpha = \lim \frac{\lambda}{\varepsilon}\text{):}$$

$$F = \lim_e F^{\varepsilon,\lambda} \text{ (cf. Theorem 1.29).}$$

This will provide both the convergence of the solutions $u_{\varepsilon,\lambda} \to u$, and of the corresponding infimum , that is "convergence of the energies". In this situation, the domain of the limit functional F, equal to $H^1(\Omega_1 \cup \Omega_2)$ contains functions with a possible jump on Σ, while the domains of the approximating functions $F^{\varepsilon,\lambda}$ are equal to $H^1(\Omega)$. Once more, this variational convergence does not reduce to pointwise convergence. Notice, too that, in this example, the two operations $\lim_{\varepsilon\to 0}$ and $\lim_{\lambda\to 0}$ do not commute: clearly taking first $\varepsilon \to 0$, then $\lambda \to 0$ yields the Dirichlet problem in Ω (i.e. complete connection, corresponding to $\lambda/\varepsilon \to +\infty$) while taking first $\lambda \to 0$, then $\varepsilon \to 0$ yields the disconnected problem along Σ (corresponding to $\lambda/\varepsilon \to 0$). □

Let us now describe a closely related limit analysis transmission problem:

B. Neumann's strainer

The two subdomains Ω_1 and Ω_2 are now connected through many small "holes" distributed on Σ. More precisely, let us assume that Σ is a hyperplane separating Ω into Ω_1 and Ω_2. On Σ let us define an ε-periodic structure: in each small cell of size ε is centred a disc T_ε^i of radius $r_\varepsilon < \varepsilon/2$. Let us denote

$$T_\varepsilon = \bigcup_i T_\varepsilon^i \text{ and } \Omega_\varepsilon = \Omega_1 \cup \Omega_2 \cup T_\varepsilon.$$

Given $f \in L^2(\Omega)$, let us consider the elliptic problem in Ω_ε:

$$\begin{cases} -\Delta u_\varepsilon = f \text{ on } \Omega_\varepsilon \\ \left.\frac{\partial u_\varepsilon}{\partial n}\right|_1 = \left.\frac{\partial u_\varepsilon}{\partial n}\right|_2 = 0 \text{ on } \Sigma \setminus T_\varepsilon \\ u_\varepsilon|_{\partial\Omega} = 0. \end{cases}$$

So, the heat flux goes only through the holes T_ε (which connect Ω_1 and Ω_2)

while $\Sigma \setminus T_\varepsilon$ is some "perfectly isolating" material. This model has its full physical significance in acoustics and was first described by Marchenko & Hruslov [1], Sanchez-Palencia [10]. Just as in the preceding situation, in the limit the two subdomains Ω_1 and Ω_2 are completely disconnected or connected, but now depending on the size of r_ε (Fig. 1.4). The critical

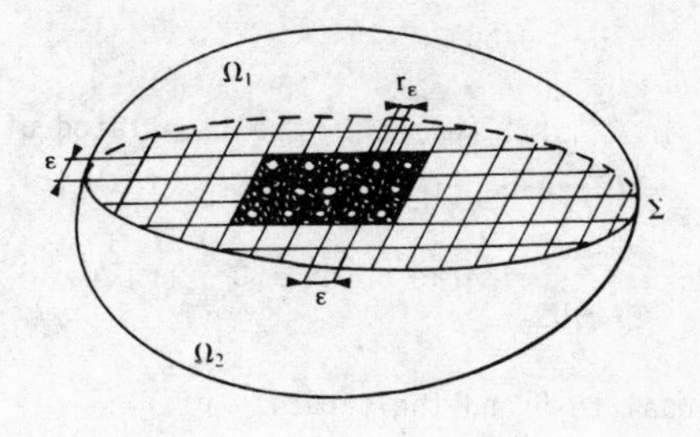

Figure 1.4

size in the present situation ($\Omega \subset \mathbb{R}^3$ and Σ a 2-dimensional manifold) is $r_\varepsilon = \varepsilon^2$. More generally, when $\Omega \subset \mathbb{R}^N$ and Σ is an (N-1)-dimensional manifold, it is equal to $r_\varepsilon = \varepsilon^{(N-1)/(N-2)}$.

<u>THEOREM 1.7</u> For every $f \in L^2(\Omega)$, the solutions $\{u_\varepsilon ; \varepsilon \to 0\}$ of

$$\begin{cases} -\Delta u_\varepsilon = f \text{ on } \Omega \\ \left.\dfrac{\partial u_\varepsilon}{\partial n}\right|_1 = \left.\dfrac{\partial u_\varepsilon}{\partial n}\right|_2 = 0 \text{ on } \Sigma \setminus T_\varepsilon \\ u_\varepsilon = 0 \text{ on } \partial\Omega \end{cases}$$

weakly converge in $V = H^1(\Omega_1 \cup \Omega_2)$ to the solution of the following limit problem:

(a) if $r_\varepsilon \ll \varepsilon^2$

$$\begin{cases} -\Delta u = f \text{ on } \Omega_1 \cup \Omega_2 \\ \left.\dfrac{\partial u}{\partial n}\right|_1 = \left.\dfrac{\partial u}{\partial n}\right|_2 = 0 \text{ on } \Sigma \\ u = 0 \text{ on } \partial\Omega \end{cases}$$

(b) if $r_\varepsilon >> \varepsilon^2$
$$\begin{cases} -\Delta u = f \text{ on } \Omega \\ u = 0 \text{ on } \partial\Omega \end{cases}$$

(c) if $r_\varepsilon = \varepsilon^2$
$$\begin{cases} -\Delta u = f \text{ on } \Omega_1 \cup \Omega_2 \\ \left.\frac{\partial u}{\partial n}\right|_1 = \left.\frac{\partial u}{\partial n}\right|_2 = \frac{C}{4}[u] \text{ on } \Sigma \\ u = 0 \text{ on } \partial\Omega \end{cases}$$

and C is the capacity in $\mathbb{R}^3$ of T (where T_ε^i are translated of $r_\varepsilon.T$).

Moreover, there is *convergence of the energies*, that is.

$$\int_{\Omega_\varepsilon} |\text{grad } u_\varepsilon|^2 dx \to \int_{\Omega_1 \cup \Omega_2} |\text{grad } u|^2 dx + C \int_\Sigma [u]^2$$

with C respectively equal to 0 and the capacity of T.

Comments: If the holes are too small, that is $r_\varepsilon << \varepsilon^2$, Ω_1 and Ω_2 are disconnected, whereas if they are too large, that is $r_\varepsilon >> \varepsilon^2$, in the limit problem Ω_1 and Ω_2 are completely connected.

The *variational approach*, once more, reveals this result: for every $\varepsilon > 0$, u_ε actually minimizes over $H^1(\Omega_\varepsilon)$ the functional $F^\varepsilon(u) - \int_\Omega fu\, dx$, where

$$F^\varepsilon(u) = \int_{\Omega_\varepsilon} |\text{grad } u|^2 dx$$

(the boundary condition on the fixed boundary of Ω does not play any role in the limit analysis).

The domain of F^ε, equal to $H^1(\Omega_\varepsilon)$, varies with respect to ε. The larger space on which all functionals F^ε are defined, and in which the solutions $\{u_\varepsilon;\ \varepsilon \to 0\}$ remain bounded, is $X = H^1(\Omega_1 \cup \Omega_2)$.

Let us now notice the following equivalence, *which provides the trick:*

$$u \in H^1(\Omega_\varepsilon) \iff u \in X \text{ and } [u] = 0 \text{ on } T_\varepsilon! \tag{1.21}$$

where, for every function u in $H^1(\Omega_1 \cup \Omega_2)$, we denote $[u] = u|_1 - u|_2$ and $u|_i$ the trace of u on Σ, when considered as an element of $H^1(\Omega_i)$. Thus, $F^\varepsilon: X \to \bar{\mathbb{R}}^+$ can be written

$$F^\varepsilon(u) = \int_{\Omega_1 \cup \Omega_2} |\text{grad } u|^2 dx + I_{K_\varepsilon}(u)$$

where $K_\varepsilon = \{u \in X/[u] = 0 \text{ on } T_\varepsilon\}$.

Introducing the function $a_\varepsilon : \Sigma \to \bar{\mathbb{R}}^+$ equal to

$$a_\varepsilon(x) = \begin{cases} +\infty & \text{if } x \in T_\varepsilon \\ 0 & \text{otherwise} \end{cases}$$

we finally get the following formulation of F^ε:

$$\text{for every } u \in H^1(\Omega_1 \cup \Omega_2),\ F^\varepsilon(u) = \int_{\Omega_1 \cup \Omega_2} |\text{grad } u|^2 dx + \int_\Sigma a_\varepsilon(x)[u]^2(x)d\sigma(x).$$

When ε goes to zero, a_ε oscillates more and more rapidly between 0 and $+\infty$, and the sequence $\{F^\varepsilon;\ \varepsilon \to 0\}$ converges in variational sense, $F = \lim_e F^\varepsilon$ (cf. Theorem 1.30). The limit functional F, described below, turns out, indeed, to be of the same nature as the approximating functionals F^ε:

$$\text{for every } u \in H^1(\Omega_1 \cup \Omega_2),\ F(u) = \int_{\Omega_1 \cup \Omega_2} |\text{grad } u|^2 dx + C\int_\Sigma [u]^2 d\sigma(x)$$

with C equal to 0 when $\lim \frac{r_\varepsilon}{\varepsilon^2} = 0$,

C equal to $+\infty$ when $\lim \frac{r_\varepsilon}{\varepsilon^2} = +\infty$,

C finite when $\lim \frac{r_\varepsilon}{\varepsilon^2}$ is finite.

Let us notice that when $C = +\infty$, $C\int_\Sigma [u]^2 d\sigma$ is equal to $+\infty$ if $[u] \neq 0$, and zero if $[u] = 0$. Thus,

$$F(u) = \int_{\Omega_1 \cup \Omega_2} |\text{grad } u|^2 dx + I_{\{[u]=0\ \Sigma\}} = \int_\Omega |\text{grad } u|^2 dx.$$

From general variational properties of epi-convergence, both convergences of the corresponding solutions and energies follow. □

1.2 EPIGRAPH-CONVERGENCE. VARIATIONAL APPROACH

In this section, we give mathematical formulations of the concepts of variational convergences introduced in the previous examples.

1.2.1 Variational approach. Definition in a general topological space

Let X be an abstract space and $(F^n)_{n\in N}$ a sequence of functions from X into $\bar{R}$.

For every $n \in N$, let us consider the minimization problem

$$\inf_{x\in X} F^n(x).$$

In general, this infimum is not achieved, but, given any sequence of strictly positive numbers $(\varepsilon_n)_{n\in N}$ decreasing to zero, we can define for every $n \in N$, an element $\bar{x}_n$, belonging to X, satisfying (1.22):

$$\left.\begin{array}{l} F^n(\bar{x}_n) \leq \inf_{x\in X} F^n(x) + \varepsilon_n \text{ if } \inf_{x\in X} F^n(x) > -\infty \\ F^n(\bar{x}_n) \leq -1/\varepsilon_n \text{ if } \inf_{x\in X} F^n(x) = -\infty. \end{array}\right\} \tag{1.22}$$

In order to study the behaviour of such a minimizing sequence $(\bar{x}_n)_{n\in N}$, a natural approach consists, first, in obtaining estimations on the sequence $(\bar{x}_n)_{n\in N}$. From these estimations, *one can derive some compactness properties:* Let us assume, for example, that the sequence $(\bar{x}_n)_{n\in N}$ is relatively compact for a topology τ on $X^{(*)}$, and extract a τ-converging subsequence:

$$\exists (n_k)_{k\in N},\ \exists \bar{x} \in X \text{ such that } \bar{x}_{n_k} \xrightarrow[k\to+\infty]{\tau} \bar{x}. \tag{1.23}$$

Let us see what limit problem $\bar{x}$ satisfies: From (1.22)

$$\liminf_{k\to+\infty} F^{n_k}(\bar{x}_k) \leq \liminf_{k\to+\infty} \inf_{x\in X} F^{n_k}(x) \tag{1.24}$$

Since we have no information on $(n_k)_{k\in N}$ and $(\bar{x}_{n_k})_{k\in N}$ except that they converge respectively to $+\infty$ and $\bar{x}$, from (1.24) we obtain

$$\inf_{\left\{\begin{array}{l}(n_k)_{k\in N}\\ x_k \xrightarrow[k]{\tau} \bar{x}\end{array}\right.} \liminf_k F^{n_k}(x_k) \leq \limsup_n \inf_{x\in X} F^n(x) \tag{1.25}$$

(*) In the previous examples, from compact embedding of $H^1(\Omega)$ into $L^2(\Omega)$, take τ = strong topology of $L^2(\Omega)$.

In order to get (1.25), we have used the fact that for any sequence $(a_n)_{n\in\mathbb{N}}$ of extended reals:

$$\sup_{(n_k)_{k\in\mathbb{N}}} \liminf_k a_{n_k} = \limsup_n a_n. \tag{1.26}$$

For any x belonging to X, let us denote by $N_\tau(x)$ the system of neighbourhoods of x with respect to topology τ. We notice that for any $V \in N_\tau(\bar{x})$, any $(n_k)_{k\in\mathbb{N}}$ and any sequence $(x_k)_{k\in\mathbb{N}}$ τ-converging to $\bar{x}$,

$$\liminf_k \inf_{x\in V} F^{n_k}(x) \leq \liminf_k F^{n_k}(x_k).$$

This implies

$$\sup_{V\in N_\tau(\bar{x})} \liminf_n \inf_{x\in V} F^n(x) \leq \inf_{\begin{cases}(n_k)_{k\in\mathbb{N}} \\ x_k \xrightarrow[k]{\tau} \bar{x}\end{cases}} \liminf_k F^{n_k}(x_k). \tag{1.27}$$

In order to get (1.27), we have used the fact that for any sequence $(a_n)_{n\in\mathbb{N}}$ of extended reals:

$$\inf_{(n_k)_{k\in\mathbb{N}}} \liminf_k a_{n_k} = \liminf_n a_n . \tag{1.28}$$

Combining (1.25) and (1.27), we obtain

$$\sup_{V\in N_\tau(\bar{x})} \liminf_n \inf_{x\in V} F^n(x) \leq \limsup_n \inf_{x\in X} F^n(x) \tag{1.29}$$

On the other hand, for every $x \in X$

$$\limsup_n \inf_{x\in X} F^n(x) \leq \sup_{V\in N_\tau(x)} \limsup_n \inf_{u\in V} F^n(u) \tag{1.30}$$

which, combined with (1.29) finally implies:

$$\forall x \in X \quad \sup_{V\in N_\tau(\bar{x})} \liminf_n \inf_{u\in V} F^n(u) \leq \sup_{V\in N_\tau(x)} \limsup_n \inf_{u\in V} F^n(u). \tag{1.31}$$

So, we have been led to introduce two new functions:

<u>DEFINITION 1.8</u> Given (X,τ) a topological space, for every $x \in X$, let us denote the system of the τ-neighbourhoods of x by $N_\tau(x)$.

With any sequence $(F^n)_{n\in N}$ of functions from X into $\bar{\mathbb{R}}$ are associated two limit functions:

(a) The τ-*epi-limit inferior* of the sequence $(F^n)_{n\in\mathbb{N}}$, denoted by $\tau\text{-li}_e F^n$ (or simply $\text{li}_e F^n$, when no confusion on the topology is possible) is defined by

$$(\tau\text{-li}_e F^n)(x) = \sup_{V\in N_\tau(x)} \liminf_n \inf_{u\in V} F^n(u) \tag{1.32}$$

(b) The τ-*epi-limit superior* of the sequence $(F^n)_{n\in N}$, denoted by $\tau\text{-ls}_e F^n$ (or simply $\text{ls}_e F^n$) is the function defined by

$$(\tau\text{-ls}_e F^n)(x) = \sup_{V\in N_\tau(x)} \limsup_n \inf_{u\in V} F^n(u). \quad \square \tag{1.33}$$

With this notation, (1.31) can be reformulated:

$$(\tau\text{-li}_e F^n)(\bar{x}) \leq \inf_{x\in X} (\tau\text{-ls}_e F^n)(x) \tag{1.34}$$

Clearly, it follows from definitions (1.32) and (1.33), that $\tau\text{-li}_e F^n$ is always less or equal than $\tau\text{-ls}_e F^n$; the interesting situation is when the equality holds; denoting by F the corresponding limit function

$$\tau\text{-li}_e F^n = \tau\text{-ls}_e F^n = F \tag{1.35}$$

(1.34) turns, in that case, into

$$F(\bar{x}) = \tau\text{-li}_e F^n(\bar{x}) \leq \inf_{x\in X} (\tau\text{-ls}_e F^n)(x) = \inf_{x\in X} F(x), \tag{1.36}$$

i.e. $\bar{x}$ minimizes F on X.

The equality (1.35) between the τ-epi-lower limit and the τ-epi upper limit of the sequence F^n implies that every τ-limit point of the minimizing sequences $(\text{Argmin } F^n)_{n\in\mathbb{N}}$ minimizes the so-defined limit function F. *In that sense,* F *is a variational limit of the sequence of functions* $(F^n)_{n\in\mathbb{N}}$:

<u>DEFINITION 1.9</u> Let (X,τ) be a general topological space and $(F^n)_{n\in\mathbb{N}}$ a sequence of functions from X into $\bar{\mathbb{R}}$. The sequence $(F^n)_{n\in\mathbb{N}}$ is said to be τ-epi convergent at x if the following equality holds:

$$\tau\text{-li}_e F^n(x) = \tau\text{-ls}_e F^n(x) \tag{1.37}$$

This common value is then denoted $\tau\text{-}\lim_e F^n(x)$ or briefly $\tau\text{-}\mathrm{lm}_e F^n(x)$:

$$\tau\text{-}\mathrm{lm}_e F^n(x) = \tau\text{-}\mathrm{li}_e F^n(x) = \tau\text{-}\mathrm{ls}_e F^n(x). \tag{1.38}$$

When equality (1.37) holds for every $x \in X$, the sequence $\{F^n; n \in \mathbb{N}\}$ is said to be τ-epi convergent and the limit function $F = \tau\text{-}\mathrm{lm}_e F^n$ defined by (1.38) is called the τ-epi-limit of the sequence F^n.

As a straight consequence of the way we introduced the notion of epi-convergence, the following theorem states the variational properties of this convergence:

<u>THEOREM 1.10</u> Let (X,τ) be a general topological space and $(F^n)_{n\in\mathbb{N}}$ a sequence of functions from X into $\bar{\mathbb{R}}$ which is τ-epi convergent:

$$F = \tau\text{-}\mathrm{lm}_e F^n.$$

Let $(x_n)_{n\in\mathbb{N}}$ be a sequence of points in X satisfying: $F^n(x_n) \leq \inf_{x\in X} F^n(x) + \varepsilon_n$ with $\{\varepsilon_n\}_{n\in\mathbb{N}}$ converging to zero. Then, for every τ-converging subsequence, $x_{n_k} \xrightarrow[k\to+\infty]{\tau} \bar{x}$, we have $F(\bar{x}) = \min_{x\in X} F(x)$ and $F(\bar{x}) = \lim_{k\to\infty} F^{n_k}(x_{n_k})$.

<u>Proof of Theorem 1.10</u> The only point which remains to prove is that $F(\bar{x}) = \lim_{k\to+\infty} F^{n_k}(x_{n_k})$. From (1.27), since $x_{n_k} \xrightarrow[k\to+\infty]{\tau} \bar{x}$,

$$F(\bar{x}) = \tau\text{-}\mathrm{li}_e F^n(\bar{x}) \leq \liminf_{k\to+\infty} F^{n_k}(x_{n_k}).$$

From (1.30), and from the definition of x_n, we get

$$F(\bar{x}) = \tau\text{-}\mathrm{ls}_e F^n(\bar{x}) \geq \limsup_{n\to+\infty} \inf_{x\in X} F^n(x) \geq \limsup_{n\to+\infty} F^n(x_n).$$

From the two preceding inequalities, it follows that

$$F(\bar{x}) = \lim F^{n_k}(x_{n_k}). \quad \square$$

<u>REMARK 1.11</u> Denoting by $\mathcal{O}_\tau$ the family of τ-open sets in X, and $\mathcal{O}_\tau(x)$ the family of the τ-open sets containing x, we notice that for every $x \in X$

$$\tau\text{-}li_e F^n(x) = \sup_{\mathcal{O}\in\mathcal{O}_\tau(x)} \liminf_n \inf_{u\in\mathcal{O}} F^n(u)$$
$$\tau\text{-}ls_e F^n(x) = \sup_{\mathcal{O}\in\mathcal{O}_\tau(x)} \limsup_n \inf_{u\in\mathcal{O}} F^n(u) \tag{1.39}$$

that is, one can replace $N_\tau(x)$ by $\mathcal{O}_\tau(x)$ in the formulation of the τ-epi-limits. Denoting by q the set function which associates $\liminf_n \inf_{u\in V} F^n(u)$ with $V \in N_\tau(x)$ noticing that $\mathcal{O}_\tau(x) \subset N_\tau(x)$, we have

$$\sup_{\mathcal{O}\in\mathcal{O}_\tau(x)} q(\mathcal{O}) \leq \sup_{V\in N_\tau(x)} q(V). \tag{1.40}$$

For every $V \in N_\tau(x)$, there exists $\mathcal{O}_V \in \mathcal{O}_\tau(x)$ such that $V \supset \mathcal{O}_V$;noticing that the set function q is decreasing

$$q(V) \leq q(\mathcal{O}_V) \leq \sup_{\mathcal{O}\in\mathcal{O}_\tau(x)} q(\mathcal{O}).$$

Thus

$$\sup_{V\in N_\tau(x)} q(V) \leq \sup_{\mathcal{O}\in\mathcal{O}_\tau(x)} q(\mathcal{O}).$$

Comparing with (1.40), we get the first equality of (1.39). The second equality of (1.39) is obtained in a similar way, taking now $q(V) = \limsup_n \inf_{u\in V} F^n(u)$, which is still a decreasing set-function. □

The *"local character"* of the epi-convergence is a straight consequence of its definition.

<u>PROPOSITION 1.12</u> Let $(F^n)_{n\in\mathbb{N}}$ and $(G^n)_{n\in\mathbb{N}}$ be two sequences of functions from (X,τ), a topological space, into $\bar{\mathbb{R}}$ and assume that there exists $\mathcal{O}$, a τ-open set of X such that

$$\forall n \in \mathbb{N} \quad \forall x \in \mathcal{O} \quad F^n(x) = G^n(x).$$

Then

$$\forall x \in \mathcal{O} \quad \tau\text{-}li_e F^n(x) = \tau\text{-}li_e G^n(x) \quad \text{and}$$
$$\tau\text{-}ls_e F^n(x) = \tau\text{-}ls_e G^n(x).$$

In other words, τ-epi convergence properties of a sequence $(F^n)_{n\in\mathbb{N}}$ at a point x are determined, since the sequence $(F^n)_{n\in\mathbb{N}}$ is known on an (arbitrarily small) τ-neighbourhood of x.

Proof of Proposition 1.12 Let us assume that $\forall n \in \mathbb{N}$ $\forall u \in \mathcal{O}_1$ $F^n(u) = G^n(u)$; take $x \in \mathcal{O}_1$. For any $\mathcal{O} \in \mathcal{O}_\tau(x)$, since $\mathcal{O} \cap \mathcal{O}_1$ is included in $\mathcal{O}$

$$\limsup_n \inf_{u \in \mathcal{O}} F^n(u) \le \limsup_n \inf_{u \in \mathcal{O} \cap \mathcal{O}_1} F^n(u).$$

Since $F^n = G^n$ on $\mathcal{O} \cap \mathcal{O}_1 \subset \mathcal{O}_1$

$$\limsup_n \inf_{u \in \mathcal{O}} F^n(u) \le \limsup_n \inf_{u \in \mathcal{O} \cap \mathcal{O}_1} G^n(u)$$

and since $\mathcal{O} \cap \mathcal{O}_1$ belongs to $\mathcal{O}_\tau(x)$,

$$\le \sup_{\mathcal{O} \in \mathcal{O}_\tau(x)} \limsup_n \inf_{u \in \mathcal{O}} G^n(u).$$

This being true for any $\mathcal{O} \in \mathcal{O}_\tau(x)$, it follows that

$$\sup_{\mathcal{O} \in \mathcal{O}_\tau(x)} \limsup_n \inf_{u \in \mathcal{O}} F^n(u) \le \sup_{\mathcal{O} \in \mathcal{O}_\tau(x)} \limsup_n \inf_{u \in \mathcal{O}} G^n(u)$$

that is

$$\tau\text{-}\mathrm{ls}_e F^n(x) \le \tau\text{-}\mathrm{ls}_e G^n(x).$$

With F^n and G^n playing symmetric roles, we finally obtain

$$\forall x \in \mathcal{O}_1 \quad \tau\text{-}\mathrm{ls}_e F^n(x) = \tau\text{-}\mathrm{ls}_e G^n(x).$$

The same type of conclusion can be derived in a similar way for the τ-epi-limit inf, instead of the τ-epi-limit sup. □

1.2.2 Epi-convergence in a first countable topological space

Let us now give, assuming (X,τ) is a first countable topological space, the sequential formulations of the epi-limits:

THEOREM 1.13 Let (X,τ) be a first countable topological space and $\{F^n, n \in \mathbb{N}\}$ a sequence of functions from X into $\bar{\mathbb{R}}$. For every $x \in X$, the following sequential formulae hold:

$$\tau\text{-}li_e F^n(x) = \min_{\{x_n \xrightarrow[n]{\tau} x\}} \liminf_{n \to +\infty} F^n(x_n) \tag{1.41}$$

$$\text{i.e.} \begin{cases} \forall x_n \xrightarrow[n]{\tau} x \quad \tau\text{-}li_e F^n(x) \leq \liminf_n F^n(x_n) \\ \exists x_n \xrightarrow[n]{\tau} x \quad \text{such that } \tau\text{-}li_e F^n(x) = \liminf_n F^n(x_n) \end{cases}$$

$$\tau\text{-}ls_e F^n(x) = \min_{\{x_n \xrightarrow[n]{\tau} x\}} \limsup_n F^n(x_n) \tag{1.42}$$

$$\begin{cases} \forall x_n \xrightarrow[n]{\tau} x \quad \tau\text{-}ls_e F^n(x) \leq \limsup_n F^n(x_n) \\ \exists x_n \xrightarrow[n]{\tau} x \quad \text{such that } F(x) = \limsup_n F^n(x_n). \end{cases}$$

Thus, if $\tau\text{-}lm_e F^n(x)$ exists,

$$\begin{aligned} \tau\text{-}lm_e F^n(x) &= \min \{\liminf_n F^n(x_n) \;/\; x_n \xrightarrow{\tau} x\} \\ &= \min \{\limsup_n F^n(x_n) \;/\; x_n \xrightarrow{\tau} x\} \\ &= \min \{\lim_n F^n(x_n) \;/\; x_n \xrightarrow{\tau} x\} \end{aligned} \tag{1.43}$$

Before proving Theorem 1.13, let us state the corresponding sequential formulation of epi-convergence:

<u>PROPOSITION 1.14</u> Let (X,τ) be a first countable topological space and $(F^n)_{n\in\mathbb{N}}$ a sequence of functions from X into $\bar{\mathbb{R}}$. The following statements are equivalent:

(i) $F(x) = \tau\text{-}lm_e F^n(x)$

$\Updownarrow$

(ii) $\begin{cases} \forall x_n \xrightarrow[n]{\tau} x \quad F(x) \leq \liminf_n F^n(x_n) \\ \exists x_n \xrightarrow[n]{\tau} x \quad \text{such that } F(x) \geq \limsup_n F^n(x_n) \end{cases}$

$\Updownarrow$

(iii) $\begin{cases} \forall x_n \xrightarrow{\tau} x \quad F(x) \leq \liminf_n F^n(x_n) \\ \exists x_n \xrightarrow{\tau} x \quad \text{such that } F(x) = \lim_n F^n(x_n). \end{cases}$

Proof of Proposition 1.14 From definition 1.9, $F(x) = \tau\text{-}lm_e F^n(x)$ is equivalent to the two equalities $F(x) = \tau\text{-}li_e F^n(x) = \tau\text{-}ls_e F^n(x)$; since $\tau\text{-}li_e F^n$ is always less or equal than $\tau\text{-}ls_e F^n$, this reduces to the system of the two inequalities

$$\tau\text{-}ls_e F^n(x) \leq F(x) \leq \tau\text{-}li_e F^n(x).$$

From characterizations (1.42) and (1.43) of $\tau\text{-}li_e F^n$ and $\tau\text{-}ls_e F^n$, we obtain formulations (ii) and (iii) of τ-epi-convergence. □

Comments: In the examples described in Section 1,1, we stressed that *epi-convergence is a concept different from pointwise convergence*. Indeed, in general, they are not comparable. This can be easily understood when examining the above sequential formulation.

Epi-convergence is expressed with the help of two sentences: The first one:

$$\text{"}\forall x \in X \quad \exists x_n \xrightarrow{\tau} x \text{ such that } F^n(x_n) \to F(x)\text{"}$$

is a *weaker* requirement than the pointwise convergence where one asks this property to hold with $x_n = x$.

The second one:

$$\text{"}\forall x \in X \quad \forall x_n \xrightarrow{\tau} x,\ F(x) \leq \liminf_n F^n(x_n)\text{"}$$

is clearly a *stronger* assumption than the pointwise convergence since one asks this property to hold for any $x_n \xrightarrow{\tau} x$ and not only for $x_n = x$!

Because of their importance, in Section 2.6.1 we shall discuss the relations between these two concepts and examine some *special situations where they coincide: monotone convergence* for example. This explains the importance of the monotone approximation schemes in optimization theory. (refer to Section 2.5). □

Let us now return to the proof of Theorem 1.13. The proof of the sequential formulae (1.41) and (1.42) relies on the following non-standard diagonalization lemmas. They turn out to be very flexible and will be used in a systematic way throughout this book. As we shall see in Section 2.1, they are related to some closure property of the epi-limit functions.

LEMMA 1.15 Let $\{a_{\nu,\mu} / \nu = 1,2,\ldots/\mu = 1,2,\ldots\}$ be a doubly indexed family in $\bar{\mathbb{R}}$. Then, there exists a mapping $\nu \to \mu(\nu)$ increasing to $+\infty$, such that:

$$\liminf_{\nu \to +\infty} a_{\nu,\mu(\nu)} \geq \liminf_{\mu \to +\infty} (\liminf_{\nu \to +\infty} a_{\nu,\mu}). \tag{1.44}$$

Proof of Lemma 1.15 Let $a_\mu = \liminf_{\nu \to +\infty} a_{\nu,\mu}$ and $a = \liminf_{\mu \to +\infty} a_\mu$. If $a = -\infty$, there is nothing to prove. So, let us assume $a > -\infty$ and take $(a_p)_{p\in\mathbb{N}}$ a sequence of real numbers strictly increasing to a:

If $a < +\infty$, take $a_p = a-2^{-p}$.

If $a = +\infty$, take $a_p = p$.

By definition of a, there exists an increasing sequence $(\mu_p)_{p\in\mathbb{N}}$, $\mu_p \to +\infty$ such that

$$a_\mu \geq a_p \text{ for all } \mu \geq \mu_p.$$

This can be condensed in:

$$a_\mu \geq \inf (a-2^{-p};p) \text{ for all } \mu \geq \mu_p. \tag{1.45}$$

In the same way, there exists an increasing sequence $(\nu_p)_{p\in\mathbb{N}}$, $\nu_p \to +\infty$ such that

$$a_{\nu,\mu_p} \geq \inf (a_{\mu_p} - 2^{-p};p) \text{ for all } \nu \geq \nu_p. \tag{1.46}$$

Set $\mu(\nu) = \mu_p$ if $\nu_p \leq \nu < \nu_{p+1}$ and verify that (1.44) is satisfied: when $\nu_p \leq \nu < \nu_{p+1}$, from (1.45) and (1.46)

$$a_{\nu,\mu(\nu)} \geq \inf (a_{\mu_p} - 2^{-p},p) \geq \inf [\inf(a-2^{-p},p)-2^{-p},p].$$

Thus, for all $\nu \geq \nu_p$

$$a_{\nu,\mu(\nu)} \geq \inf [\inf (a-2^{-p},p)-2^{-p},p].$$

It follows that

$$\liminf_{\nu \to +\infty} a_{\nu,\mu(\nu)} \geq \inf [\inf(a-2^{-p},p)-2^{-p},p].$$

This being true for any $p \in \mathbb{N}$, using the fact that for any $a \in \bar{\mathbb{R}}$,

$$\inf\,[\inf(a-2^{p},p) - 2^{p},p]$$

increases to a as p goes to $+\infty$, we get:

$$\liminf_{\nu \to +\infty} a_{\nu,\mu(\nu)} \geq a = \liminf_{\mu \to +\infty} (\liminf_{\nu \to +\infty} a_{\nu,\mu}). \quad \square$$

COROLLARY 1.16 Let $\{a_{\nu,\mu} / \nu = 1,2\ldots / \mu = 1,2,\ldots\}$ be a doubly indexed family in $\bar{\mathbb{R}}$. Then, there exists a mapping $\nu \to \mu(\nu)$, increasing to $+\infty$, such that:

$$\limsup_{\nu \to +\infty} a_{\nu,\mu(\nu)} \leq \limsup_{\mu \to +\infty} (\limsup_{\nu \to +\infty} a_{\nu,\mu}). \tag{1.47}$$

LEMMA 1.17 Let $\{a_{\nu,\mu} / \nu = 1,2,\ldots / \mu = 1,2,\ldots\}$ be a doubly indexed family in $\bar{\mathbb{R}}$. Then, there exists a mapping $\nu \to \mu(\nu)$ increasing to $+\infty$, such that:

$$\limsup_{\mu \to +\infty} (\liminf_{\nu \to +\infty} a_{\nu,\mu}) \geq \liminf_{\nu \to +\infty} a_{\nu,\mu(\nu)}. \tag{1.48}$$

Proof of Lemma 1.17 The idea of the proof is similar to that of Lemma 1.15. Let us denote $a_{\mu} = \liminf_{\nu \to +\infty} a_{\nu,\mu}$ and $a = \limsup_{\mu \to +\infty} a_{\mu}$. If $a = +\infty$, there is nothing to prove; so, let us assume $a < +\infty$.

By definition of a, there exists an increasing sequence $(\mu_p)_{p\in\mathbb{N}}$, $\mu_p \xrightarrow[p\to+\infty]{} +\infty$ such that:

$$\sup\,(-p,a + 2^{-p}) \geq a_{\mu} \quad \text{for all } \mu \geq \mu_p. \tag{1.49}$$

By definition of a_{μ}, there exists an increasing sequence $(\nu_p)_{p\in\mathbb{N}}$, $\nu_p \to +\infty$ such that:

$$\sup\,(-p,a_{\mu_p} + 2^{-p}) \geq a_{\nu_p,\mu_p} \quad \text{for all } p \in \mathbb{N}. \tag{1.50}$$

Let us define $\mu(\nu) = \mu_p$ if $\nu_p \leq \nu < \nu_{p+1}$. Then,

$$\liminf_{\nu \to +\infty} a_{\nu,\mu(\nu)} \leq \liminf_{p \to +\infty} a_{\nu_p,\mu(\nu_p)} = \liminf_{p \to +\infty} a_{\nu_p,\mu_p},$$

from the definition of $\mu(\nu_p) = \mu_p$. From (1.48) and (1.50), we get

$$\liminf_{\nu \to +\infty} \; a_{\nu,\mu(\nu)} \leq \liminf_{p \to +\infty} [\sup\,(-p, \sup(-p,a+2^{-p}) + 2^{-p})]$$

$\leq a$ which ends the proof of the Lemma 1.17. $\square$

We are now able to give the proof of Theorem 1.13:

Proof of Theorem 1.13

(a) Let us recall that

$$\tau\text{-}li_e F^n(x) = \sup_{V \in N_\tau(x)} \liminf_{n \to +\infty} \inf_{u \in V} F^n(u)$$

and prove that

$$\tau\text{-}li_e F^n(x) = \min_{\{x_n \xrightarrow{\tau} x\}} \liminf_{n \to +\infty} F^n(x_n).$$

One inequality is true in a general topological setting: for any $x_n \xrightarrow{\tau} x$, for any $V \in N_\tau(x)$, $\liminf_n F^n(x_n) \geq \liminf_n \inf_{u \in V} F^n(u)$; thus,

$$\inf_{\{x_n \xrightarrow{\tau} x\}} \liminf_n F^n(x_n) \geq \sup_{V \in N_\tau(x)} \liminf_n \inf_{u \in V} F^n(u). \tag{1.51}$$

In order to prove the opposite inequality, we use that (X,τ) is a first countable topological space: let $(V_k)_{k \in \mathbb{N}}$ be a countable base of (open) neighbourhoods of x, decreasing with k and such that $\bigcap_k V_k = \{x\}$. By definition,

$$\tau\text{-}li_e F^n(x) = \lim_{k \to +\infty} \liminf_{n \to +\infty} \inf_{u \in V_k} F^n(u). \tag{1.52}$$

For each $k \in \mathbb{N}$ and each $n \in \mathbb{N}$, let us choose an element $x_{n,k}$ in X satisfying

$$x_{n,k} \in V_k \text{ and } F^n(x_{n,k}) \leq \sup \{\inf_{u \in V_k} F^n(u) + 2^{-n}, -n\}. \tag{1.53}$$

Let us prove that

$$\liminf_{n \to +\infty} F^n(x_{n,k}) = \liminf_{n \to +\infty} \inf_{u \in V_k} F^n(u). \tag{1.54}$$

Since $x_{n,k}$ belongs to V_k, the left-hand side is larger than the right-hand side in (1.54). Let us prove the opposite inequality:

$$\liminf_{n \to +\infty} F^n(x_{n,k}) \leq \liminf_{n \to +\infty} \inf_{u \in V_k} F^n(u). \tag{1.55}$$

Let us assume $\liminf_{n \to +\infty} F^n(x_{n,k}) > -\infty$; otherwise (1.55) is automatically satisfied. By definition of the lower limit, there exists a real number r_o

and an index n_o such that:

$$\forall n > n_o \quad F^n(x_{n,k}) > r_o.$$

From (1.53), this implies

$$\forall n > n_o \quad r_o < \sup\{\inf_{u\in V_k} F^n(u) + 2^{-n}, -n\}.$$

So,

$$\forall n > \sup\{n_o, -r_o+1\}, \ \sup\{\inf_{u\in V_k} F^n(u)+2^{-n}, -n\} = \inf_{u\in V_k} F^n(u)+2^{-n}$$

and

$$F^n(x_{n,k}) < \inf_{u\in V_k} F^n(u) + 2^{-n}.$$

Taking the lower limit with respect to n, we obtain (1.55).

From (1.52) and (1.54), we obtain:

$$\tau\text{-}li_e F^n(x) = \lim_{k\to+\infty} \liminf_{n\to+\infty} F^n(x_{n,k}), \quad x_{n,k} \in V_k.$$

From diagonalization Lemma 1.17, there exists a mapping $n \to k(n)$ increasing to $+\infty$ such that:

$$\tau\text{-}li_e F^n(x) > \liminf_{n\to+\infty} F^n(x_{n,k(n)}).$$

Taking $x_n = x_{n,k(n)}$, we have that $x_n \in V_{k(n)}$; so $x_n \xrightarrow[n\to+\infty]{\tau} x$ and

$$\tau\text{-}li_e F^n(x) > \liminf_{n\to+\infty} F^n(x_n). \qquad (1.56)$$

From (1.51) and (1.56) we obtain

$$\tau\text{-}li_e F^n(x) = \min_{\{x_n \xrightarrow{\tau} x\}} \liminf_{n\to+\infty} F^n(x_n).$$

(b) Let us now prove that, for every $x \in X$

$$\tau\text{-}ls_e F^n(x) = \min_{\{x_n \xrightarrow{\tau} x\}} \limsup_{n\to+\infty} F^n(x_n).$$

We recall that

$$\tau\text{-}ls_e F^n(x) = \sup_{V\in N_\tau(x)} \limsup_{n\to+\infty} \inf_{u\in V} F^n(u).$$

One inequality is true in a general topological setting: for any -converging sequence $x_n \longrightarrow x$ and any $V \in N_\tau(x)$

$$\limsup_{n \to +\infty} F^n(x_n) \geq \limsup_{n \to +\infty} \inf_{u \in V} F^n(u).$$

This implies

$$\inf_{\{x_n \xrightarrow{\tau} x\}} \limsup_{n \to +\infty} F^n(x_n) \geq \sup_{V \in N_\tau(x)} \limsup_{n \to +\infty} \inf_{u \in V} F^n(u). \quad (1.57)$$

In order to prove the opposite inequality, we use the fact that (X,τ) is a first countable topological space and denote by $(V_k)_{k \in \mathbb{N}}$ a countable base of neighbourhoods of x, decreasing with k and such that $\bigcap_{k \in \mathbb{N}} V_k = \{x\}$.

By definition of $\tau\text{-}ls_e F^n(x)$,

$$\tau\text{-}ls_e F^n(x) = \lim_{k \to +\infty} \limsup_{n \to +\infty} \inf_{u \in V_k} F^n(u). \quad (1.58)$$

For each $k \in \mathbb{N}$ and $n \in \mathbb{N}$, let choose $x_{n,k}$ in V_k satisfying (1.53), and prove that

$$\limsup_{n \to +\infty} F^n(x_{n,k}) = \limsup_{n \to +\infty} \inf_{u \in V_k} F^n(u) \text{ for every } k \in \mathbb{N}. \quad (1.59)$$

Since x_{n_k} belongs to V_k, the right-hand member of (1.59) is less than or equal to the left-hand one. Let us prove the opposite inequality:

$$\limsup_{n \to +\infty} F^n(x_{n,k}) \leq \limsup_{n \to +\infty} \inf_{u \in V_k} F^n(u).$$

If $\limsup_{n \to +\infty} F^n(x_{n,k}) = -\infty$, there is nothing to prove. So, let us assume that this quantity is strictly greater than $-\infty$, i.e. there exist $a > -\infty$ and a subsequence $\{n_p; p = 1,2,\dots\}$ such that:

$$-\infty < \limsup_{n \to +\infty} F^n(x_{n,k}) = a = \lim_{p \to +\infty} F^{n_p}(x_{n_p,k}).$$

There exists $p_o \in \mathbb{N}$ such that

$$\forall p \geq p_o \quad F^{n_p}(x_{n_p,k}) \geq a-1;$$

from definition (1.53) of $x_{n,k}$, this implies that:

$$\forall n_p > \sup\{n_{p_o}, -a+2\} \quad F^{n_p}(x_{n_p,k}) < \inf_{u\in V_k} F^{n_p}(u) + 2^{-n_p}$$

and consequently

$$\limsup_{n\to+\infty} F^n(x_{n,k}) = \lim_{p\to+\infty} F^{n_p}(x_{n_p,k}) < \limsup_{p\to+\infty} \inf_{u\in V_k} F^{n_p}(u)$$

$$< \limsup_{n\to+\infty} \inf_{u\in V_k} F^n(u).$$

So, (1.59) is proved, and combined with (1.58)

$$\tau\text{-}ls_e F^n(x) = \lim_{k\to+\infty} \limsup_{n\to+\infty} F^n(x_{n,k}).$$

By the diagonalization Lemma 1.15 and its Corollary 1.16, there exists a map $n \to k(n)$, increasing to $+\infty$ such that

$$\tau\text{-}ls_e F^n(x) > \limsup_{n\to+\infty} F^n(x_{n,k(n)}).$$

Taking $x_n = x_{n,k(n)}$, noticing that $x_n \in V_{k(n)}$, it follows that $x_n \xrightarrow{\tau} x$ as $n \to +\infty$, and

$$\tau\text{-}ls_e F^n(x) > \limsup_{n\to+\infty} F^n(x_n).$$

Combining (1.57) with the above inequality, we obtain

$$\tau\text{-}ls_e F^n(x) = \min_{\{x_n \xrightarrow{\tau} x\}} \limsup_{n\to+\infty} F^n(x_n). \quad \square$$

Let us give a useful corollary of the diagonalization formula (1.47):

COROLLARY 1.18 Let (X,τ) be a metrizable space and $\{x_{\nu,\mu} / \nu \in \mathbb{N}; \mu \in \mathbb{N}\}$ a double indexed sequence in X such that:

$$x_{\nu,\mu} \xrightarrow[\nu\to+\infty]{\tau} x_\mu$$

and

$$x_\mu \xrightarrow[\mu\to+\infty]{\tau} x.$$

Then, there exists a mapping $\nu \to \mu(\nu)$ increasing to $+\infty$ such that

$$x_{\nu,\mu(\nu)} \xrightarrow[\nu\to+\infty]{\tau} x.$$

Proof of Corollary 1.18 Let us denote by d a distance on X inducing the topology τ, and $a_{\nu,\mu} = d(x_{\nu,\mu},x)$; the double indexed family $\{a_{\nu,\mu}/(\nu,\mu) \in \mathbb{N} \times \mathbb{N}\}$, belongs to $\mathbb{R}$, and from the Corollary 1.16, there exists an increasing map $\nu \to \mu(\nu)$ such that

$$\limsup_{\nu \to +\infty} a_{\nu,\mu(\nu)} \leq \limsup_{\mu \to +\infty} (\limsup_{\nu \to +\infty} a_{\nu,\mu}).$$

By definition of $a_{\nu,\mu}$,

$$\limsup_{\nu \to +\infty} a_{\nu,\mu} = \lim_{\nu \to +\infty} d(x_{\nu,\mu},x) = d(x_{\mu},x)$$

and

$$\limsup_{\mu \to +\infty} (\limsup_{\nu \to +\infty} a_{\nu,\mu}) = \lim_{\mu \to +\infty} d(x_{\mu},x) = 0.$$

So,

$$\limsup_{\nu \to +\infty} d(x_{\nu,\mu(\nu)},x) = 0,$$

which means that

$$x_{\nu,\mu(\nu)} \xrightarrow[\nu \to +\infty]{\tau} x. \quad \square$$

Let us end this section with the following remark:

REMARK 1.19 Let (X,τ) be a general topological space; then

$$\inf_{\{x_n \xrightarrow[n]{\tau} x\}} \liminf_{n} F^n(x_n) = \inf_{\left\{\begin{array}{l}(n_k)_{k \in \mathbb{N}} \\ x_k \xrightarrow{\tau} x\end{array}\right\}} \liminf_{k} F^{n_k}(x_k)$$

which gives us an equivalent formulation of the $\tau\text{-}li_e F^n$ in a first countable topological space. Clearly, the right-hand side is less than or equal to the left-hand one. On the other hand, given a subsequence $(n_k)_{k\in\mathbb{N}}$ and $x_k \xrightarrow{\tau} x$ let us consider the sequence $(y_n)_{n\in\mathbb{N}}$ defined by

$$y_n = \begin{cases} x_k & \text{if for some } k \in \mathbb{N},\ n = n_k \\ x & \text{otherwise.} \end{cases}$$

Then, $y_n \xrightarrow{\tau} x$ and $\liminf_k F^{n_k}(x_k) \geq \liminf_n F^n(y_n)$, which gives the desired inequality

1.2.3 Epigraph-Convergence. A variational notion of convergence

When introducing the concept of epi-convergence, our main motivation was to find a minimal notion of convergence

$$F^n \to F \text{ as } n \text{ goes to } +\infty$$

which allows us to go to the limit on the corresponding minimization problems. We have introduced this concept in a general topological framework. Indeed, in a metrizable setting, it can be very simply formulated (Corollary 1.14) and its properties obtained as straight consequences of the definition.

For example, let us return to the proof of Theorem 1.10:

Let $\{x_n ; n \in \mathbb{N}\}$ a sequence of ε_n-minimizers ($\varepsilon_n \to 0$)

$$F^n(x_n) \leq F^n(y) + \varepsilon_n \text{ for every } y \in X,\ n \in \mathbb{N}$$

and assume that

$$\left| \begin{array}{l} F = \tau\text{-}\lim_e F^n \\ \\ \text{For some subsequence } \{n_k\},\ x_{n_k} \xrightarrow{\tau} x. \end{array} \right.$$

By definition of epi-convergence,

$$\begin{cases} \text{(i) for every } y \in X, \text{ there exists a sequence } y_n \xrightarrow{\tau} y \text{ s.t } F^n(y_n) \to F(y). \\ \\ \text{(ii) for every } x_k \xrightarrow{\tau} x,\ \ F(x) \leq \liminf_k F^{n_k}(x_k). \end{cases}$$

Returning to the definition of x_n,

$$F^{n_k}(x_{n_k}) \leq F^{n_k}(y_{n_k}) + \varepsilon_{n_k} \quad \text{for every } k \in \mathbb{N}, \tag{1.60}$$

property (i) of epi-convergence allows us to go to the limit on the right-hand member of (1.60), while property (ii) allows us to go to the limit on the left-hand member. We thus obtain

$$F(x) \leq F(y) \text{ for every } y \in X, \text{ i.e. } F(x) = \min F.$$

One should notice, by the way, that we have not used all the information contained in the concept of epi-convergence in order to pass to the limit:

more precisely, we have not used part of the sentence (i): "$y_n \xrightarrow{\tau} y$".
So, one might be tempted to introduce a weaker notion of convergence, that is

$$\left.\begin{array}{l} \text{for every } x \in X, \text{ there exists a sequence } \{x_n; n \in \mathbb{N}\} \\ \qquad\qquad \text{such that } F^n(x_n) \to F(x) \\ \text{for every } x_n \xrightarrow{\tau} x, \ \liminf_n F^n(x_n) \geq F(x): \end{array}\right\} \quad (1.61)$$

But there is another property a convergence notion has to satisfy in order to deserve *"variational convergence"* status: The guarantee of convergence of the infimum is really significant only if *this property still holds for a rather large class of perturbations*. For example, in calculus of variations, F^ε usually describes the internal energy of some system, ε is a physical parameter and the sentence

$$F = \lim_{\varepsilon \to 0} F^\varepsilon$$

has a physical meaning only if it does guarantee the convergence of the minimization problems

$$\inf_{u \in X} \{F^\varepsilon(u) - \langle f,u \rangle\} \xrightarrow[\varepsilon \to 0]{} \inf_{u \in X} \{F(u) - \langle f,u \rangle\}$$

for every external action f on the system. In this example, natural perturbations for which this stability property is required are the linear continuous ones.

Epi-convergence does satisfy such a stability property. As a direct consequence of the definition of epi-convergence, one obtains the following implication:

$$F = \tau\text{-}\mathrm{lm}_e F^n \Rightarrow \left| \begin{array}{l} \text{for every } G: X \to \mathbb{R} \ \tau\text{-continuous}, \\ \quad F + G = \tau\text{-}\mathrm{lm}_e(F^n + G). \end{array}\right.$$

So, *epi-convergence is a variational notion of convergence*. But the convergence notion (1.61), which does not satisfy any stability property, is not. We like to stress the importance of this perturbation feature. We could have indeed introduced the epi-convergence $F^n \to F$ as the convergence of

$$\inf_{u \in X} \{F^n(u) + G(u)\} \to \inf_{u \in X} \{F(u) + G(u)\}$$

for a class $\mathfrak{G}$ of perturbations G. This will be made precise in Section 2.7, the Moreau-Yosida approximation being precisely designed in the above way, taking G as quadratic functions of the distance.

Variational properties of the epi-convergence will be discussed in detail in Sections 2.2 and 2.3.

1.3 DIRECT APPROACH BY EPI-CONVERGENCE. PROOF OF EXAMPLES 1.1

1.3.1 Proof of epi-convergence results: various methods

Given a sequence of minimization problems (take $\varepsilon = 1/n$, $n \to +\infty$)

$$\min_{u \in X} F^\varepsilon(u), \ \varepsilon \to 0,$$

we have explained in the preceding section that, since one has obtained some relative compactness τ on a corresponding minimizing sequence $\{u_\varepsilon; \varepsilon \to 0\}$, the limit analysis, as ε goes to zero, reduces to an epi-convergence problem:

$$\tau\text{-}\mathrm{lm}_e F^\varepsilon = F.$$

More precisely, we have to study the existence of the τ-epi limit of the sequence $\{F^\varepsilon, \varepsilon \to 0\}$ and compute (when it exists!) this limit function F. Throughout this book, we shall describe various methods, of independent interest, which allow us to obtain such epi-convergence results.

A. Direct method

1. This is the more natural one. It is described in the present section. It just consists in exhibiting the limit function and verifying the (sequential) definition of epi-convergence. The proof is divided into two parts corresponding to each of the two sentences of Definition - Proposition 1.14.

(a) for every $u \in X$, one has to exhibit a sequence $\{u_\varepsilon; \varepsilon \to 0\}$ τ-converging to u such that $F^\varepsilon(u_\varepsilon) \xrightarrow[\varepsilon \to 0]{} F(u)$.

(b) for every $u \in X$ and every -converging sequence $\{u_\varepsilon; \varepsilon \to 0\}$ τ-converging to u one has to verify that $\liminf_{\varepsilon \to 0} F^\varepsilon(u_\varepsilon) \geq F(u)$.

Noticing that when it exists

$$F(u) = (\tau\text{-}lm_e F^{\varepsilon})(u) = \min \{\lim_{\varepsilon \to 0} F^{\varepsilon}(u_{\varepsilon}) / u_{\varepsilon} \xrightarrow[\varepsilon \to 0]{\tau} u\}$$

the decisive step in the direct proof of the epi-convergence consists in finding, for any $u \in X$, a sequence $\{u_{\varepsilon}; \varepsilon \to 0\}$ τ-converging to u which realizes the above minimum. This is often difficult to achieve for a general u. It can usually be done for a dense class of u, the argument being completed by a density argument. The class of the *piecewise affine continuous functions* will often provide such a dense class (refer to the homogenization example, Section 1.3.2).

2. One of the major advantages of the epi-convergence, from a technical point of view, is its *flexibility:* it can easily be combined with tools or arguments of optimization theory. The simplest one relies on the following *comparison argument:*

Given $\{F^{\varepsilon}, \varepsilon \to 0\}$, $\{G^{\varepsilon}; \varepsilon \to 0\}$ two sequences from (X,τ) into $\bar{\mathbb{R}}$ such that: $F^{\varepsilon}(u) \leq G^{\varepsilon}(u)$ for every $\varepsilon > 0$ and $u \in X$, then

$$\tau\text{-}li_e F^{\varepsilon} \leq \tau\text{-}li_e G^{\varepsilon}$$

$$\tau\text{-}ls_e F^{\varepsilon} \leq \tau\text{-}ls_e G^{\varepsilon},$$

these last inequalities being direct consequences of definition 1.8 of the epi-limits.

This allows us to obtain direct *majorization or minorization of the epi-limits:* In some situations it is interesting (refer to the elasto-plastic torsion problem, 1.6.1) to approach the sequence $\{F^{\varepsilon}; \varepsilon \to 0\}$ by a monotone sequence (increasing or decreasing), $\{F^{\varepsilon}_{\lambda}; \varepsilon > 0, \lambda > 0\}$; for instance, for every $\varepsilon > 0$

$$F^{\varepsilon}_{\lambda} \uparrow F^{\varepsilon} \text{ as } \lambda\text{-decreases to zero.}$$

Taking advantage of the fact that the epi-convergence problems $\{F^{\varepsilon}_{\lambda}; \varepsilon \to 0\}$ may be simpler to study (for example when obtained after penalization of a constraint), we obtain

$$F_{\lambda} = \tau\text{-}lm_e F^{\varepsilon}_{\lambda} \leq \tau\text{-}li_e F^{\varepsilon} \text{ for every } \lambda > 0,$$

and hence

$$\sup_{\lambda > 0} F_{\lambda} \leq \tau\text{-}li_e F^{\varepsilon}.$$

Similarly, a decreasing approximation yields a bound from above. In Chapter 3, in the case of convex functions, we shall explain how, by *duality transform*, one can obtain new epi-convergence results for the conjugate functions so obtained.

3. Another interesting feature of epi-convergence is that, by application of Theorem 1.10, it provides simultaneously the convergence of the solutions and of the corresponding minimum values, that's what we called in Theorems 1.1, 1.2, 1.3, 1.6, 1.7 *"convergence of the energies"*.

4. Let us mention, finally, that, when attacking a given problem by the direct epi-convergence method, the big difficulty is to exhibit the functional candidate to be the epi-limit and guess the possible various situations and corresponding critical parameters. This can be done usually by heuristic arguments relying on asymptotic developments (cf. homogenization, singular perturbation problems,...), and computation of apriori estimates on the solutions. These considerations play in turn a decisive role in the development of the epi-convergence proof.

So, we can summarize these considerations by saying that epi-convergence fits well with other mathematical analyses of approaches to a limit such as asymptotic developments and energy methods. In several situations, it is the joint utilization of all these techniques which finally provides a successful treatment (refer to homogenization of fissured elastic bodies 1.3.3, Neumann strainer 1.3.5, elasto plastic torsion 1.6.1).

B. Compactness method

This method, which will be developed in Section 2.4, relies on compactness theorems in the epi-convergence sense for classes of functionals: They are usually obtained by axiomatization of the properties of functions of the studied class $\mathcal{F}$, which are preserved by epi-convergence, such as local properties, homogeneity, invariance properties, increasing or decreasing properties,...

The proof is then completed by a representation theorem. Following the works of Serrin [1], [2], De Giorgi & Spagnolo [1], it is a major contribution of De Giorgi [1], [3], then of Carbone & Sbordone [2], Buttazzo & Dal Maso [2], to show how, in calculus of variations, such a programme can be realized by marrying epi-convergence and measure theory: functionals F are considered not only depending on u but also on the set of observations Ω:

$$F(u,\Omega) = \int_\Omega j(x, \text{grad } u(x),\dots,)dx.$$

Given a sequence of functions $\{F^\varepsilon; \varepsilon \to 0\}$ of the class $\mathcal{F}$ under consideration, epi-convergence follows then by identification of the limit function. Let us finally notice that compactness results have nice applications, too, in optimal control theory and stochastic optimization (cf. Section 2.8.3). □

1.3.2 Homogenization of elliptic operators

The following theorem describes the epi-convergence approach to Example 1.1.1 (Theorem 1.1).

THEOREM 1.20 Let Ω be a regular bounded open set in $\mathbb{R}^N$ and $\{F^\varepsilon; \varepsilon > 0\}$ the sequence of convex continuous functionals from $H^1(\Omega)$ into $\mathbb{R}^+$ defined by

$$F^\varepsilon(u) = \int_\Omega j(\frac{x}{\varepsilon}, \text{grad } u(x))dx$$

where

$$j:\mathbb{R}^N \times \mathbb{R}^N \to \mathbb{R}^+$$

$(y,z) \to j(y,z)$ is Y-periodic in y, convex continuous with respect to z and satisfies $(0 < \lambda_o \leq \Lambda_o < +\infty)$

$$\lambda_o|z|^2 \leq j(y,z) \leq \Lambda_o(1 + |z|^2)$$

for a.e. $y \in \mathbb{R}^N$, for any $z \in \mathbb{R}^N$.

Then, for every $u \in H^1(\Omega)$, taking τ equal to the strong topology of $L^2(\Omega)$

$$\tau\text{-}\mathrm{lm}_e F^\varepsilon(u) = F^{hom}(u) \text{ exists}$$

where

$$F^{hom}(u) = \int_\Omega j^{hom}(\text{grad } u(x))dx,$$

and

$$\text{for every } z \in \mathbb{R}^N, \quad j^{hom}(z) = \min_{\{w \text{ Y-periodic}\}} \int_Y j(y,\text{grad } w(y) + z)dy. \tag{1.62}$$

Proof of Theorem 1.20 For any $z \in \mathbb{R}^N$, let us denote by w_z an element in $H^1(Y)$ minimizing (1.62):

$$\left.\begin{aligned} &-\operatorname{div} \partial j(y, \operatorname{grad} w_z(y) + z) = 0 \text{ on } Y \\ &w_z \quad Y\text{-periodic} \end{aligned}\right\} \tag{1.63}$$

where ∂j denotes the subdifferential of the convex continuous function j (when j is differentiable, $\partial j = \operatorname{grad} j$). Then,

$$j^{hom}(z) = \int_Y j(y, \operatorname{grad} w_z(y) + z)dy.$$

1. Let us first prove the inequality $F^{hom} \geq \tau\text{-}ls_e F^\varepsilon$.

Equivalently, let us prove that, for every u belonging to $H^1(\Omega)$, there exists a sequence $\{u_\varepsilon ; \varepsilon > 0\}$ satisfying

$$u_\varepsilon \to u \text{ in } s\text{-}L^2(\Omega), \text{ as } \varepsilon \text{ goes to zero.}$$

$$F^{hom}(u) \geq \limsup_{\varepsilon \to 0} F^\varepsilon(u_\varepsilon) \text{ as } \varepsilon \text{ goes to zero.}$$

A first approach would consist to take

$$u_\varepsilon(x) = u(x) + \varepsilon w_{\operatorname{grad}u(x)}(\tfrac{x}{\varepsilon}). \tag{1.64}$$

If u is not very smooth (at least C^1), take for example u a continuous piecewise affine function ($\operatorname{grad} u(x) = z_1$ on Ω_1, $\operatorname{grad} u(x) = z_2$ on Ω_2 with $z_1 \neq z_2$), u_ε does not belong to $H^1(\Omega)$! So, as a first step, one has to assume u regular in order to be able to derive regularity properties of the function $x \to w_{\operatorname{grad}u(x)}(y)$: that is a difficult point, because of the non-linear way w_z depends on z (refer to 1.63). Then, denoting $w_1(x,y) = w_{\operatorname{grad}u(x)}(y)$ and noticing that

$$\operatorname{grad}u_\varepsilon(x) = \operatorname{grad}u(x) + \operatorname{grad}_y w_1(x,\tfrac{x}{\varepsilon}) + \varepsilon \operatorname{grad}_x w_1(x,\tfrac{x}{\varepsilon})$$

a formal computation yields

$$\int_\Omega j(\tfrac{x}{\varepsilon}, \operatorname{grad} u_\varepsilon(x))dx \simeq \int_\Omega j(\tfrac{x}{\varepsilon}, \operatorname{grad}u(x) + \operatorname{grad}_y w_{\operatorname{grad}u(x)}(\tfrac{x}{\varepsilon}))dx$$

$$\underset{(\varepsilon \to 0)}{\longrightarrow} \int_\Omega dx \int_Y j(y, \operatorname{grad}u(x) + \operatorname{grad}_y w_{\operatorname{grad}u(x)}(y))dy$$

$$= \int_\Omega j^{hom}(\text{grad } u(x))dx.$$

Then, one would complete this proof by a density argument. But it is very technical indeed to make this argument rigorous. So we overcome the difficulty and proceed by steps using the density in $H^1(\Omega)$ of *the piecewise affine continuous functions* and taking advantage of the local character of the functionals F^ε.

Step one Take first u an affine function:

$$u(x) = \langle z,x\rangle + \alpha, \quad z \in \mathbb{R}^N, \quad \alpha \in \mathbb{R}.$$

Then, grad $u(x) \equiv z$. The above argument can now be made rigorous: the sequence $\{u_\varepsilon; \varepsilon > 0\}$ defined by (1.64)

$$u_\varepsilon(x) = u(x) + \varepsilon w_z(\frac{x}{\varepsilon})$$

converges to u in s-$L^2(\Omega)$, as ε goes to zero, and satisfies:

$$F^\varepsilon(u_\varepsilon) = \int_\Omega j(\frac{x}{\varepsilon}, \text{grad } u_\varepsilon(x))dx$$

$$= \int_\Omega j(\frac{x}{\varepsilon}, z + \text{grad } w_z(\frac{x}{\varepsilon}))dx$$

$$\xrightarrow[\varepsilon\to 0]{} \int_\Omega dx \int_Y j(y,z + \text{grad } w_z(y))dy.$$

From the definition (1.62) and (1.63) of j^{hom} and w_z,

$$\lim_{\varepsilon\to 0} F^\varepsilon(u_\varepsilon) = \int_\Omega j^{hom}(z)dx$$

$$= \int_\Omega j^{hom}(\text{grad } u(x))dx.$$

Step two Then, use the local character of the functionals $\{F^\varepsilon; \varepsilon > 0\}$ and take u a piecewise continuous affine function:

$$u(x) = u^i(x) = \langle z_i,x\rangle + \alpha_i \text{ on } \Omega_i, \ i \in I \text{ finite},$$

where the Ω_i form a partition of by polyhedral sets. On each Ω_i, take

$$u^i_\varepsilon(x) = u(x) + \varepsilon w_{z_i}(\frac{x}{\varepsilon}) \text{ as given in step one.}$$

The problem, as already mentioned, is that one cannot take u_ε to be equal to u_ε^i on each Ω_i: on each interface Σ_{ij} between Ω_i and Ω_j this would result in a jump, due to the fact that $w_{z_i} \neq w_{z_j}$. Taking advantage of the fact that this jump is of size ε, one can overcome this difficulty, recollecting u_ε^i and u_ε^j in a smooth way. Let us make this precise and for simplicity of notation take i = 1,2. The argument can be easily extended to any finite number of sets Ω_i. Let Σ be the interface between Ω_1 and Ω_2, and $\Sigma_\delta = \{x \in \Omega / \text{dist}\,(x,\Sigma) \leq \delta\}$, $\delta > 0$ (Fig. 1.5). Introducing $\phi_\delta \in W^{1,\infty}(\Omega)$, $0 \leq \phi_\delta \leq 1$

$$\phi_\delta = \begin{cases} 1 \text{ on } \Sigma_\delta \\ 0 \text{ on } \Omega \setminus \Sigma_{2\delta} \end{cases}$$

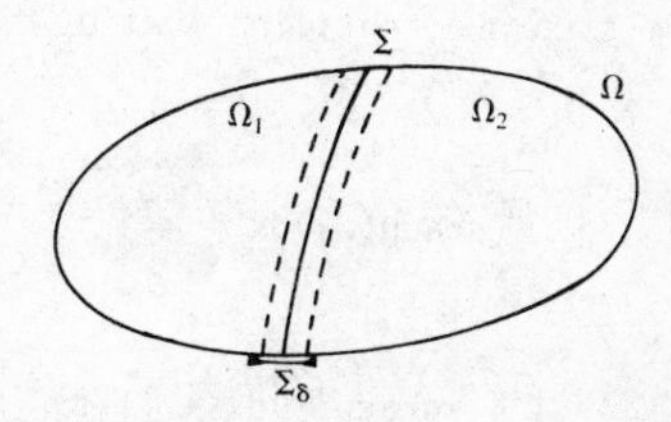

Figure 1.5

let us define

$$u_\varepsilon^\delta = \begin{cases} (1-\phi_\delta)u_\varepsilon^1 + \phi_\delta u \text{ on } \Omega_1 \\ (1-\phi_\delta)u_\varepsilon^2 + \phi_\delta u \text{ on } \Omega_2 \end{cases} \tag{1.65}$$

The function u_ε^δ so defined belongs to $H^1(\Omega)$ since, on Σ_δ,

$$(1-\phi_\delta)u_\varepsilon^1 + \phi_\delta u = (1-\phi_\delta)u_\varepsilon^2 + \phi_\delta u = u!$$

Given $0 < t < 1$,

$$F^\varepsilon(tu_\varepsilon^\delta) = \sum_i \int_{\Omega_i} j(\frac{x}{\varepsilon}, t(1-\phi_\delta)\text{grad}u_\varepsilon^i + t\phi_\delta \text{grad}u + (1-t)\frac{t}{1-t}(u-u_\varepsilon^i)\text{grad}\phi_\delta)dx.$$

From the convexity of j, noticing that $t(1-\phi_\delta) + t\phi_\delta + (1-t) = 1$,

$$F^\varepsilon(tu_\varepsilon^\delta) \leq \sum_i \int_{\Omega_i} j(\frac{x}{\varepsilon}, \text{grad}u_\varepsilon^i)dx + \int_{\Sigma_{2\delta}} j(\frac{x}{\varepsilon}, \text{grad}u)dx$$

$$+ \sum_i (1-t) \int_{\Omega_i} j(\frac{x}{\varepsilon}, \frac{t}{1-t}(u-u_\varepsilon^i)\text{grad}\,\phi_\delta)dx.$$

Using the majorization from above on j, $j(y,z) \leq \Lambda_o(1 + |z|^2)$,

$$F^\varepsilon(tu_\varepsilon^\delta) \leq \sum_i \int_{\Omega_i} j(\frac{x}{\varepsilon}, \text{grad}u_\varepsilon^i)dx + \int_{\Sigma_{2\delta}} \Lambda_o(1 + |\text{grad}u|^2)dx$$

$$+ \sum_i (1-t) \int_{\Omega_i} \Lambda_o(1 + (\frac{t}{1-t})^2(u-u_\varepsilon^i)^2|\text{grad}\phi_\delta|^2)dx.$$

Now, let ε go to zero. From step one, noticing that $u_\varepsilon^i \to u$ in $L^2(\Omega_i)$ as ε goes to zero, we obtain

$$\limsup_{\varepsilon\to 0} F^\varepsilon(tu_\varepsilon^\delta) \leq \sum_i \int_{\Omega_i} j^{hom}(\text{grad}u(x))dx$$

$$+ \int_{\Sigma_{2\delta}} \Lambda_o(1 + |\text{grad}u|^2)dx + (1-t)\Lambda_o \text{ meas}(\Omega).$$

Letting δ tend to zero and t to one, one obtains

$$\limsup_{\substack{\delta\to 0\\ t\to 1}} \limsup_{\varepsilon\to 0} F^\varepsilon(tu_\varepsilon^\delta) \leq F^{hom}(u).$$

From diagonalization result (1.47) (refer to Corollary 1.16), there exists a map $\varepsilon \to (\delta(\varepsilon), t(\varepsilon))$ such that

$$\lim_{\varepsilon\to 0} \delta(\varepsilon) = 0$$

$$\lim_{\varepsilon\to 0} t(\varepsilon) = 1$$

and

$$\limsup_{\varepsilon\to 0} F^\varepsilon(t(\varepsilon))u_\varepsilon^{\delta(\varepsilon)}) \leq \limsup_{\substack{t\to 1\\ \delta\to 0}} \limsup_{\varepsilon\to 0} F^\varepsilon(tu_\varepsilon^\delta).$$

Taking $u_\varepsilon = t(\varepsilon)u_\varepsilon^{\delta(\varepsilon)}$ and noticing that u_ε converges to u in $L^2(\Omega)$ as ε goes to zero, we finally obtain

$$\limsup_{\varepsilon \to 0} F^\varepsilon(u_\varepsilon) \leq F^{hom}(u).$$

Step three The proof is now completed by a *density argument*. First, let us notice that F^{hom} is a convex continuous function on $H^1(\Omega)$. The convexity of j^{hom} is immediate: take $z_1, z_2 \in \mathbb{R}^N$

$$j^{hom}(z_i) = \int_Y j(y, \text{grad } w_{z_i}(y) + z_i)dy \quad i = 1,2.$$

Hence,

$$\frac{1}{2}(j^{hom}(z_1) + j^{hom}(z_2)) = \frac{1}{2}\int_Y \{j(y, \text{grad } w_{z_1} + z_1)$$

$$+ j(y, \text{grad } w_{z_2} + z_2)\}dy$$

$$\geq \int_Y j(y, \text{grad } (\frac{w_{z_1} + w_{z_2}}{2}) + \frac{z_1 + z_2}{2})dy.$$

Noticing that $\frac{1}{2}(w_{z_1} + w_{z_2})$ is still Y-periodic

$$\geq j^{hom}(\frac{z_1 + z_2}{2}).$$

Noticing that

$$j^{hom}(z) \leq \int_Y j(y,z)dy \qquad \text{(take } w = 0 \text{ in (1.62))}$$

$$\leq \Lambda_o(1 + |z|^2)$$

it follows that F^{hom} is a convex finitely valued function on $H^1(\Omega)$ and hence continuous.

Given u belonging to $H^1(\Omega)$, let $\{u_k; k \in \mathbb{N}\}$ be a sequence of piecewise affine continuous functions such that

$$u_k \xrightarrow[k\to+\infty]{} u$$

in the strong topology of $H^1(\Omega)$, (refer to Ekeland & Teman [1]).

By continuity of F^{hom} on $H^1(\Omega)$

$$F^{hom}(u_k) \xrightarrow[k\to+\infty]{} F^{hom}(u).$$

From step two, for every $k \in \mathbb{N}$, there exists a sequence $\{u_{\varepsilon,k};\ \varepsilon \to 0\}$ such that

$$u_{\varepsilon,k} \xrightarrow[\varepsilon\to 0]{} u_k \text{ in s-}L^2(\Omega)$$

$$F^{hom}(u_k) \geq \limsup_{\varepsilon\to 0} F^{\varepsilon}(u_{\varepsilon,k}).$$

Thus,

$$\begin{cases} F^{hom}(u) \geq \limsup\limits_{k\to+\infty}\ \limsup\limits_{\varepsilon\to 0} F^{\varepsilon}(u_{\varepsilon,k}) \\ 0 \qquad\quad \geq \limsup\limits_{k\to+\infty}\ \limsup\limits_{\varepsilon\to 0} |u_{\varepsilon,k}-u|_{L^2(\Omega)} . \end{cases}$$

Applying diagonalization formula (1.16) to the above situation, we derive existence of a strictly increasing mapping $\varepsilon \to k(\varepsilon)$ such that, denoting $u_{\varepsilon} = u_{\varepsilon,k(\varepsilon)}$, the sequence $\{u_{\varepsilon};\ \varepsilon \to 0\}$ tends to u in s-$L^2(\Omega)$ and

$$F^{hom}(u) \geq \limsup_{\varepsilon\to 0} F^{\varepsilon}(u_{\varepsilon}). \qquad \square$$

2. Let us now prove the inequality $\tau\text{-li}_e F^{\varepsilon} \geq F^{hom}$.

Equivalently, we have to prove that, for every sequence $\{u_{\varepsilon};\ \varepsilon \to 0\}$ which strongly converges in $L^2(\Omega)$ to some u

$$\liminf_{\varepsilon\to 0} F^{\varepsilon}(u_{\varepsilon}) \geq F^{hom}(u).$$

The idea of the proof is to minorize, via a convex subdifferential inequality, $F^{\varepsilon}(u_{\varepsilon})$ by some $F^{\varepsilon}(v_{\varepsilon})$, which, from the preceding step, is known to be converging to $F^{hom}(v)$. Once more, we proceed by steps, taking v first as a piecewise affine function, then completing the proof by a density argument: one lets v tend to u, in order to kill the correcting terms.

For any such piecewise affine continuous function v, following (1.65), let us denote, on each Ω_i (grad $v \equiv z_i$ on Ω_i), $\{v_{\varepsilon}^i;\ \varepsilon > 0\}$, the approximating sequence

$$v_\varepsilon^i(x) = v(x) + \varepsilon w_{z_i}(\frac{x}{\varepsilon}),\ x \in \Omega_i.$$

For each Ω_i, let us introduce $\phi_i \in D(\Omega_i)$ (indefinitely differentiable with compact support) such that $0 \leqslant \phi_i \leqslant 1$ (then, we shall let ϕ_i tend to one). Let us write the subdifferential inequality

$$j(\frac{x}{\varepsilon}, \mathrm{grad} u_\varepsilon(x))\phi_i(x) \geqslant j(\frac{x}{\varepsilon}, \mathrm{grad} v_\varepsilon(x))\phi_i(x)$$
$$+ \langle \partial j(\frac{x}{\varepsilon}, \mathrm{grad} v_\varepsilon(x)), \mathrm{grad} u_\varepsilon - \mathrm{grad} v_\varepsilon \rangle \cdot \phi_i(x).$$

Integrating over Ω_i and summing over i, we obtain

$$\int_\Omega j(\frac{x}{\varepsilon}, \mathrm{grad}\ u_\varepsilon)dx \geqslant \sum_i \int_{\Omega_i} j(\frac{x}{\varepsilon}, \mathrm{grad}\ u_\varepsilon(x))\phi_i(x)dx$$
$$\geqslant \sum_i \int_{\Omega_i} j(\frac{x}{\varepsilon}, \mathrm{grad} v_\varepsilon^i)\phi_i(x)dx + \sum_i \int_{\Omega_i} \langle \partial j(\frac{x}{\varepsilon}, \mathrm{grad} v_\varepsilon^i), \mathrm{grad}(u_\varepsilon - v_\varepsilon^i)\rangle \phi_i dx. \quad (1.66)$$

The same computation as before yields

$$\lim_{\varepsilon \to 0} \sum_i \int_{\Omega_i} j(\frac{x}{\varepsilon}, \mathrm{grad}\ v_\varepsilon^i)\phi_i dx = \sum_i \int_{\Omega_i} j^{hom}(\mathrm{grad}\ v(x))\phi_i(x)dx. \quad (1.67)$$

Integrating by parts and using the fact that $-\mathrm{div}\ \partial j(\frac{x}{\varepsilon}, \mathrm{grad}\ v_\varepsilon^i) = 0$ on Ω_i,

$$\sum_i \int_{\Omega_i} \langle \partial j(\frac{x}{\varepsilon}, \mathrm{grad}\ v_\varepsilon^i), \mathrm{grad}(u_\varepsilon - v_\varepsilon^i)\rangle \phi_i dx$$
$$= - \sum_i \int_{\Omega_i} \langle \partial j(\frac{x}{\varepsilon}, \mathrm{grad}\ v_\varepsilon^i), (u_\varepsilon - v_\varepsilon^i)\mathrm{grad}\phi_i \rangle dx. \quad (1.68)$$

Going to the limit in (1.66) and using (1.67) and (1.68),

$$\liminf_{\varepsilon \to 0} F^\varepsilon(u_\varepsilon) \geqslant \sum_i \int_{\Omega_i} j^{hom}(\mathrm{grad} v(x))\phi_i(x)dx - \sum_i \int_{\Omega_i} \langle \partial j^{hom}(\mathrm{grad} v), u-v \rangle \mathrm{grad}\,\phi_i. \quad (1.69)$$

We have used the fact that

$$\partial j(\frac{x}{\varepsilon}, z_i + \mathrm{grad} w_{z_i}(\frac{x}{\varepsilon})) \longrightarrow \mathfrak{m}_Y\ \partial j(\cdot, z_i + \mathrm{grad} w_{z_i}(\cdot)) \text{ in weak-}L^2(\Omega)$$

and

$$\partial j^{hom}(z) = \mathbb{m}_Y \, \partial j(\cdot, z + \text{grad } w_z(\cdot))$$

(we shall comment on this equality at the end of this paragraph, Proposition 1.21). Noticing that $\partial j^{hom}(\text{grad } v) \equiv \partial j^{hom}(z_i)$ on Ω_i, it follows that

$$- \sum_i \int_{\Omega_i} \langle \partial j^{hom}(\text{grad}v), u-v \rangle \text{grad}\phi_i dx = \sum_i \int_{\Omega_i} \langle \partial j^{hom}(\text{grad}v), \text{grad}(u-v) \rangle \phi_i dx.$$

Returning to (1.69), we obtain

$$\liminf_{\varepsilon \to 0} F^{\varepsilon}(u_{\varepsilon}) \geq \sum_i \int_{\Omega_i} j^{hom}(\text{grad}v)\phi_i dx + \sum_i \int_{\Omega_i} \langle \partial j^{hom}(\text{grad}v), \text{grad}(u-v) \rangle \phi_i dx. \tag{1.70}$$

The above inequality is true for any $\phi_i \in D(\Omega_i)$, $0 \leq \phi_i \leq 1$. Letting ϕ_i converge increasingly to one on Ω_i, from the Beppo-Levy theorem the sequence

$$\sum_i \int_{\Omega_i} j^{hom}(\text{grad}v)\phi_i \, dx$$

converges increasingly to

$$\int_{\Omega} j^{hom}(\text{grad } v)dx.$$

On the other hand, as already noticed,

$$0 \leq j^{hom}(z) \leq \Lambda_o(1 + |z|^2).$$

This majorization, combined with the convexity of j^{hom}, yields:

$$\forall z \in \mathbb{R}^N \quad |\partial j^{hom}(z)| \leq M(1 + |z|), \text{ for some } M \in \mathbb{R}^+. \tag{1.71}$$

In order to obtain the above inequality, just write the subdifferential inequality

$$\forall v \in B(0,2|z|) \quad j^{hom}(v) \geq j^{hom}(z) + \langle \partial j^{hom}(z), v-z \rangle;$$

thus

$$\Lambda_o(1 + 4|z|^2) \geq \sup_{|\xi| \leq |z|} \langle \partial j^{hom}(z), \xi \rangle = |\partial j^{hom}(z)| \cdot |z|.$$

Hence, $|\partial j^{hom}(z)| \leq \Lambda_o(1 + 4|z|)$ for $|z| \geq 1$.

Since u and v belong to $H^1(\Omega)$, using (1.71) and the Lebesgue dominated convergence theorem, we obtain

$$\sum_i \int_{\Omega_i} \langle \partial j^{hom}(\text{grad}v), \text{grad}(u-v)\rangle \phi_i dx \to \int_\Omega \langle \partial j^{hom}(\text{grad}v), \text{grad}(u-v)\rangle dx.$$

So, for every v piecewise affine continuous function,

$$\liminf_{\varepsilon \to 0} F^\varepsilon(u_\varepsilon) \geq \int_\Omega j^{hom}(\text{grad}v)dx + \int_\Omega \langle \partial j^{hom}(\text{grad}v), \text{grad}(u-v)\rangle dx. \quad (1.72)$$

Using once more the density of the piecewise affine continuous functions in $H^1(\Omega)$, let v converge to u in $H^1(\Omega)$. From the continuity of F^{hom} on $H^1(\Omega)$ and the majorization (1.71) on $|\partial j^{hom}(z)|$, we finally get

$$\liminf_{\varepsilon \to 0} F^\varepsilon(u_\varepsilon) \geq F^{hom}(u). \quad \square \quad (1.73)$$

PROPOSITION 1.21 The function $j^{hom}: \mathbb{R}^N \to \mathbb{R}^+$ is a convex continuous function satisfying, like j, the following minorization and majorization:

$$\lambda_0 |z|^2 \leq j^{hom}(z) \leq \Lambda_0(1 + |z|^2) \text{ for every } z \in \mathbb{R}^N. \quad (1.74)$$

Its subdifferential $A^{hom} = \partial j^{hom}$ is the maximal monotone operator (everywhere defined) from $\mathbb{R}^N$ into $\mathbb{R}^N$ given by

$$A^{hom}(z) = \mathfrak{m}_Y\, \partial j(\cdot, z + \text{grad}w_z(\cdot)). \quad (1.75)$$

When j is quadratic,

$$j(y,z) = \frac{1}{2} \sum_{i,j=1}^N a_{ij}(y) z_i z_j$$

so is j^{hom}

$$j^{hom}(z) = \frac{1}{2} \sum_{i,j=1}^N a_{ij}^{hom} z_i z_j,$$

and

$$a_{ij}^{hom} = \mathfrak{m}_Y(a_{ij}) - \mathfrak{m}_Y\left(\sum_{k=1}^N a_{ik} \frac{\partial \chi^j}{\partial y_k}\right). \quad (1.76)$$

where χ^j is the solution of

$$\begin{cases} A\chi^j = Ay^j \\ \chi^j \text{ Y-periodic} \end{cases}$$

and

$$Av = -\sum_{i,j} \frac{\partial}{\partial y_i}\left(a_{ij}(y)\frac{\partial v}{\partial y_j}\right).$$

Proof of Proposition 1.21 We have already verified that j^{hom} is a convex continuous function from $\mathbb{R}^N$ into $\mathbb{R}^+$, and

$$0 \leq j^{hom}(z) \leq \Lambda_o(1 + |z|^2) \text{ for every } z \in \mathbb{R}^N.$$

On the other hand, for every Y-periodic function w,

$$\int_Y j(y,\text{grad}w(y) + z)dy \geq \lambda_o \int_Y \{|\text{grad}w|^2 + 2\langle z,\text{grad}w(y)\rangle + |z|^2\}dy$$

$$\geq \lambda_o|z|^2.$$

Hence, $j^{hom}(z) \geq \lambda_o|z|^2$ for any $z \in \mathbb{R}^N$.

Let us now verify that $\partial j^{hom} = A^{hom}$ where A^{hom} is given by (1.75). First, let us prove that A^{hom} is a maximal monotone operator: for every z_1 and $z_2 \in \mathbb{R}^N$,

$$\langle A^{hom}(z_1) - A^{hom}(z_2), z_1 - z_2\rangle$$

$$= \mathfrak{m}_Y \langle \partial j(y,z_1+\text{grad}w_{z_1}(y)-\partial j(y,z_2+\text{grad}w_{z_2}(y),z_1-z_2\rangle$$

$$= \mathfrak{m}_Y \langle \partial j(y,z_1+\text{grad}w_{z_1}-\partial j(y,z_2+\text{grad}w_{z_2}),$$
$$(z_1+\text{grad}w_{z_1}) - (z_2+\text{grad}w_{z_2})\rangle$$

$$+ \mathfrak{m}_Y \langle \partial j(y,z_1+\text{grad}w_{z_1})-\partial j(y,z_2+\text{grad}w_{z_2}),\text{grad}w_{z_2}-\text{grad}w_{z_1}\rangle.$$

From the monotonicity of ∂j

$$\langle A^{hom}(z_1)-A^{hom}(z_2),z_1-z_2\rangle \geq \int_Y \langle \partial j(y,z_1+\text{grad}w_{z_1})-\partial j(y,z_2-\text{grad}w_{z_2}),$$
$$\text{grad}(w_{z_2}-w_{z_1})\rangle . \qquad (1.77)$$

From the definition of w_z (1.63), $-\text{div}\, \partial j(y,\text{grad}w_z(y)+z) = 0$ and w_z is a

Y-periodic function. Integrating by parts the second member of (1.77), we obtain

$$\forall z_1, z_2 \in \mathbb{R}^N \quad \langle A^{hom}(z_1) - A^{hom}(z_2), z_1 - z_2 \rangle \geq 0.$$

The maximality of A^{hom} follows from its continuity properties (refer to Brezis [1] for a complete survey on maximal monotone operators): for simplicity, let us assume ∂j to be strongly coercive and Lipschitz, i.e.

$$\begin{aligned} \forall z_1, z_2 \in \mathbb{R}^N \quad & \langle \partial j(z_1) - j(z_2), z_1 - z_2 \rangle \geq \lambda_o |z_1 - z_2|^2 \\ & |\partial j(z_1) - j(z_2)| \leq \Lambda_o |z_1 - z_2|. \end{aligned} \tag{1.78}$$

Noticing that

$$A^{hom}(z) = \lim_{\varepsilon \to 0} \partial j(\frac{\cdot}{\varepsilon}, z + \operatorname{grad} w_z(\frac{\cdot}{\varepsilon}))$$

in the weak topology of $L^2(\Omega)$ and,

$$\begin{aligned} &\int_Y |\partial j(\frac{x}{\varepsilon}, z_1 + \operatorname{grad} w_{z_1}(\frac{x}{\varepsilon})) - \partial j(\frac{x}{\varepsilon}, z_2 + \operatorname{grad} w_{z_2}(\frac{x}{\varepsilon})|^2 dx \\ &\leq \frac{\Lambda_o^2}{\lambda_o} \int_Y \langle \partial j(\frac{x}{\varepsilon}, z_1 + \operatorname{grad} w_{z_1}(\frac{x}{\varepsilon})) - \partial j(\frac{x}{\varepsilon}, z_2 + \operatorname{grad} w_{z_2}(\frac{x}{\varepsilon})), \\ &\qquad\qquad z_1 - z_2 + \operatorname{grad}(w_{z_1} - w_{z_2})(\frac{x}{\varepsilon}) \rangle dx \\ &\leq \frac{\Lambda_o^2}{\lambda_o} \int_Y \langle \partial j(\frac{x}{\varepsilon}, z_1 + \operatorname{grad} w_{z_1}(\frac{x}{\varepsilon})) - \partial j(\frac{x}{\varepsilon}, z_2 + \operatorname{grad} w_{z_2}(\frac{x}{\varepsilon})), z_1 - z_2 \rangle dx \end{aligned}$$

from the lower semicontinuity for the weak topology of $L^2(Y)$ of the $L^2(Y)$-norm, it follows

$$|A^{hom}(z_1) - A^{hom}(z_2)|^2 \leq \frac{\Lambda_o^2}{\lambda_o} \langle A^{hom}(z_1) - A^{hom}(z_2), z_1 - z_2 \rangle.$$

Thus,

$$|A^{hom}(z_1) - A^{hom}(z_2)| \leq \frac{\Lambda_o^2}{\lambda_o} \; |z_1 - z_2|.$$

We now verify that $A^{hom} \subset \partial j^{hom}$. From the maximal monotony of A^{hom} the equality follows. So, let us verify that

$$\forall \xi \in \mathbb{R}^N, \quad \forall z \in \mathbb{R}^N \quad j^{hom}(\xi) \geq j^{hom}(z) + \langle A^{hom}(z), \xi - z \rangle. \tag{1.79}$$

From the convex subdifferential inequality

$$j(y, \text{grad}w_\xi(y)+\xi) \geq j(y, \text{grad}w_z(y)+z)$$

$$+ \langle \partial j(y, \text{grad}w_z(y)+z), \text{grad}(w_\xi - w_z)(y) + \xi - z \rangle$$

integrating on Y, by the same type of argument as above, we obtain

$$j^{hom}(\xi) \geq j^{hom}(z) + \langle \mathfrak{m}_Y(\partial j(y, \text{grad}w_z(y)+z), \xi - z \rangle,$$

that is (1.79).

Let us now examine the quadratic case. First, we can observe that the quadratic property $F^\varepsilon(\lambda u) = \lambda^2 F^\varepsilon(u)$ is preserved by the epi-limit process. Thus, j^{hom} is a quadratic, convex, positive, i.e.

$$\forall z \in \mathbb{R}^N \quad j^{hom}(z) = \frac{1}{2} \sum_{i,j} a_{ij}^{hom} z_i z_j$$

with

$$a_{ij}^{hom} = a_{ji}^{hom},$$

and

$$A^{hom}(z)_i = \sum_j a_{ij}^{hom} z_j.$$

On the other hand, from (1.75)

$$A^{hom}(z)_i = \mathfrak{m}_Y \left(\sum_k a_{ik}(y) (z + \text{grad}w_z(y))_k\right).$$

Comparing these two last formulae, and taking $z = e_j = (\delta_{kj})_{k=1,\ldots,N}$, we obtain

$$a_{ij}^{hom} = \mathfrak{m}_Y(a_{ij}) - \mathfrak{m}_Y\left(\sum_{k=1}^N a_{ik} \frac{\partial \chi^j}{\partial y_k}\right)$$

where $\chi^j = -w_{e_j}$. This is (1.76). □

Combining Theorem 1.20, the variational property of the epi-convergence (Theorem 1.10) and the above proposition, we can formulate the following result, which extends to a non-linear setting the Theorem 1.1 announced in Section 1.1.1.

COROLLARY 1.22 For any $f \in H^{-1}(\Omega)$ the sequence of solutions $\{u_\varepsilon ; \varepsilon \to 0\}$ of the following problems (where we use the notation $A = \partial j$)

$$\begin{cases} - \operatorname{div} A(\frac{x}{\varepsilon}, \operatorname{grad} u_\varepsilon) = f \text{ on } \Omega \\ u_{\varepsilon|\partial\Omega} = 0 \end{cases}$$

weakly converges in $H_0^1(\Omega)$ to the solution u of

$$\begin{cases} - \operatorname{div} A^{hom}(\operatorname{grad} u) = f \text{ on } \Omega \\ u|\partial\Omega = 0 \end{cases}$$

where A^{hom} is given by (1.75):

$$A^{hom}(z) = \mathfrak{m}_Y A(\cdot, z + \operatorname{grad} w_z(\cdot)).$$

Moreover,

$$\int_\Omega j(\frac{x}{\varepsilon}, \operatorname{grad} u_\varepsilon)dx \to \int_\Omega j^{hom}(\operatorname{grad} u)dx \text{ as } \varepsilon \to 0.$$

1.3.3 Domains with many small holes. Neumann boundary conditions

In the previous situation, we used in an essential way that the integrand $j(y,z)$ is bounded from above, let us say

$$j(y,z) \leq \Lambda_0(1 + |z|^2)$$

and takes, for $z \neq 0$, strictly positive values

$$\lambda_0|z|^2 \leq j(y,z), \text{ for some } 0 < \lambda_0 \leq \Lambda_0 < +\infty.$$

These two assumptions, in the case of an isotropic material, i.e.

$$j(y,z) = a(y)|z|^2$$

constrain the conductivity coefficient a to take only finite and strictly positive values.

It is of physical interest to consider the following limiting cases:

a = 0 corresponds to "perfectly" isolating inclusions

while

$a = +\infty$ corresponds to highly conducting inclusions.

A general approach to such limit analysis problems will be described in Chapter 2. The following results describe the homogenization of materials with many small isolating inclusions. Corresponding Euler equations involve Neumann boundary conditions.

A. Homogenization of strongly connected domains (cf. Theorem 1.2)

THEOREM 1.23 Let T be a "hole" with *smooth* boundary, strongly included in $Y = (0,1)^N$. For every $\varepsilon > 0$, let us consider the periodic structure in $\mathbb{R}^N$ generated by an ε-homothetic of this basic cell. For any bounded open set Ω in $\mathbb{R}^N$, let us denote $T_\varepsilon = \bigcup_{i\in I(\varepsilon)} (\varepsilon T)_i$ the union the "holes" included in Ω and $\Omega_\varepsilon = \Omega \setminus T_\varepsilon$ the "material" (Fig. 1.6).

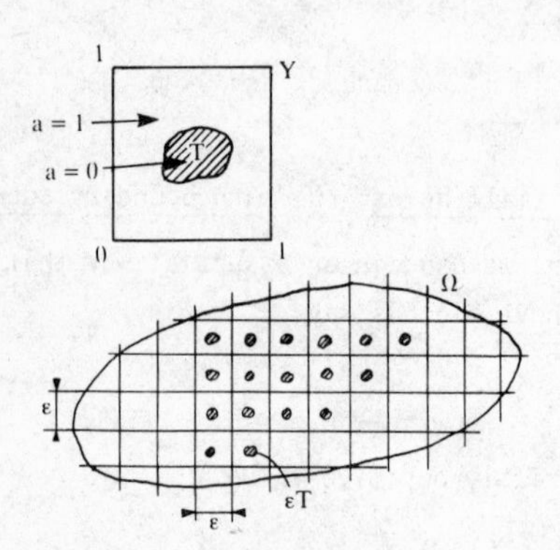

Figure 1.6

Take $F^\varepsilon(u) = \int_\Omega a(\frac{x}{\varepsilon})|\text{grad}u|^2 dx$ where a is the Y-periodic function equal to one in $Y \setminus T$ and zero in T.

In other words, for every $u \in H^1(\Omega)$

$$F^\varepsilon(u) = \int_{\Omega_\varepsilon} |\text{grad}u|^2 dx.$$

Then, taking $X = H^1(\Omega)$, τ = strong topology of $L^2(\Omega)$, for every $u \in H^1(\Omega)$

$$\tau\text{-}\lim_e F^\varepsilon(u) = F^{hom}(u) \text{ exists}$$

with

$$F^{hom}(u) = \int_\Omega j^{hom}(\text{grad}u)dx$$

and, for every $z \in \mathbb{R}^N$

$$j^{hom}(z) = \min_{\{w \text{ Y-periodic}\}} \int_{Y\setminus T} |\text{grad}w(y) + z|^2 dy. \tag{1.80}$$

Proof of Theorem 1.23

(a) The inequality $\tau\text{-}\text{ls}_e F^\varepsilon \leq F^{hom}$ is obtained in a similar way as in Theorem 1.20, since j satisfies

$$j(y,z) = a(y)|z|^2 \leq |z|^2$$

and the majorization from above is the only inequality used in this part of the proof.

(b) The difficulty comes from the inequality

$$\tau\text{-}\text{li}_e F^\varepsilon \geq F^{hom}$$

since, because of the degeneracy of the coefficients, we no longer have

$$j(y,z) \geq \lambda_o|z|^2 \text{ for some } \lambda_o > 0!$$

We have to adapt the argument developed in the proof of Theorem 1.20: in a similar way, for every $z \in \mathbb{R}^N$, let us introduce w_z a solution of the minimization problem (1.80). It satisfies: $w_z \in H^1(Y\setminus T)$ and

$$\begin{cases} -\Delta w_z = 0 \text{ on } Y\setminus T \\ w_z \quad \text{Y-periodic} \\ \dfrac{\partial w_z}{\partial n} + \langle z,n\rangle = 0 \text{ on } \partial T. \end{cases}$$

Taking $v(x) = \langle x,z\rangle + \alpha$ an affine function and $\phi \in D(\Omega)$, $0 \leq \phi \leq 1$, let us consider the approximating sequence

$$v_\varepsilon(x) = v(x) + \varepsilon w_z(\frac{x}{\varepsilon}).$$

For any weakly converging sequence $u_\varepsilon \longrightarrow u$ in $w\text{-}H^1(\Omega)$,

$$\int_{\Omega_\varepsilon} |\mathrm{grad}u_\varepsilon|^2 dx \geqslant \int_{\Omega_\varepsilon} \phi(x)|\mathrm{grad}v_\varepsilon|^2 dx$$

$$+ 2\int_{\Omega_\varepsilon} \phi(x)\ \mathrm{grad}v_\varepsilon(x)\cdot\mathrm{grad}(u_\varepsilon - v_\varepsilon)dx. \qquad (1.81)$$

In a similar way, as in Theorem 1.20,

$$\int_{\Omega_\varepsilon} \phi(x)|\mathrm{grad}v_\varepsilon|^2 dx \xrightarrow[\varepsilon\to 0]{} \int_\Omega \phi(x)\ j^{hom}(\mathrm{grad}v)dx.$$

Noticing that the sequence $\{v_\varepsilon;\ \varepsilon > 0\}$ satisfies

$$\begin{cases} -\ \Delta v_\varepsilon = 0 \text{ on } \Omega_\varepsilon \\ \dfrac{\partial v_\varepsilon}{\partial n} = 0 \text{ on } \partial T_\varepsilon \end{cases}$$

by integration by parts

$$\int_{\Omega_\varepsilon} \phi(x)\ \mathrm{grad}v_\varepsilon\cdot\mathrm{grad}(u_\varepsilon - v_\varepsilon)dx = -\int_{\Omega_\varepsilon} \mathrm{grad}\ \phi\cdot\mathrm{grad}v_\varepsilon\ (u_\varepsilon - v_\varepsilon)dx. \qquad (1.82)$$

By assumption, the sequence $\{u_\varepsilon;\ \varepsilon \to 0\}$ weakly converges in $H^1(\Omega)$ and hence strongly converges to u in $L^2(\Omega)$. On the other hand, for each $\varepsilon > 0$, the function v_ε is only, a priori, defined on Ω_ε. Let us extend it to Ω, take

$$\tilde{v}_\varepsilon(x) = v(x) + \varepsilon\tilde{w}_z(\tfrac{x}{\varepsilon})$$

where

$\tilde{w}_z$ is an $H^1(Y)$ extension of w_z to Y,

in such a way that

$$\tilde{v}_\varepsilon(x) \xrightarrow[\varepsilon\to 0]{} v(x) \text{ weakly in } H^1(\Omega).$$

Going to the limit on (1.81), by using (1.82) and the above remarks, we get

$$\liminf_{\varepsilon\to 0} \int_{\Omega_\varepsilon} |\mathrm{grad}u_\varepsilon|^2 dx \geqslant \int_\Omega \phi(x)\cdot j^{hom}(\mathrm{grad}v)dx$$

$$- 2\int_\Omega \mathrm{grad}\ \phi\cdot\partial j^{hom}(\mathrm{grad}v)(u-v)dx.$$

Integrating by parts again

$$\geq \int_\Omega \phi(x) j^{hom}(\text{grad}v)dx + 2\int_\Omega \phi(x)\cdot \partial j^{hom}(\text{grad}v)\cdot \text{grad}(u-v)dx.$$

The proof is then completed as in Theorem 1.20. □

Application to Theorem 1.3: Let us now explain how to derive from the above theorem the convergence of the solutions of the corresponding minimization problems. In the preceding non-degenerate case, there was no difficulty: because of the uniform coerciveness of functionals $\{F^\varepsilon; \varepsilon \to 0\}$ the minimizing sequence $\{u_\varepsilon; \varepsilon \to 0\}$ was bounded in $H^1(\Omega)$ and hence relatively compact in $L^2(\Omega)$. But now, functionals F^ε are naturally defined on $H^1(\Omega_\varepsilon)$ and corresponding solutions u_ε only satisfy

$$\sup_{\varepsilon>0} \|u_\varepsilon\|_{H^1(\Omega_\varepsilon)} < +\infty \;!$$

In order to conclude, one has to use an *extension theorem:* following Tartar [2], one can prove the existence, for any $\varepsilon > 0$, of an extension operator

$$\mathbb{P}^\varepsilon : H^1(\Omega_\varepsilon) \to H^1(\Omega)$$

such that the family $(\mathbb{P}_\varepsilon)_{\varepsilon\to 0}$ satisfies

$$\sup_{\varepsilon>0} \|\mathbb{P}^\varepsilon\|_{\mathcal{L}(H^1(\Omega_\varepsilon),H^1(\Omega))} < +\infty .$$

Thus, denoting a_ε the characteristic function of Ω_ε, $a_\varepsilon = 1$ on Ω_ε, $a_\varepsilon = 0$ on T_ε

$$F^\varepsilon(\mathbb{P}^\varepsilon(u_\varepsilon)) - \langle fa_\varepsilon, \mathbb{P}^\varepsilon(u_\varepsilon)\rangle = \min_{v\in H^1(\Omega)} \{F^\varepsilon(v) - \langle fa_\varepsilon, v\rangle\}.$$

We can now complete our argument with the help of Theorem 1.23, since the sequence $\mathbb{P}^\varepsilon(u_\varepsilon)$ is bounded in $H^1(\Omega)$ and $a_\varepsilon \rightharpoonup \theta = |T|$ weakly in $L^2(\Omega)$.

So, we have obtained conclusion (b) of Theorem 1.3. In order to derive the result in the case where the size r_ε of the holes is of order less than ε, we can proceed by comparison: If $r_\varepsilon \ll \varepsilon$ then $r_\varepsilon \leq k\varepsilon$ for any $k > 0$ and ε sufficiently small. Thus, $r_\varepsilon T \subset \varepsilon(kT)$ and

$$F^\varepsilon(u) = \int_{\Omega_\varepsilon} |\text{grad}u|^2 dx \geq G_k^\varepsilon(u) = \int_{\Omega_\varepsilon^k} |\text{grad}u|^2 dx$$

where $\Omega_\varepsilon^k = \Omega \setminus \cup\, \varepsilon(kT)$: we have enlarged the holes in order to reduce to the above situation. It follows that, for every $k > 0$ and $u \in H^1(\Omega)$,

$$\tau\text{-}li_e F^\varepsilon(u) \geq \tau\text{-}li_e G_k^\varepsilon(u) = \int_\Omega j_k^{hom}(\text{grad}u)dx$$

where

$$j_k^{hom}(z) = \min_{\{wY \text{ periodic}\}} \int_{Y\setminus kT} |\text{grad}w(y) + z|^2 dy.$$

Letting k go to zero, we easily obtain that

$$j_k^{hom}(z) \uparrow \min_{\{w\ Y \text{ periodic}\}} \int_Y |\text{grad}w(y) + z|^2 dy = |z|^2.$$

Thus,

$$\tau\text{-}li_e F^\varepsilon(u) \geq \int_\Omega |\text{grad}u|^2 dx.$$

Since, on the other hand

$$F^\varepsilon(u) \leq \int_\Omega |\text{grad}u|^2 dx$$

we obtain

$$\tau\text{-}ls_e F^\varepsilon(u) \leq \int_\Omega |\text{grad}u|^2 dx$$

and

$$\tau\text{-}lm_e F^\varepsilon(u) = \int_\Omega |\text{grad}u|^2 dx$$

exists for every $u \in H^1(\Omega)$.

We complete the proof after noticing that the limit θ of a_ε is identically equal to one (in fact the sequence $\{a_\varepsilon;\ \varepsilon \to 0\}$ is strongly convergent in any $L^p(\Omega)$, $1 \leq p < +\infty$, to one).

REMARK 1.24 In fact, in the real physical situation, the conductivity coefficient a is not equal to zero on T: it is equal to some δ, with $\delta > 0$, very small, and the isolating inclusions are separated by some length $\varepsilon > 0$ which is small too. These two physical parameters are fixed, strictly positive. We have described above the following scheme:

first $\delta \to 0$, then $\varepsilon \to 0$.

It is a natural question to ask: what is the limit problem when taking

$\delta = \delta(\varepsilon)$ and $\varepsilon \to 0$? We are going to prove that indeed the limit problem is the same, as in the previous situation.

For any $\delta > 0$, let us introduce $a_\delta : \mathbb{R}^N \to \mathbb{R}^+$ the Y-periodic function equal to δ on T and 1 on $Y \setminus T$. Let us denote

$$F^{\varepsilon,\delta}(u) = \int_\Omega a_\delta(\frac{x}{\varepsilon}) \ |\mathrm{grad} u|^2 dx.$$

$$F^{\varepsilon,\delta(\varepsilon)}(u) = \int_\Omega a_{\delta(\varepsilon)}(\frac{x}{\varepsilon}) \ |\mathrm{grad} u|^2 dx$$

$$F^\varepsilon(u) = \int_{\Omega_\varepsilon} |\mathrm{grad} u|^2 dx.$$

Given any $\varepsilon > 0$, for ε sufficiently small, since $\delta(\varepsilon)$ goes to zero as $\varepsilon \to 0$

$$F^\varepsilon \leqslant F^{\varepsilon,\delta(\varepsilon)} \leqslant F^{\varepsilon,\delta}.$$

Hence,

$$\tau\text{-}\mathrm{lm}_e F^\varepsilon \leqslant \tau\text{-}\mathrm{li}_e F^{\varepsilon,\delta(\varepsilon)} \leqslant \tau\text{-}\mathrm{ls}_e F^{\varepsilon,\delta(\varepsilon)} \leqslant \tau\text{-}\mathrm{lm}_e F^{\varepsilon,\delta} \tag{1.83}$$

From Theorems 1.20 and 1.23

$$\tau\text{-}\mathrm{lm}_e F^{\varepsilon,\delta} = F^\delta_{hom}$$

where

$$F^\delta_{hom}(u) = \int_\Omega j_\delta(\mathrm{grad} u(x)) dx$$

and

$$j_\delta(z) = \min_{\{w \text{ Y-periodic}\}} \int_Y a_\delta(y) |\mathrm{grad} w(y) + z|^2 dy \tag{1.84}$$

and

$\tau\text{-}\mathrm{lm}_e F^\varepsilon = F^{hom}$ is given by (1.80).

The inequalities (1.83) being valid for any $\delta > 0$, let δ decrease to zero. From

$$\lim_{\delta \to 0} j_\delta(z) = \min_{\{w \text{ Y-periodic}\}} \int_{Y \setminus T} |\mathrm{grad} w(y) + z|^2 dy = j_{hom}(z)$$

and

$$\lim_{\delta \to 0} F^\delta_{hom}(u) = F_{hom}(u)$$

(Beppo-Levy monotone convergence theorem) we finally get that for any $u \in H^1(\Omega)$

$$F^{hom}(u) \leq \tau\text{-}li_e F^{\varepsilon,\delta(\varepsilon)}(u) \leq \tau\text{-}ls_e F^{\varepsilon,\delta(\varepsilon)}(u) \leq F^{hom}(u)$$

i.e.

$$F^{hom}(u) = \tau\text{-}lm_e F^{\varepsilon,\delta(\varepsilon)}(u) \text{ for every } u \in H^1(\Omega). \quad \square$$

The above argument illustrates the flexibility of the epi-convergence theory: one can obtain rather easily information on the epi-limits functionals by comparison with other functionals whose epi-limits are already known. We shall systematically use this type of technique. □

REMARK 1.25 The proof of Theorem 1.23 relies essentially on the existence of a sequence of extension operators $\{\mathbb{P}_\varepsilon ; \varepsilon \to 0\}$ from $H^1(\Omega_\varepsilon)$ into $H^1(\Omega)$ which remains uniformly bounded. This property, called *"strong connectivity"* of the Ω_ε by Hruslov [1], indeed, guarantees as we shall see in Section 2.4, that the limit functional F^{hom} "lives" only on $H^1(\Omega)$. In "Neumann's strainer" example (Section 1.1.3) this condition fails to be satisfied and the limit functional F^{hom} "lives" on a larger space than $H^1(\Omega)$. But this condition is only sufficient, not necessary: it may happen, as described in the example below, that the domains Ω_ε are not strongly connected and the limit functional F^{hom} does live only on H^1!

B. Homogenization of thin isolating inclusions

THEOREM 1.26 Take T to be a smooth (N-1) manifold strongly included in Y. As in Theorem 1.23, let us consider the periodic structure on $\mathbb{R}^N$ generated by an ε-homothetic of this basic cell. We denote by $T_\varepsilon = \cup(\varepsilon T)_i$ the "holes" included in Ω and by $\Omega_\varepsilon = \Omega \setminus T_\varepsilon$ the "material" (Fig. 1.7). Let $F^\varepsilon : L^2(\Omega) \to \bar{\mathbb{R}}^+$

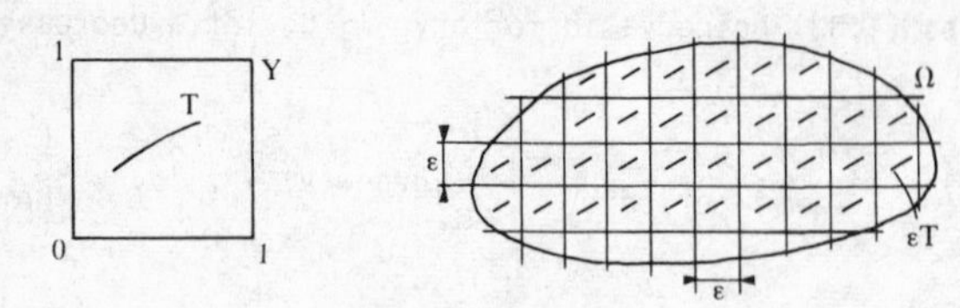

Figure 1.7

be the functional with domain $H^1(\Omega_\varepsilon)$ (i.e. including functions with jumps across T_ε)

$$F^\varepsilon(u) = \int_{\Omega_\varepsilon} |\mathrm{grad}u|^2 dx.$$

Then, taking τ = the strong topology of $L^2(\Omega)$, the sequence $\{F^\varepsilon; \varepsilon \to 0\}$, τ-epi converges to F^{hom} *the functional with domain equal to* $H^1(\Omega)$ given by:

$$F^{hom}(u) = \int_\Omega j^{hom}(\mathrm{grad}u)dx,$$

$$j^{hom}(z) = \min_{\{w\ Y\text{-periodic}\}} \int_{Y\setminus T} |\mathrm{grad}w(y) + z|^2 dy. \qquad (1.85)$$

Proof of Theorem 1.26 We could attack this problem directly by the same technique as in Theorem 1.20. But we would like to show how one can quite simply reduce it to the preceding situation.

The idea is to "enlarge" the hole. To that end, *we introduce another parameter* $\eta > 0$ *which describes the thickness of the enlarged hole* T^η (Fig. 1.8).

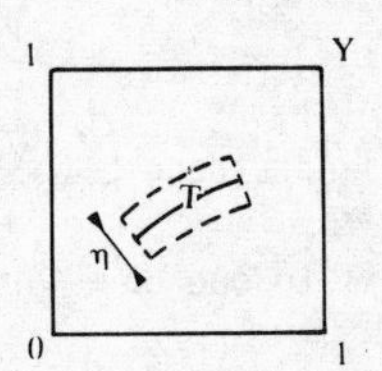

Figure 1.8

We use the notation $\Omega_\varepsilon^\eta = \Omega\setminus T_\varepsilon^\eta$, where $T_\varepsilon^\eta = \bigcup_i (\varepsilon T^\eta)_i$ and

$$F_\eta^\varepsilon(u) = \int_{\Omega_\varepsilon^\eta} |\mathrm{grad}u|^2 dx.$$

We can now assert the existence, for every $\eta > 0$, of extension operators

$$\mathbb{P}_\eta^\varepsilon : H^1(\Omega_\varepsilon^\eta) \to H^1(\Omega)$$

with

$$\sup_{\varepsilon>0} \|\mathbb{P}^\varepsilon_\eta\| = C(\eta) < +\infty \qquad (\lim_{\eta\to 0} C(\eta) = +\infty!).$$

Taking advantage of the inequality

$$F^\varepsilon \geq F^\varepsilon_\eta \tag{1.86}$$

one obtains the following conclusions:

(a) <u>The limit function F^{hom} "lives" on $H^1(\Omega)$</u>: let $u_\varepsilon \to u$ in s-$L^2(\Omega)$ such that

$$\liminf_{\varepsilon\to 0} F^\varepsilon(u_\varepsilon) < +\infty.$$

Let us prove that $u \in H^1(\Omega)$. From (1.86),

$$\liminf_{\varepsilon\to 0} F^\varepsilon_\eta(u_\varepsilon) < +\infty,$$

where we have fixed some $\eta > 0$.

It follows that (at least for a subsequence still denoted ε)

$$\sup_{\varepsilon>0} \|\mathbb{P}^\varepsilon_\eta(u_\varepsilon)\|_{H^1(\Omega)} < +\infty.$$

Let us extract a subsequence

$$\mathbb{P}^\varepsilon_\eta(u_\varepsilon) \longrightarrow u^* \text{ in weak-}H^1(\Omega) \text{ and strong } L^2(\Omega)$$

and introduce X^η_ε the function equal to one on Ω^η_ε and zero elsewhere. Noticing that

$$\begin{aligned} &\mathbb{P}^\varepsilon_\eta(u_\varepsilon)\, X^\eta_\varepsilon = u_\varepsilon X^\eta_\varepsilon \\ &X^\eta_\varepsilon \longrightarrow \theta = \text{meas } T^\eta \text{ weakly in } L^2(\Omega), \text{ with meas } T^\eta > 0, \end{aligned} \tag{1.87}$$

we get

$$u^*\theta = u\theta,$$

i.e. $u = u^*$ and u belongs to $H^1(\Omega)$; in other words, $\tau\text{-lm}_e F^\varepsilon = +\infty$ on $L^2(\Omega)\setminus H^1(\Omega)$!

(b) $\tau\text{-li}_e F^\varepsilon \geq F^{hom}$. From (1.86), for every $u \in H^1(\Omega)$, for every $\eta > 0$

$$\tau\text{-li}_e F^\varepsilon(u) \geq \tau\text{-li}_e F^\varepsilon_\eta(u) = \int_\Omega j^{hom}_\eta(\text{grad}u)dx \tag{1.88}$$

with

$$j^{hom}_\eta(z) = \min_{\{w \text{ Y-periodic}\}} \int_{Y \setminus T^\eta} |\text{grad}w(y) + z|^2 dy.$$

We have applied the conclusion of Theorem 1.26 to $\{F^\varepsilon_\eta;\ \varepsilon \to 0\}$ in order to get the above equality.

The inequality (1.88) being valid for any $\eta > 0$, taking the supremum as η decreases to zero on its right member, we get the required conclusion. The other inequality

$$F^{hom} \geq \tau\text{-ls}_e F^\varepsilon$$

is obtained in a quite similar way to that used in the proof of Theorem 1.20.

When applying Theorem 1.26 to prove convergence of the solutions $\{u_\varepsilon;\ \varepsilon \to 0\}$ of the corresponding minimization problems

$$\min \{F^\varepsilon(u) - \int fu\}$$

one has to verify, too, that these solutions are relatively compact in $L^2(\Omega)$. This can be proved, using the same trick as above: for every $\eta > 0$, the sequence $\{\mathbb{P}^\varepsilon_\eta(u_\varepsilon);\ \varepsilon \to 0\}$ is bounded in $H^1(\Omega)$ and hence relatively compact in $L^2(\Omega)$. Then apply a diagonalization argument and build some $u \in H^1(\Omega)$ such that

$$\forall \eta > 0 \quad \mathbb{P}^{\varepsilon_\nu}_\eta(u_{\varepsilon_\nu}) \to u \text{ strongly in } L^2(\Omega).$$

From

$$\|u_\varepsilon - u\|_{L^2} \leq \|u_\varepsilon - \mathbb{P}^\varepsilon_\eta(u_\varepsilon)\|_{L^2} + \|\mathbb{P}^\varepsilon_\eta(u_\varepsilon) - u\|$$

using that $\{u_\varepsilon;\ \varepsilon > 0\}$ is in fact bounded in some $L^p(\Omega)$ with $p > 2$, and hence enjoys a suitable equi-integrability property, we derive

$$u_{\varepsilon_\nu} \to u \text{ in strong } L^2(\Omega).$$

Preceding considerations lead naturally to the following variational inequalities introduced by E. Sanchez-Palencia [9] in order to describe fissured elastic materials.

C. Homogenization of fissured elastic materials

The situation and notations are the same as in Theorem 1.26. But now, we consider an oriented normal to T, $\vec{n}$ (which is directed outward from a side of T labelled 1, the other side being labelled 2 - see Fig. 1.9). For any

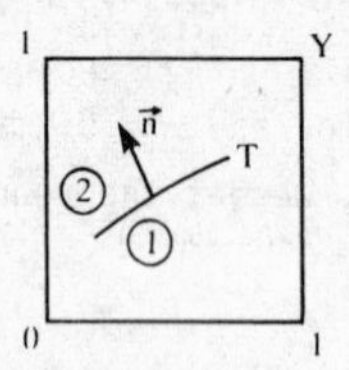

Figure 1.9

function u with a possible jump across T, we use the notation

$$[u] = u_2 - u_1$$

where u_2 and u_1 are respective traces of u from side 2 and 1, on T. The constraint involved in the description of fissured material is: $[\vec{u}\cdot\vec{n}] \geqslant 0$ on T. At equilibrium, the material tends to minimize the internal energy with respect to all displacements satisfying this constraint. Physically, this means that the two lips of the fissure cannot interpenetrate! At equilibrium this implies that the force is normal to T, that there is compression but no traction on T, and that, if the fissure is open at one point, the force is zero at that point.

The study of *homogenization of such material* leads to the following epi-convergence problem (for simplicity, we consider the corresponding scalar case).

$$F^{\varepsilon}(u) = \int_{\Omega_{\varepsilon}} |\text{grad}\, u|^2 dx + I_{K^{\varepsilon}}(u)$$

where $I_{K^{\varepsilon}}$ is the indicator functional of the constraint K^{ε}

$$K^{\varepsilon} = \{u \in H^1(\Omega_{\varepsilon}) / [u] \geqslant 0 \text{ on } T_{\varepsilon}\}.$$

A quite similar approach as in Theorem 1.20 and comparison with Theorem 1.26 yields (cf. Attouch & Murat [2]) the epi-convergence of the sequence $\{F^\varepsilon;\ \varepsilon \to 0\}$ to F^{hom}, described as follows:

$$F^{hom}(u) = \int_\Omega j^{hom}(\text{grad}u)dx$$

with for every $z \in \mathbb{R}^N$

$$j^{hom}(z) = \min \{ \int_{Y\setminus T} |\text{grad}w(y) + z|^2 dy / \begin{smallmatrix} w \text{ Y-periodic} \\ [w] \geq 0 \text{ on } T \end{smallmatrix} \}.$$

1.3.4 Domains with many small holes. Dirichlet boundary conditions

Let us describe the epi-convergence analysis of example 1.1.2(B).

THEOREM 1.27 Let $T_\varepsilon = \bigcup_{i \in I(\varepsilon)} T^i_\varepsilon$ be the union of small holes of size r_ε which are periodically distributed with period εY in $\mathbb{R}^N$. Each hole T^i_ε is obtained by translating an r_ε-homothetic of a "fixed" hole $T \subset Y$. Let us consider the sequence of functionals $\{F^\varepsilon;\ \varepsilon \to 0\}$, from $H^1_0(\Omega)$ into $\mathbb{R}^+$ defined by

$$F^\varepsilon(u) = \int_\Omega |\text{grad}u|^2 dx + I_{K^\varepsilon}(u)$$

where I_{K^ε} is the indicator functional of the convex set $K^\varepsilon = \{u \in H^1_0(\Omega)/u = 0 \text{ on } T_\varepsilon\}$. Equivalently,

$$F^\varepsilon(u) = \begin{cases} \int_\Omega |\text{grad}u|^2 dx & \text{if } u = 0 \text{ on } T_\varepsilon \\ +\infty \text{ otherwise.} \end{cases}$$

The description of the results depends on the dimension of the space, so take $N = 3$. Then, depending on the behaviour of $r_\varepsilon/\varepsilon^3$ as ε goes to zero, three different situations can occur. Denoting s (resp. w) the strong (resp. weak) topology of $H^1_0(\Omega)$,

(a) When $r_\varepsilon \ll \varepsilon^3$

$$s\text{-}\mathrm{lm}_e F^\varepsilon = w\text{-}\mathrm{lm}_e F^\varepsilon = \Phi \text{ where } \Phi(u) = \int_\Omega |\text{grad}u|^2 dx \quad \forall u \in H^1_0(\Omega).$$

(b) When $r_\varepsilon \gg \varepsilon^3$

$$s\text{-}\lim_e F^\varepsilon = w\text{-}\lim_e F^\varepsilon = I_{\{0\}} \text{ is the indicator function of } \{0\}.$$

(c) When $r_\varepsilon \sim \varepsilon^3$

$$w\text{-}\lim_e F^\varepsilon = F$$

where

$$\text{for every } u \in H_0^1(\Omega) \quad F(u) = \int_\Omega |\text{grad}u|^2 dx + C \int_\Omega u^2(x)dx. \tag{1.89}$$

C is equal to the capacity of the hole T in $\mathbb{R}^N$:

$$C = \inf \{ \int_{\mathbb{R}^3} |\text{grad}w|^2 dx; \ w \in H^1(\mathbb{R}^3), \ \tilde{w} \geq 1 \text{ q.e on } T\}.$$

<u>Proof of Theorem 1.27</u> Let us introduce the "test functions" $\{w_\varepsilon; \varepsilon \to 0\}$ which are solutions of the corresponding variational problems on the microscopic cells. On $Y_\varepsilon = \varepsilon Y$, w_ε is defined as the solution of the minimization problem

$$\min \{ \int_{Y_\varepsilon} |\text{grad}w|^2 dx / w = 0 \text{ on } T_\varepsilon, \ w = 1 \text{ on } Y_\varepsilon \setminus B_\varepsilon \}, \tag{1.90}$$

where B_ε denotes the ball of radius $\frac{\varepsilon}{2}$ included in Y_ε. Then, w_ε is extended by the periodicity εY to all of $\mathbb{R}^3$.

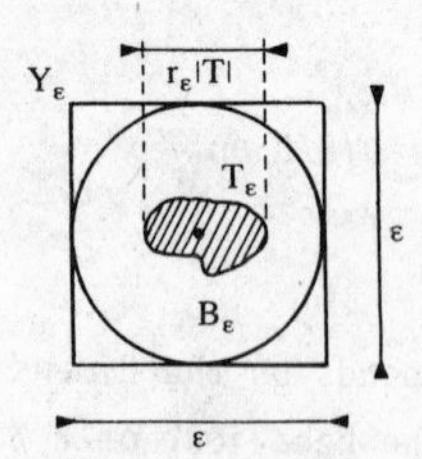

<u>Figure 1.10</u>

Let us compute the $H^1(\Omega)$ norm of w_ε:

$$\int_\Omega |\text{grad}w_\varepsilon|^2 dx \simeq \frac{\text{meas}\Omega}{\varepsilon^3} \int_{Y_\varepsilon} |\text{grad}w_\varepsilon|^2 dx,$$

since the number of microscopic cells Y_ε^i included in Ω is equivalent to $\text{meas}\Omega/\varepsilon^3$. Hence,

$$\int_\Omega |\text{grad}w_\varepsilon|^2 dx \simeq \frac{\text{meas}\Omega}{\varepsilon^3} \min\{\int_{B_\varepsilon} |\text{grad}w|^2 dx / w = 0 \text{ on } T_\varepsilon,\ w = 1 \text{ on } \partial B_\varepsilon\}.$$

By changing the scale, $x = r_\varepsilon y$,

$$\int_\Omega |\text{grad}w_\varepsilon|^2 dx \simeq \frac{\text{meas}\Omega}{\varepsilon^3} \min\{\int_{B_{\varepsilon/r_\varepsilon}} \frac{|\text{grad}w|^2}{r_\varepsilon^2} r_\varepsilon^3 dy / w=0 \text{ on } T, w=1 \text{ on } \partial B_{\varepsilon/r_\varepsilon}\},$$

where we notice that the hole $r_\varepsilon T = T_\varepsilon$ has been transformed into the initial hole T. Therefore

$$\int_\Omega |\text{grad}w_\varepsilon|^2 dx \simeq \text{meas}\ \Omega . \frac{r_\varepsilon}{\varepsilon^3} \min\{\int_{B_{\varepsilon/r_\varepsilon}} |\text{grad}w|^2 dy / w=0 \text{ on } T,\ w=1 \text{ on } \partial B_{\varepsilon/r_\varepsilon}\}$$

When ε goes to zero this last minimum converges to $\text{Cap}_{\mathbb{R}^3} T$ and finally

$$\int_\Omega |\text{grad}w_\varepsilon|^2 dx \simeq \text{Cap}_{\mathbb{R}^3} T.\ \text{meas}\ \Omega . \frac{r_\varepsilon}{\varepsilon^3}. \tag{1.91}$$

If $\limsup_{\varepsilon \to 0} \frac{r_\varepsilon}{\varepsilon^3} < +\infty$, then the sequence $\{w_\varepsilon ; \varepsilon \to 0\}$ is bounded in $H^1(\Omega)$ and hence relatively compact in $L^2(\Omega)$. Let us introduce X_ε the function equal to 1 on $\cup_i Y_\varepsilon^i \backslash B_\varepsilon^i$ and 0 elsewhere. Then,

$$(w_\varepsilon - 1)X_\varepsilon = 0 \text{ on } \mathbb{R}^3. \tag{1.92}$$

Noticing that the sequence $\{X_\varepsilon ; \varepsilon \to 0\}$ w-$L^2(\Omega)$ converges to some strictly positive constant θ, $0 < \theta < 1$, every s-$L^2(\Omega)$ limit value w of the sequence $\{w_\varepsilon ; \varepsilon \to 0\}$, from (1.92), satisfies

$$(w-1)\theta = 0 \text{ on } \mathbb{R}.$$

Hence, the whole sequence $\{w_\varepsilon ; \varepsilon \to 0\}$ strongly converges in $L^2(\Omega)$ to one.

Let us now examine, depending upon the behaviour of $r_\varepsilon/\varepsilon^3$, the three cases:

1. <u>Let us first examine the critical case $r_\varepsilon \simeq \varepsilon^3$</u>

(a) $F \geqslant$ w-$H_o^1(\Omega)\text{ls}_e F^\varepsilon$, i.e. for every $u \in H_o^1(\Omega)$ there exists a sequence $\{u_\varepsilon ; \varepsilon \to 0\}$ satisfying

$$\begin{cases} u_\varepsilon = 0 \text{ on } T_\varepsilon \\ u_\varepsilon \xrightarrow[(\varepsilon\to 0)]{} u \text{ in } w\text{-}H_o^1(\Omega) \text{ and } F(u) \geq \limsup_{\varepsilon\to 0} \Phi(u_\varepsilon). \end{cases}$$

A natural approach would consist of taking directly

$$u_\varepsilon = u \cdot w_\varepsilon .$$

Then, $u_\varepsilon = 0$ on T_ε and

$$\begin{aligned} \int_\Omega |\text{grad} u_\varepsilon|^2 dx &= \int_\Omega |w_\varepsilon \text{grad} u + u \text{ grad} w_\varepsilon|^2 dx \\ &= \int_\Omega \{w_\varepsilon^2 |\text{grad} u|^2 + 2w_\varepsilon . \text{grad} w_\varepsilon . u . \text{grad} u + u^2 |\text{grad} w_\varepsilon|^2\} dx \end{aligned} \tag{1.93}$$

One can pass to the limit, as ε goes to zero, assuming u to be regular. Denoting $\mu_\varepsilon = |\text{grad} w_\varepsilon|^2 dx$, from (1.91) the sequence of positive Radon measures μ_ε on Ω is uniformly bounded,

$$\sup_{\varepsilon\to 0} \int_\Omega d\mu_\varepsilon < + \infty,$$

and for any step function

$$v = \sum_i \alpha_i X_{\omega_i}$$

$$\int_\Omega v \, d\mu_\varepsilon \xrightarrow[(\varepsilon\to 0)]{} \text{CapT.} \sum_i \alpha_i \text{ meas } \omega_i = \text{CapT} \int_\Omega v \cdot dx.$$

Therefore

$$\mu_\varepsilon \xrightarrow[(\varepsilon\to 0)]{} C \, dx \text{ for the topology } \sigma(\mathfrak{m}^b(\Omega), \mathbb{C}(\Omega)). \tag{1.94}$$

So, take first u a $\mathbb{C}^1$ function. Using (1.94), noticing that $\text{grad} w_\varepsilon$ weakly converges to zero in $L^2(\Omega)$ and w_ε strongly converges to one on $L^2(\Omega)$, from (1.93)

$$\lim_{\varepsilon\to 0} \int_\Omega |\text{grad} u_\varepsilon|^2 dx = \int_\Omega |\text{grad} u|^2 dx + C \int_\Omega u^2 dx = F(u).$$

We now complete the demonstration by a diagonalization argument: Noticing that F is continuous on $H_o^1(\Omega)$ for the norm topology, given $u \in H_o^1(\Omega)$ let

$$u_k \xrightarrow[k \to \infty]{} u \text{ in } s\text{-}H_o^1(\Omega), \text{ then } F(u_k) \xrightarrow[k \to \infty]{} F(u).$$

From the preceding argument, for every $k \in \mathbb{N}$ there exists an approximating sequence $\{u_{k,\varepsilon} : \varepsilon \to 0\}$ such that

$$(u_{k,\varepsilon}, F^\varepsilon(u_{k,\varepsilon})) \xrightarrow[\varepsilon \to 0]{s\text{-}L^2 \times \mathbb{R}} (u_k, F(u_k)) \underset{k\to+\infty}{\downarrow} s\text{-}L^2 \times \mathbb{R} \quad (u, F(u)).$$

By diagonalization (Corollary 1.18), there exists an increasing map $\varepsilon \to k(\varepsilon)$ such that

$$(u_{k(\varepsilon),\varepsilon}, F^\varepsilon(u_{k(\varepsilon),\varepsilon})) \xrightarrow[(\varepsilon \to 0)]{s\text{-}L^2 \times \mathbb{R}} (u,F(u)).$$

Denoting $u_\varepsilon = u_{k(\varepsilon),\varepsilon}$, from the uniform coerciveness property of the F^ε,

$$u_\varepsilon \xrightarrow[\varepsilon \to 0]{} u \text{ weakly in } H_o^1(\Omega) \text{ and } F^\varepsilon(u_\varepsilon) = \Phi(u_\varepsilon) \xrightarrow[\varepsilon \to 0]{} F(u).$$

(b) Let us now prove (this is the difficult point) that $F \leqslant w\text{-}H_o^1(\Omega)\text{-}li_e F^\varepsilon$, i.e. for every weakly converging sequence in $H_o^1(\Omega)$, $u_\varepsilon \to u$,

$$F(u) \leqslant \liminf_{\varepsilon \to 0} F^\varepsilon(u_\varepsilon).$$

As in the preceding homogenization argument, the idea is to compare $F^\varepsilon(u_\varepsilon)$ with $F^\varepsilon(w_\varepsilon v)$, via a subdifferential inequality. We first take v regular in order to justify the computations, then let v tend to u: Given $v \in D(\Omega)$, a $\mathbb{C}^\infty$ function with compact support, let us denote $v_\varepsilon = v \cdot w_\varepsilon$. Let us write the subdifferential inequality

$$\int_\Omega |\text{grad}u_\varepsilon|^2 dx \geqslant \int_\Omega |\text{grad}v_\varepsilon|^2 dx + 2\int_\Omega \langle \text{grad}v_\varepsilon, \text{grad}u_\varepsilon - \text{grad}v_\varepsilon \rangle dx.$$

From the first part of the proof,

$$\lim_{\varepsilon\to 0} \int_\Omega |\text{grad}v_\varepsilon|^2 dx = \int_\Omega |\text{grad}v|^2 dx + C \int_\Omega v^2 dx = F(v).$$

Thus,

$$\liminf_{\varepsilon \to 0} F^\varepsilon(u_\varepsilon) \geqslant F(v) + 2 \liminf_{\varepsilon \to 0} I_\varepsilon \tag{1.95}$$

where we denote

$$I_\varepsilon = \int_\Omega \langle \mathrm{grad} v_\varepsilon, \mathrm{grad} u_\varepsilon \rangle dx - \int_\Omega |\mathrm{grad} v_\varepsilon|^2 dx.$$

The difficult point is to go to the limit on the duality bracket $\langle v_\varepsilon, u_\varepsilon \rangle_{(H_0^1, H_0^1)}$, since we know these two sequences to be convergent only in $w\text{-}H_0^1(\Omega)$!

$$\int_\Omega \langle \mathrm{grad} v_\varepsilon, \mathrm{grad} u_\varepsilon \rangle dx = \int_\Omega w_\varepsilon \langle \mathrm{grad} v, \mathrm{grad} u_\varepsilon \rangle dx + \int_\Omega v \langle \mathrm{grad} w_\varepsilon, \mathrm{grad} u_\varepsilon \rangle dx.$$

Since

$$w_\varepsilon \xrightarrow[(\varepsilon \to 0)]{} 1 \text{ in } s\text{-}L^2(\Omega),\ \mathrm{grad} u_\varepsilon \to \mathrm{grad} u \text{ in } w\text{-}L^2(\Omega), \text{ and } v \in D(\Omega),$$

$$\int_\Omega w_\varepsilon \langle \mathrm{grad} v, \mathrm{grad} u_\varepsilon \rangle dx \xrightarrow[(\varepsilon \to 0)]{} \int_\Omega \langle \mathrm{grad} v, \mathrm{grad} u \rangle dx.$$

Thus,

$$\liminf_{\varepsilon \to 0} I_\varepsilon = \int_\Omega \langle \mathrm{grad} v, \mathrm{grad} u - \mathrm{grad} v \rangle dx - C \int_\Omega v^2 dx + \liminf_{\varepsilon \to 0} \int_\Omega v \langle \mathrm{grad} w_\varepsilon, \mathrm{grad} u_\varepsilon \rangle dx.$$

Let us integrate by parts this last expression

$$\int_\Omega \langle \mathrm{grad} u_\varepsilon, v\,\mathrm{grad} w_\varepsilon \rangle dx = -\int_\Omega u_\varepsilon \langle \mathrm{grad} v, \mathrm{grad} w_\varepsilon \rangle dx + \langle -\Delta w_\varepsilon, u_\varepsilon v \rangle_{(H^{-1}, H_0^1)}.$$

Noticing that $u_\varepsilon \to u$ in $s\text{-}L^2(\Omega)$ and $\mathrm{grad} w_\varepsilon \longrightarrow 0$ in $w\text{-}L^2$,

$$\int_\Omega u_\varepsilon \langle \mathrm{grad} v, \mathrm{grad} w_\varepsilon \rangle dx \xrightarrow[\varepsilon \to 0]{} 0.$$

Thus,

$$\liminf_{\varepsilon \to 0} I_\varepsilon = \int_\Omega \langle \mathrm{grad} v, \mathrm{grad}(u-v) \rangle dx - C \int_\Omega v^2 dx + \liminf_{\varepsilon \to 0} J_\varepsilon, \quad (1.96)$$

where

$$J_\varepsilon = \langle -\Delta w_\varepsilon, u_\varepsilon v\rangle_{(H^{-1},H_0^1)}.$$

We are going to prove that

$$\lim_{\varepsilon\to 0} J_\varepsilon = C \int_\Omega uv\, dx. \tag{1.97}$$

For the moment let us assume (1.97) and complete the proof: from (1.95), (1.96) and (1.97),

$$\liminf_{\varepsilon \to 0} F^\varepsilon(u_\varepsilon) \geq F(v) + 2[\int_\Omega \langle \text{grad} v, \text{grad}(u-v)\rangle dx + C \int_\Omega v(u-v)dx]. \tag{1.98}$$

Since this last inequality is valid for any $v \in D(\Omega)$, letting v tend to u in the norm topology of $H_0^1(\Omega)$ and using the continuity of F for this topology, we finally get

$$\liminf_{\varepsilon \to 0} F^\varepsilon(u_\varepsilon) \geq F(u).$$

In order to prove (1.97) we follow the argument developed by Cioranescu & Murat [1].

Let us write the Euler equation of the minimization problem (1.80) defining w_ε:

$$\begin{cases} -\Delta w_\varepsilon = 0 \text{ on } B_\varepsilon \setminus T_\varepsilon \\ w_\varepsilon = 0 \text{ on } T_\varepsilon \\ w_\varepsilon = 1 \text{ on } Y_\varepsilon \setminus B_\varepsilon \end{cases} \tag{1.99}$$

and $w_\varepsilon \in H^1(Y_\varepsilon)$.

Thus, $-\Delta w_\varepsilon$ is equal to zero except on the boundary of the sets ∂B_ε^i and ∂T_ε^i:

$$-\Delta w_\varepsilon = \mu_\varepsilon - \gamma_\varepsilon$$

where

$\mu^\varepsilon = \left.\dfrac{\partial w^\varepsilon}{\partial \mathbf{n}}\right|_{\partial B_\varepsilon^i}$ is supported by the spheres ∂B_ε^i

$\gamma_\varepsilon = \left.\dfrac{\partial w^\varepsilon}{\partial n}\right|_{\partial T_\varepsilon^i}$ is supported by the boundaries ∂T_ε^i of the holes T_ε^i.

Noticing that u_ε is equal to zero on ∂T^i_ε,

$$J_\varepsilon = \langle -\Delta w_\varepsilon, u_\varepsilon v\rangle = \sum_{i\in I(\varepsilon)} \int_{\partial B^i_\varepsilon} u_\varepsilon v \frac{\partial w^\varepsilon}{\partial n} d\sigma$$

$$= \int_\Omega u_\varepsilon v \, d\mu_\varepsilon \quad \text{where } \mu_\varepsilon = \sum_{i\in I(\varepsilon)} \frac{\partial w^\varepsilon}{\partial n}\Big|_{\partial B^i_\varepsilon}$$

is a positive measure of finite energy. We are going to prove that

$$\mu^\varepsilon \xrightarrow[(\varepsilon\to 0)]{} Cdx \text{ for the norm topology of } H^{-1}(\Omega). \tag{1.100}$$

This is the important point of the proof: the sequence of signed measures $\{-\Delta w_\varepsilon;\ \varepsilon \to 0\}$ converges only weakly in $H^{-1}(\Omega)$, but their positive parts $\{(-\Delta w_\varepsilon)^+; \varepsilon\to 0\}$ which are the only ones to consider in the limit process (since $u_\varepsilon = 0$ on the support of $(-\Delta w_\varepsilon)^-$!) do converge strongly in $H^{-1}(\Omega)$. Then, one can go to the limit on the bracket $J_\varepsilon = \langle u_\varepsilon v, \mu_\varepsilon\rangle_{(H^1_o, H^{-1})}$ and get (1.97).

We first notice that, from (1.99),

$$0 = \int_{B_\varepsilon} -\Delta w_\varepsilon . w_\varepsilon \, dx = \int_{B_\varepsilon} |\text{grad} w_\varepsilon|^2 dx - \int_{\partial B_\varepsilon} \frac{\partial w_\varepsilon}{\partial n} d\sigma.$$

Thus, the same computation as (1.91) yields that, for any step function ϕ

$$\int_\Omega \phi d\mu_\varepsilon \simeq \int_\Omega \phi \ |\text{grad} w_\varepsilon|^2 dx \to C \int_\Omega \phi \, dx.$$

This implies that

$$\mu_\varepsilon \longrightarrow Cdx \text{ in } \sigma(\mathfrak{m}^b(\Omega), \mathfrak{C}(\Omega)) \text{ and weakly in } H^{-1}(\Omega). \tag{1.101}$$

In order to get the strong $H^{-1}(\Omega)$ convergence of the sequence $\{\mu_\varepsilon;\ \varepsilon \to 0\}$ we proceed by comparison.

(a) Let us first assume T to be a ball of radius r, $0 < r < \frac{1}{2}$. Then, an explicit computation yields

$$\mu_\varepsilon \simeq k(\varepsilon) \sum_{i\in I(\varepsilon)} \varepsilon \, \delta^i_\varepsilon$$

where δ^i_ε is the superficial measure supported by the sphere $S^i_\varepsilon = \partial B^i_\varepsilon$, and $k(\varepsilon)$ is a positive constant, $k(\varepsilon) \underset{\varepsilon\to 0}{\to} 1$. Introducing q_ε the auxiliary function which is a solution on each Y_ε^i of

$$\begin{cases} \Delta q_\varepsilon = N \text{ on } B_\varepsilon^i \quad (N = 3) \\ q_\varepsilon = 0 \text{ on } Y_\varepsilon^i \setminus B_\varepsilon^i \\ \dfrac{\partial q_\varepsilon}{\partial n} = \dfrac{\varepsilon}{2} \text{ on } \partial B_\varepsilon^i \end{cases} \tag{1.102}$$

and noticing that these conditions are compatible, we get

$$-\Delta q_\varepsilon = -N\chi_B^\varepsilon + \frac{1}{2} \Sigma\varepsilon\delta_\varepsilon^i$$

where χ_B^ε is the function equal to one on $\cup B_\varepsilon^i$ and zero elsewhere. Since $|\text{grad} q_\varepsilon| < \varepsilon$, $q_\varepsilon \underset{(\varepsilon\to 0)}{\to} 0$ in the norm topology of $W^{1,\infty}(\Omega)$. On the other hand, the sequence $\{-N\chi_B^\varepsilon;\ \varepsilon \to 0\}$ weakly converges in $L^2(\Omega)$ to some constant and hence strongly converges in $H^{-1}(\Omega)$.

Thus, $\sum_{i\in I(\varepsilon)} \varepsilon\,\delta_\varepsilon^i$ is strongly convergent in $H^{-1}(\Omega)$ (in fact to a constant function) and so is $\{\mu_\varepsilon;\ \varepsilon \to 0\}$.

(b) Then, we complete the proof by a comparison argument which relies on the maximum principle:

Given $T \subset Y$, let $T \subset B_\rho \subset Y$, where B_ρ denotes a ball of radius ρ with the same centre as Y. Let us denote by $\bar{w}_\varepsilon$ the corresponding test function, that is obtained by replacing in (1.90) $T_\varepsilon = r_\varepsilon T$ by $r_\varepsilon B_\rho = \bar{T}_\varepsilon$. From the maximum principle (see Fig. 1.11),

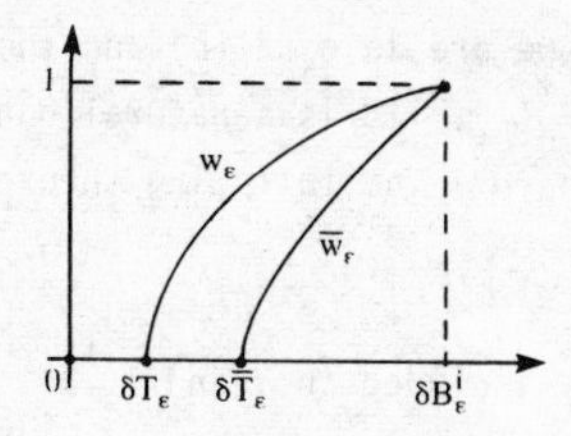

Figure 1.11

$$0 < \bar{w}_\varepsilon < w_\varepsilon$$

and

$$0 < \frac{\partial w_\varepsilon}{\partial n} < \frac{\partial \bar{w}_\varepsilon}{\partial n} \text{ on } \partial B_\varepsilon^i .$$

By the preceding argument, the sequence $\bar{\mu}_\varepsilon = \sum_i \frac{\partial \bar{w}_\varepsilon}{\partial n}\Big|_{\partial B^i_\varepsilon}$ strongly converges in $H^{-1}(\Omega)$ to some measure $\bar{\mu}$.

We know that

$$\mu_\varepsilon = \sum_i \frac{\partial w_\varepsilon}{\partial n}\Big|_{\partial B^i_\varepsilon}$$

weakly converges in $H^{-1}(\Omega)$ to $\mu = Cdx$ and by the above inequality

$$0 \leq \mu_\varepsilon \leq \bar{\mu}_\varepsilon.$$

This implies (Cioranescu & Murat [1], Lemma 2.8) that the sequence $\{\mu_\varepsilon; \varepsilon \to 0\}$ is also strongly convergent in $H^{-1}(\Omega)$.

REMARKS

1. The conclusion of Theorem 1.27 can be condensed in a single formula: for every $u \in H^1_0(\Omega)$

$$(w\text{-}\lim_e F^\varepsilon)(u) = \int_\Omega |\text{grad} u|^2 dx + C \int_\Omega u^2 dx$$

with C equal respectively to 0, CapT, $+\infty$ when $r_\varepsilon \ll \varepsilon^3$, $r_\varepsilon \sim \varepsilon^3$, $r_\varepsilon \gg \varepsilon^3$. Variational properties of the epi-convergence (Theorem 1.10) yield both convergence of solutions and energies of the corresponding minimization problems (Theorem 1.3).

2. When the size of the "holes" is comparable to the size of the microscopic cells (that is r_ε *of order* ε), we are in case (b) and the sequence $\{u_\varepsilon; \varepsilon \to 0\}$ converges to zero strongly in $H^1_0(\Omega)$. It is a natural question to ask, *at what rate does* u_ε *converge to zero*? The following theorem answers this question:

THEOREM 1.28 Let T be strongly included in Y and $T_\varepsilon = \bigcup_{i \in I(\varepsilon)} T^i_\varepsilon$ be the union of ε-homothetic of T periodically distributed with period εY in $\mathbb{R}^N$.

For every $f \in L^2(\Omega)$, for every $\varepsilon > 0$, let us denote by u_ε the solution of

$$\begin{cases} -\Delta u_\varepsilon = f \text{ on } \Omega_\varepsilon & \Omega_\varepsilon = \Omega \setminus T_\varepsilon \\ u_\varepsilon = 0 \text{ on } \partial\Omega_\varepsilon & \partial\Omega_\varepsilon = \partial T_\varepsilon \cup \partial\Omega. \end{cases} \tag{1.103}$$

Then,

$$u_\varepsilon \to 0 \text{ strongly in } H_0^1(\Omega), \text{ as } \varepsilon \to 0$$

$$\frac{u_\varepsilon}{\varepsilon} \rightharpoonup 0 \quad \text{weakly in } H_0^1(\Omega), \text{ as } \varepsilon \to 0 \tag{1.104}$$

$$\frac{u_\varepsilon}{\varepsilon^2} \rightharpoonup \quad f.\mathfrak{m}_Y(w) \text{ weakly in } L^2(\Omega), \text{ as } \varepsilon \to 0,$$

where w is the solution of

$$\begin{cases} -\Delta w = 1 \text{ in } Y \setminus T \\ w = 0 \text{ on } T \\ w \quad Y\text{-periodic and } \mathfrak{m}_Y(w) = \int_Y w(y)dy. \end{cases} \tag{1.105}$$

In terms of epi-convergence, this last result can be interpreted as: $v_\varepsilon = u_\varepsilon/\varepsilon^2$ minimizes over $H_0^1(\Omega)$ the functional $F^\varepsilon(v) - \int_\Omega fv$, where

$$F^\varepsilon(v) = \frac{1}{2}\,\varepsilon^2 \int_\Omega |\text{grad} v|^2 dx + \int_\Omega a(\frac{x}{\varepsilon})v^2(x)dx$$

and a(y) is the Y-periodic function equal to $+\infty$ on T, zero on $Y \setminus T$. Then, for every $v \in L^2(\Omega)$

$$w\text{-}L^2(\Omega)\text{lm}_e F^\varepsilon(v) = F(v) \text{ exists, with } F(v) = \frac{1}{2}\,\frac{1}{\mathfrak{m}_Y(w)} \int_\Omega v^2(x)dx. \tag{1.106}$$

Moreover, there is convergence of the corresponding minimum values, that is:

$$\int_\Omega |\text{grad} u_\varepsilon|^2 \sim \varepsilon^2.\mathfrak{m}(w) \int_\Omega f^2 dx.$$

<u>Proof of Theorem 1.28</u> Let us apply the Poincaré inequality on $Y \setminus T$: there exists a strictly positive constant C such that, for every $u \in H^1(Y)$ which satisfies u = 0 on T,

$$\int_{Y\setminus T} u^2(y)dy \leq C \int_{Y\setminus T} |\text{grad} u(y)|^2 dy.$$

By changing the scale, $x = \varepsilon y$, one obtains that, for every $v \in H^1(\Omega)$, v = 0 on T_ε

$$\int_{\Omega_\varepsilon} v^2(x)dx \leq \varepsilon^2.C \int_{\Omega_\varepsilon} |\text{grad} v|^2 dx. \tag{1.107}$$

Noticing that the solution u_ε of (1.103) satisfies

$$\int_{\Omega_\varepsilon} |\mathrm{grad} u_\varepsilon|^2 dx = \int_{\Omega_\varepsilon} f.u_\varepsilon \leq (\int_\Omega f^2 dx)^{\frac{1}{2}} (\int_{\Omega_\varepsilon} u_\varepsilon^2 dx)^{\frac{1}{2}},$$

by taking $v = u_\varepsilon$ in (1.107), we obtain

$$\|\mathrm{grad} u_\varepsilon\|^2_{L^2} \leq \|f\|_{L^2} \|u_\varepsilon\|_{L^2}$$

$$\leq \|f\|_{L^2} \sqrt{C}.\ \varepsilon.\ \|\mathrm{grad} u_\varepsilon\|_{L^2}$$

that is,

$$\|\mathrm{grad} u_\varepsilon\|_{L^2(\Omega)} \leq \sqrt{C}.\varepsilon.\ \|f\|_{L^2(\Omega)}$$

$$\|u_\varepsilon\|_{L^2(\Omega)} \leq C.\varepsilon^2\ \|f\|_{L^2(\Omega)}.$$

This clearly yields the two first conclusions of (1.104).

The sequence $\{v_\varepsilon = u_\varepsilon/\varepsilon^2;\ \varepsilon \to 0\}$ is weakly relatively compact in $L^2(\Omega)$. The last conclusion of (1.104) will follow from the epi-convergence result (1.106), since the minimum point of $F(v) - \int fv$ is precisely $v = \mathfrak{m}_Y(w).f$: Let us define $w_\varepsilon(x) = w(x/\varepsilon)$ where w is the solution of (1.105). Then,

$$w_\varepsilon \longrightarrow \mathfrak{m}_Y(w) \text{ in weak-}L^2(\Omega),$$

as ε goes to zero.

(i) Given $v \in H^1(\Omega) \cap \mathcal{C}(\Omega)$, take $v_\varepsilon = v.\ \frac{1}{\mathfrak{m}_Y(w)}\ w_\varepsilon$.

Then, $v_\varepsilon \longrightarrow v$ weakly in $L^2(\Omega)$, $v_\varepsilon = 0$ on T_ε and

$$F^\varepsilon(v_\varepsilon) = \frac{\varepsilon^2}{2} \int_{\Omega_\varepsilon} |\mathrm{grad} v_\varepsilon|^2 dx$$

$$= \frac{\varepsilon^2}{2}\ \frac{1}{\mathfrak{m}(w)^2} \int_{\Omega_\varepsilon} \{|\mathrm{grad} v|^2.w_\varepsilon^2 + 2v.w_\varepsilon \langle \mathrm{grad} v, \mathrm{grad} w_\varepsilon \rangle + v^2 |\mathrm{grad} w_\varepsilon|^2\} dx.$$

On the other hand, for every open set $A \subset \Omega$,

$$\varepsilon^2.\int_A |\mathrm{grad} w_\varepsilon|^2 dx \sim \varepsilon^2.\ \frac{\mathrm{meas} A}{\varepsilon^N}\ .\ \int_{Y_\varepsilon \setminus T_\varepsilon} |\mathrm{grad} w_\varepsilon(x)|^2 dx.$$

Changing the scale $x = \varepsilon y$, we obtain that

$$\varepsilon^2 \int_A |\text{grad} w_\varepsilon|^2 \xrightarrow[(\varepsilon \to 0)]{} \text{meas } A. \int_{Y \setminus T} |\text{grad} w(y)|^2 dy.$$

Multiplying (1.105) by w and integrating over Y we get

$$\int_{Y \setminus T} |\text{grad} w(y)|^2 dy = \int_{Y \setminus T} w(y) dy = \mathfrak{m}(w).$$

Thus, the sequence of measures $\{\varepsilon^2 |\text{grad} w_\varepsilon|^2 dx;\ \varepsilon \to 0\}$ w*-converges to $\mathfrak{m}(w)dx$ and

$$F^\varepsilon(v_\varepsilon) \to F(v) \text{ as } \varepsilon \to 0.$$

(ii) For any sequence $\{v_\varepsilon;\ \varepsilon \to 0\}$ of functions v_ε equal to zero on T_ε and weakly converging to some v in $L^2(\Omega)$, let us prove that

$$\liminf_{\varepsilon \to 0} F^\varepsilon(v_\varepsilon) \geq F(v).$$

Take some $z \in D(\Omega)$, and introduce $z_\varepsilon = \frac{1}{\mathfrak{m}(w)} \cdot w_\varepsilon . z$ given by the preceding argument. Writing

$$F^\varepsilon(v_\varepsilon) + F^\varepsilon(z_\varepsilon) \geq \varepsilon^2 \int_{\Omega_\varepsilon} \text{grad} v_\varepsilon . \text{grad} z_\varepsilon = \frac{\varepsilon^2}{\mathfrak{m}(w)} \int_{\Omega_\varepsilon} \{\text{grad} v_\varepsilon . w_\varepsilon \text{ grad} z + \text{grad} v_\varepsilon \cdot \text{grad} w_\varepsilon \cdot z\} dx$$

and going to the limit as $\varepsilon \to 0$,

$$\liminf_{\varepsilon \to 0} F^\varepsilon(v_\varepsilon) + \frac{1}{2\mathfrak{m}(w)} \int_\Omega z^2(x) dx \geq \liminf_{\varepsilon \to 0} \frac{\varepsilon^2}{\mathfrak{m}(w)} \int_{\Omega_\varepsilon} \text{grad} v_\varepsilon \cdot \text{grad} w_\varepsilon \cdot z \, dx.$$

Once more, we have to go to the limit on the product of two weakly convergent sequences $\{\varepsilon \text{ grad } v_\varepsilon\}$ and $\{\varepsilon \text{ grad } w_\varepsilon\}$! In order to remove the difficulty, we integrate by parts this last expression, taking advantage of the fact that

$$-\Delta(\varepsilon^2 w_\varepsilon) = 1 \text{ in } \Omega_\varepsilon .$$

(This follows from (1.105) and the Y-periodicity of w). We obtain

$$\liminf_{\varepsilon \to 0} F^{\varepsilon}(v_{\varepsilon}) + \frac{1}{2\mathfrak{m}(w)} \int_{\Omega} z^2(x)dx \geq \frac{1}{\mathfrak{m}(w)} \liminf_{\varepsilon \to 0} \int_{\Omega_{\varepsilon}} v_{\varepsilon} z \, dx$$

$$= \frac{1}{\mathfrak{m}(w)} \int_{\Omega} v \cdot z \, dx.$$

Letting z go to v in $L^2(\Omega)$ (by density of $D(\Omega)$ in $L^2(\Omega)$) we finally obtain

$$\liminf_{\varepsilon \to 0} F^{\varepsilon}(v_{\varepsilon}) \geq \frac{1}{2\mathfrak{m}(w)} \int_{\Omega} v^2 dx,$$

which completes the proof of (1.106).

1.3.5 Transmission problems through thin layers

Let us describe the epi-convergence approach to transmission problems (Examples 1.1.3).

A. Thin isolating layer (refer to Theorem 1.6)

We recall that a layer Σ_{ε} of thickness ε and of conductivity λ is included in a material Ω of conductivity 1. Corresponding energy functional $F^{\varepsilon,\lambda}: H_0^1(\Omega) \to \mathbb{R}^+$, is given by

$$F^{\varepsilon,\lambda}(u) = \frac{1}{2} \int_{\Omega_{\varepsilon}} |\text{grad}\, u|^2 dx + \frac{1}{2} \int_{\Sigma_{\varepsilon}} \lambda \, |\text{grad}\, u|^2 dx,$$

where $\Omega_{\varepsilon} = \Omega \setminus \Sigma_{\varepsilon}$. As ε goes to zero, Σ_{ε} shrinks to Σ (a smooth manifold which splits Ω into two subdomains Ω_1 and Ω_2).

Because of the lack of uniform coercivity of functionals $\{F^{\varepsilon,\lambda}; \varepsilon \to 0, \lambda \to 0\}$, the topology τ for which we have to compute the epi-limit of the sequence $\{F^{\varepsilon,\lambda}\}$ is not given a priori. We have, first, to establish estimations on solutions $\{u_{\varepsilon,\lambda}; \begin{smallmatrix}\varepsilon \to 0\\ \lambda \to 0\end{smallmatrix}\}$ of the minimization problems

$$\min_{u \in H_0^1(\Omega)} \{F^{\varepsilon,\lambda}(u) - \int_{\Omega} fu \, dx\}.$$

Following Brezis et al. [1], Lemma 5.1, by standard estimations and change of scale (just as in the proof of Theorem 1.28) one obtains existence of a constant M such that, for all ε,λ sufficiently small

$$\left.\begin{aligned}&\int_{\Omega}|u_{\varepsilon,\lambda}(x)|^2dx \leq M[(\tfrac{\varepsilon}{\lambda})^2 + 1]\\&\int_{\Omega_\varepsilon}|\text{grad}u_{\varepsilon,\lambda}(x)|^2dx + \lambda\int_{\Sigma_\varepsilon}|\text{grad}u_{\varepsilon,\lambda}(x)|^2dx \leq M[\tfrac{\varepsilon}{\lambda} + 1].\end{aligned}\right\} \quad (1.108)$$

Thus, when $\lim\limits_{\substack{\varepsilon\to 0\\ \lambda\to 0}} (\frac{\lambda}{\varepsilon}) < +\infty$, the sequence $\{u_{\varepsilon,\lambda};\ \varepsilon \to 0,\ \lambda \to 0$ remains bounded in $L^2(\Omega)$, and the limit analysis problem can be reduced to the τ-epi convergence of the sequence $\{F^{\varepsilon,\lambda};\ \varepsilon \to 0,\ \lambda \to 0\}$ on $X = L^2(\Omega)$, with τ = weak topology of $L^2(\Omega)$. (Let us notice that, since we work on an apriori bounded set of $L^2(\Omega)$, this topology is metrizable).

THEOREM 1.29

(a) When $\lambda/\varepsilon \to \alpha$, $0 < \alpha < +\infty$, as ε and λ go to zero, the following convergence holds:

$$F = \tau\text{-lm}_e F^{\varepsilon,\lambda} \text{ exists,} \quad (\tau = w\text{-}L^2(\Omega))$$

where F, for any $u \in L^2(\Omega)$ is given by

$$F(u) = \begin{cases} \frac{1}{2}\int_{\Omega_1\cup\Omega_2}|\text{grad}u|^2dx + \frac{\alpha}{2}\int_{\Sigma}[u]^2dx & \text{if } u \in H^1(\Omega_1 \cup \Omega_2)\\ +\infty & \text{otherwise.}\end{cases}$$

(b) When $\lambda/\varepsilon \to +\infty$, $F = \tau\text{-lm}_e\, F^{\varepsilon,\lambda}$ is equal to $(\alpha = +\infty)$

$$F(u) = \begin{cases} \frac{1}{2}\int_{\Omega}|\text{grad}u|^2dx & \text{if } u \in H^1(\Omega)\\ +\infty & \text{otherwise.}\end{cases}$$

(c) When $\lambda/\varepsilon \to 0$, taking τ = topology of s-L^2 convergence on compact subsets of $\Omega_1 \cup \Omega_2$ then $F = \tau\text{-lm}_e F^{\varepsilon,\lambda}$ exists and is equal to $(\alpha = 0)$

$$F(u) = \begin{cases} \frac{1}{2}\int_{\Omega_1\cup\Omega_2}|\text{grad}u|^2dx & \text{if } u \in H^1(\Omega_1 \cup \Omega_2)\\ +\infty & \text{otherwise.}\end{cases}$$

Proof of Theorem 1.29 Let us sketch the proof and take, for simplicity, Σ the hyperplane $x_3 = 0$ in $\mathbb{R}^3$. Let us treat the case, $\lim_{\{\varepsilon\to 0, \lambda\to 0\}} \lambda/\varepsilon = \alpha$, $0 < \alpha < +\infty$, the limiting cases $\alpha = 0$ and $\alpha = +\infty$ being then easily deduced by a comparison argument. From estimations (1.108) it clearly follows that

$$\mathrm{lm}_e F^{\varepsilon,\lambda}(u) = +\infty \text{ on } L^2(\Omega)\setminus H^1(\Omega_1 \cup \Omega_2).$$

Denoting $\lambda = \lambda(\varepsilon) \sim \varepsilon\,\alpha$, and $F^{\varepsilon} = F^{\varepsilon,\lambda(\varepsilon)}$, we have to find for any $u \in H^1(\Omega_1 \cup \Omega_2)$ on approximating sequence $u_\varepsilon \longrightarrow u$ (weakly in $L^2(\Omega)$) which realizes the minimum of $\{\lim F^{\varepsilon}(u_\varepsilon)/u_\varepsilon \longrightarrow u\}$.

(a) Take

$$u_\varepsilon(x_1,x_2,x_3) = \begin{cases} u(x_1,x_2,x_3) \text{ if } |x_3| > \frac{\varepsilon}{2} \\ \frac{1}{2}[u(x_1,x_2,\frac{\varepsilon}{2})+u(x_1,x_2,-\frac{\varepsilon}{2})]+\frac{x_3}{\varepsilon}[u(x_1,x_2,\frac{\varepsilon}{2})-u(x_1,x_2,-\frac{\varepsilon}{2})] \\ \qquad \text{if } |x_3| < \varepsilon/2. \end{cases}$$

Clearly, $u_\varepsilon \in H_0^1(\Omega)$, $u_\varepsilon \xrightarrow[(\varepsilon\to 0)]{} u$ and

$$F^{\varepsilon}(u_\varepsilon) = \frac{1}{2}\int_{\Omega_\varepsilon} |\mathrm{grad}\,u|^2 dx + \frac{\varepsilon\alpha}{2}\int_{|x_3|<\varepsilon/2} |\frac{1}{2}\,\mathrm{grad}(u(x_1,x_2,\frac{\varepsilon}{2})+u(x_1,x_2,-\frac{\varepsilon}{2}))$$

$$+ \mathrm{grad}(\frac{x_3}{\varepsilon}[u(x_1,x_2,\frac{\varepsilon}{2}) - u(x_1,x_2,-\frac{\varepsilon}{2})])|^2 dx$$

$$\sim \frac{1}{2}\int_{\Omega_\varepsilon} |\mathrm{grad}\,u|^2 dx + \frac{\alpha}{2\varepsilon}\int_{|x_3|<\varepsilon/2} |\mathrm{grad}[x_3(u(\cdot,\frac{\varepsilon}{2})-u(\cdot,-\frac{\varepsilon}{2})]|^2 dx$$

$$\sim \frac{1}{2}\int_{\Omega_\varepsilon} |\mathrm{grad}\,u|^2 dx + \frac{\alpha}{2\varepsilon}\int_{|x_3|<\varepsilon/2} |u(\cdot,\frac{\varepsilon}{2})-u(\cdot,-\frac{\varepsilon}{2})|^2 dx$$

$$+ \frac{\alpha}{2\varepsilon}\int_{|x_3|<\varepsilon/2} |x_3.\mathrm{grad}[u(\cdot,\frac{\varepsilon}{2})-u(\cdot,-\frac{\varepsilon}{2})]|^2 dx$$

$$+ \frac{\alpha}{\varepsilon}\int_{|x_3|<\varepsilon/2} x_3\cdot(u(\cdot,\frac{\varepsilon}{2})-u(\cdot,-\frac{\varepsilon}{2}))\cdot\frac{\partial}{\partial x_3}[u(\cdot,\frac{\varepsilon}{2}) - u(\cdot,-\frac{\varepsilon}{2})]dx.$$

These two last integrals go to zero, being the average of quantities which go to zero with ε (since $x_3 \to 0$ appears under the integral sign). Therefore,

$$F^{\varepsilon}(u_\varepsilon) \sim \frac{1}{2}\int_{\Omega_\varepsilon} |\mathrm{grad}u|^2 dx + \frac{\alpha}{2\varepsilon}\int_{|x_3|<\varepsilon/2} |u(\cdot,\varepsilon/2)-u(\cdot,-\varepsilon/2)|^2 dx$$

and for u sufficiently smooth

$$F^{\varepsilon}(u_\varepsilon) \xrightarrow[(\varepsilon \to 0)]{} \frac{1}{2}\int_{\Omega_1 \cup \Omega_2} |\mathrm{grad}u|^2 dx + \frac{\alpha}{2}\int_{\Sigma} [u]^2 d\sigma = F(u).$$

This first part of the proof is then completed by a density argument.

(b) We have now to prove, in order to complete the proof, that, for any sequence $v_\varepsilon \rightharpoonup u$ in $w\text{-}L^2(\Omega)$, $\liminf_{\varepsilon \to 0} F^{\varepsilon}(v_\varepsilon) \geq F(u)$. Once more, we compare $F^{\varepsilon}(v_\varepsilon)$ to $F^{\varepsilon}(u_\varepsilon)$, where u_ε is constructed as above (which in fact amounts to proving that u_ε realizes such an infimum):

$$\frac{\alpha\varepsilon}{2}\int_{\Sigma_\varepsilon} |\mathrm{grad}v_\varepsilon|^2 + \frac{\alpha\varepsilon}{2}\int_{\Sigma_\varepsilon} |\mathrm{grad}u_\varepsilon|^2 dx \geq \alpha\varepsilon\int_{\Sigma_\varepsilon} \langle \mathrm{grad}v_\varepsilon, \mathrm{grad}u_\varepsilon\rangle dx$$

$$\geq \alpha\varepsilon\int_{\Sigma_\varepsilon} \langle \frac{1}{2}\,\mathrm{grad}(u(\cdot,\tfrac{\varepsilon}{2}) + u(\cdot,\tfrac{-\varepsilon}{2})) + \frac{x_3}{\varepsilon}\mathrm{grad}(u(\cdot,\tfrac{\varepsilon}{2})-u(\cdot,\tfrac{-\varepsilon}{2})) + \frac{\vec{e}_3}{\varepsilon}(u(\cdot,\tfrac{\varepsilon}{2})-u(\cdot,\tfrac{-\varepsilon}{2})), \mathrm{grad}v_\varepsilon\rangle.$$

The same computation as above yields (we can, without restriction, assume u to be regular. Otherwise (cf. Theorem 1.20) take u_ε associated to some regular v, then let $v \to u$!):

$$\liminf_{\varepsilon \to 0} \frac{\alpha\varepsilon}{2}\int_{\Sigma_\varepsilon} |\mathrm{grad}v_\varepsilon|^2 dx + \frac{\alpha}{2}\int_{\Sigma} [u]^2 d\sigma$$

$$\geq \liminf_{\varepsilon \to 0} \alpha\int_{\Sigma_\varepsilon} \langle \mathrm{grad}(u(\cdot,\tfrac{\varepsilon}{2})-u(\cdot,\tfrac{-\varepsilon}{2})), \mathrm{grad}v_\varepsilon\rangle\cdot x_3 \, dx$$

$$+ \liminf_{\varepsilon \to 0} \alpha\int_{\Sigma_\varepsilon} \langle \vec{e}_3, \mathrm{grad}v_\varepsilon\rangle(u(\cdot,\tfrac{\varepsilon}{2})-u(\cdot,\tfrac{-\varepsilon}{2}))\, dx.$$

Without restriction, we can assume $\liminf_{\varepsilon \to 0} F^{\varepsilon}(v_\varepsilon) < +\infty$, otherwise the inequality is clear. Hence,

$$\sup_{\varepsilon>0} \varepsilon\int_{\Sigma_\varepsilon} |\mathrm{grad}v_\varepsilon|^2 dx < +\infty,$$

which implies, for example, that

$$\lim_{\varepsilon\to 0} \alpha\varepsilon \int_{\Sigma_\varepsilon} \langle \frac{1}{2}\,(\mathrm{grad}u(\cdot,\tfrac{\varepsilon}{2}) + \mathrm{grad}u(\cdot,-\tfrac{\varepsilon}{2})), \mathrm{grad}v_\varepsilon \rangle dx$$

$$\leq C\alpha\varepsilon.(\int_{\Sigma_\varepsilon} |\mathrm{grad}v_\varepsilon|^2 dx)^{\frac{1}{2}} \leq \frac{C\alpha\varepsilon}{\sqrt{\varepsilon}} = C\alpha\sqrt{\varepsilon} \to 0, \text{ as } \varepsilon \to 0.$$

This type of inequality has been used in the above computation. Similarly,

$$\limsup_{\varepsilon\to 0} \alpha \int_{\Sigma_\varepsilon} \langle \mathrm{grad}[u(\cdot,\tfrac{\varepsilon}{2})-u(\cdot,-\tfrac{\varepsilon}{2})],\ \mathrm{grad}v_\varepsilon \rangle x_3\ dx$$

$$\leq \limsup_{\varepsilon\to 0} \alpha\ .(\int_{\Sigma_\varepsilon} |\mathrm{grad}v_\varepsilon|^2 dx)^{\frac{1}{2}}(\int_{\Sigma_\varepsilon} |\mathrm{grad}[u(\cdot,\tfrac{\varepsilon}{2})-u(\cdot,-\tfrac{\varepsilon}{2})]|^2 x_3^2 dx)^{\frac{1}{2}}$$

$$\leq \limsup_{\varepsilon\to 0} \alpha\ .\ \frac{C}{\sqrt{\varepsilon}}.\varepsilon(\int_{\Sigma_\varepsilon} |\mathrm{grad}[u(\cdot,-\tfrac{\varepsilon}{2})-u(\cdot,\tfrac{\varepsilon}{2})]|^2 dx)^{\frac{1}{2}} \to 0 \text{ as } \varepsilon \to 0.$$

(We have majorized this last integral by using: $|x_3| \leq \varepsilon$ on Σ_ε). Therefore,

$$\liminf_{\varepsilon\to 0} \frac{\alpha\varepsilon}{2} \int_{\Sigma_\varepsilon} |\mathrm{grad}v_\varepsilon|^2 dx + \frac{\alpha}{2}\int_\Sigma [u]^2 d\sigma$$

$$\geq \liminf_{\varepsilon\to 0} \alpha \int_{\Sigma_\varepsilon} \langle \vec{e}_3, \mathrm{grad}v_\varepsilon \rangle \cdot (u(\cdot,\tfrac{\varepsilon}{2})-u(\cdot,-\tfrac{\varepsilon}{2}))dx. \tag{1.109}$$

By integration (since $u(\cdot,\frac{\varepsilon}{2})$ and $u(\cdot,-\frac{\varepsilon}{2})$ do not depend on x_3!),

$$\alpha\int_{\Sigma_\varepsilon} \langle \vec{e}_3, \mathrm{grad}v_\varepsilon \rangle (u(\cdot,\tfrac{\varepsilon}{2})-u(\cdot,-\tfrac{\varepsilon}{2}))dx$$

$$=\alpha\int_{(\partial\Sigma_\varepsilon)^+} v_\varepsilon\ (u(\cdot,\tfrac{\varepsilon}{2})-u(\cdot,-\tfrac{\varepsilon}{2}))d\sigma - \alpha\int_{(\partial\Sigma_\varepsilon)^-} v_\varepsilon.(u(\cdot,\tfrac{\varepsilon}{2})-u(\cdot,-\tfrac{\varepsilon}{2}))d\sigma.$$

Since $v_{\varepsilon_{(\partial\Sigma_\varepsilon)^+}} \longrightarrow u(\cdot,0^+)$, $v_{\varepsilon_{(\partial\Sigma_\varepsilon)^-}} \longrightarrow u(\cdot,0^-)$, we obtain

$$\lim_{\varepsilon\to 0} \alpha\int_{\Sigma_\varepsilon} \langle \vec{e}_3, \mathrm{grad}v_\varepsilon \rangle (u(\cdot,\tfrac{\varepsilon}{2})-u(\cdot,-\tfrac{\varepsilon}{2}))dx = \alpha\int_\Sigma [u]^2 d\sigma.$$

Returning to (1.109)

$$\liminf_{\varepsilon} \frac{\alpha\varepsilon}{2}\int_{\Sigma_\varepsilon} |\mathrm{grad}v_\varepsilon|^2 dx + \frac{\alpha}{2}\int_\Sigma [u]^2 d\sigma \geq \alpha\int_\Sigma [u]^2 d\sigma$$

which finally implies

$$\liminf_{\varepsilon \to 0} F^{\varepsilon}(v_{\varepsilon}) \geq \int_{\Omega_1 \cup \Omega_2} |\text{grad} u|^2 dx + \frac{\alpha}{2} \int_{\Sigma} [u]^2 d\sigma. \quad \square$$

B. Neumann's strainer (refer to Example 1.1.3, Theorem 1.7)

We recall that Σ, a hyperplane in $\mathbb{R}^3$, splits an open bounded set Ω into two subdomains Ω_1 and Ω_2. These subdomains Ω_1 and Ω_2 are, in fact, connected by many small "holes" T_{ε} periodically distributed on Σ, while $\Sigma \setminus T_{\varepsilon}$ can be imagined as an isolating material. The holes $T_{\varepsilon} = \cup T_{\varepsilon}^{i}$ are obtained by taking an r_{ε}-homothetic of a fixed hole T strongly included in $Y =]-\frac{1}{2}, +\frac{1}{2}[^{N-1}$ which is then translated in order to obtain an ε-periodic configuration in all directions of Σ. The "density of holes" is described by r_{ε} (see Fig. 1.12).

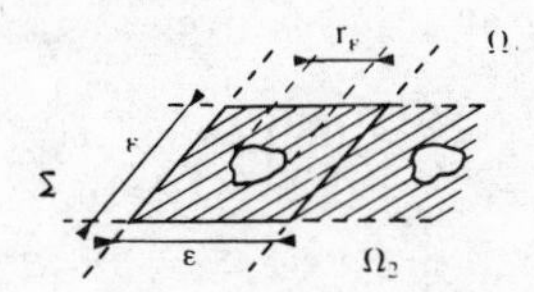

Figure 1.12

The epi-convergence approach to such limit analysis problem ($\varepsilon \to 0$) relies on the following variational formulation introduced by Attouch, Damlamian et al. [1]. The heat temperature u_{ε}, for a given source f, is a solution of

$$\min_{u \in H_0^1(\Omega_1 \cup \Omega_2)} \{F^{\varepsilon}(u) - \int_{\Omega} fu\}$$

$F^{\varepsilon} : H_0^1(\Omega_1 \cup \Omega_2) \to \bar{\mathbb{R}}^+$ is equal to

$$F^{\varepsilon}(u) = \int_{\Omega_1 \cup \Omega_2} |\text{grad} u|^2 dx + I_{K^{\varepsilon}}(u)$$

where

$$K^{\varepsilon} = \{u \in H^1(\Omega_1 \cup \Omega_2) / [u] = 0 \text{ on } T_{\varepsilon}\}.$$

The property $u \in H_0^1(\Omega_{\varepsilon})$, where $\Omega_{\varepsilon} = \Omega_{\varepsilon} = \Omega_1 \cup \Omega_2 \cup T_{\varepsilon}$ is interpreted as a constraint imposed on the jump [u] of u across Σ ($[u] = u|_1 - u|_2$), that is, to be equal to zero on T_ε! The space on which functionals $\{F^\varepsilon; \varepsilon \to 0\}$ are naturally defined and uniformly coercive is $X = H^1(\Omega_1 \cup \Omega_2)$. Hence, the limit analysis problem is equivalent to the study of τ-epi convergence of the sequence of functionals $\{F^\varepsilon : X \to \bar{\mathbb{R}}^+; \varepsilon \to 0\}$, τ-weak topology of X.

THEOREM 1.30 When ε goes to zero, the sequence $\{F^\varepsilon; \varepsilon \to 0\}$ τ-epi converges. Depending on $\lim_{\varepsilon \to 0} r_\varepsilon/\varepsilon^2$, its epi-limit $F = \tau\text{-lm}_e F^\varepsilon$ is equal to:

(a) $r_\varepsilon \ll \varepsilon^2$ $\quad F(u) = \int_{\Omega_1 \cup \Omega_2} |\text{grad}u|^2 dx$ for every $u \in H^1(\Omega_1 \cup \Omega_2)$

(b) $r_\varepsilon = \varepsilon^2$ $\quad F(u) = \int_{\Omega_1 \cup \Omega_2} |\text{grad}u|^2 dx + \frac{C}{4} \int_\Sigma [u]^2 d\sigma$ for every $u \in H^1(\Omega_1 \cup \Omega_2)$

where C is equal to the capacity of T in $\mathbb{R}^3$, i.e.

$$C = \inf \{\int_{\mathbb{R}^3} |\text{grad}w(x)|^2 dx;\ w \in H^1(\mathbb{R}^3),\ w = 1 \text{ on } T\}$$

(c) $r_\varepsilon \gg \varepsilon^2$

$$F(u) = \begin{cases} \int_\Omega |\text{grad}u|^2 dx & \text{if } u \in H^1(\Omega) \\ +\infty & \text{otherwise.} \end{cases}$$

As we have already noticed, these results can be formulated in a unified way:

$$F = \tau\text{-lm}_e F^\varepsilon \text{ with } F(u) = \int_{\Omega_1 \cup \Omega_2} |\text{grad}u|^2 dx + C \int_\Sigma [u]^2 dx$$

with C equal to zero, $\frac{\text{CapT}}{4}$, $+\infty$ respectively as $\lim_{\varepsilon \to 0} r_\varepsilon/\varepsilon^2$ is equal to 0,1, $+\infty$. Moreover, the approximating and limit functionals turn out to be of the same nature when noticing that

$$F^\varepsilon(u) = \int_{\Omega_1 \cup \Omega_2} |\text{grad}u|^2 dx + \int_\Sigma a_\varepsilon(x)[u]^2 dx \text{ with } a_\varepsilon(x) = \begin{cases} 0 & \text{if } x \in \Sigma \setminus T_\varepsilon \\ +\infty & \text{if } x \in T_\varepsilon. \end{cases}$$

Proof of Theorem 1.30 We just sketch the proof and refer to Attouch, Damlamian et al. [1] for details. In order to justify the introduction of the test functions, let us assume for a moment that $F^{\varepsilon} \xrightarrow{e} F$ and examine the form of this limit functional:

$$F(u) = \min \{\lim_{\varepsilon \to 0} \int_{\Omega_1 \cup \Omega_2} |\mathrm{grad} u_\varepsilon|^2 dx;\ u_\varepsilon \longrightarrow u,\ [u_\varepsilon] = 0 \text{ on } T_\varepsilon\}.$$

Writing $u_\varepsilon = u - z_\varepsilon$, $z_\varepsilon \longrightarrow 0$, we obtain

$$F(u) = \int_{\Omega_1 \cup \Omega_2} |\mathrm{grad} u|^2 dx + G([u])$$

where

$$G([u]) = \min \{\lim_{\varepsilon \to 0} \int_{\Omega_1 \cup \Omega_2} |\mathrm{grad} z_\varepsilon|^2 dx;\ z_\varepsilon \longrightarrow 0;\ [z_\varepsilon] = [u] \text{ on } T_\varepsilon\}.$$

Because of the local character of the problems, invariance with respect to translations of the limit problem and quadratic properties of the functionals, G is necessarily of the type

$$G(u) = C \int_\Sigma [u]^2 d\sigma.$$

Thus, an heuristic argument tells us that the limit functional is of the type

$$F(u) = \int_{\Omega_1 \cup \Omega_2} |\mathrm{grad} u|^2 dx + C \int_\Sigma [u]^2 d\sigma.$$

Taking $u \equiv +1/2$ on Ω_1, $u \equiv -1/2$ on Ω_2, we derive

$$C.\mathrm{meas}\ \Sigma = \min \{\lim_{\varepsilon \to 0} \int_{\Omega_\varepsilon} |\mathrm{grad} u_\varepsilon|^2 dx / u_\varepsilon \to u,\ [u_\varepsilon] = 0 \text{ on } T_\varepsilon\}.$$

But because of the odd property of the function u, to approach it with minimal energy by (a) functions with jump equal to zero on T_ε, or by (b) functions equal to zero on T_ε are two equivalent problems!

This makes the connection with the problem $\{u = 0 \text{ on } T_\varepsilon\}$ studied in Section 1.3.4 (Theorem 1.27), explains why we find the "same" constants ($C = (\mathrm{Cap}T)/4$), and introduces the good test functions $\{w_\varepsilon; \varepsilon \to 0\}$:

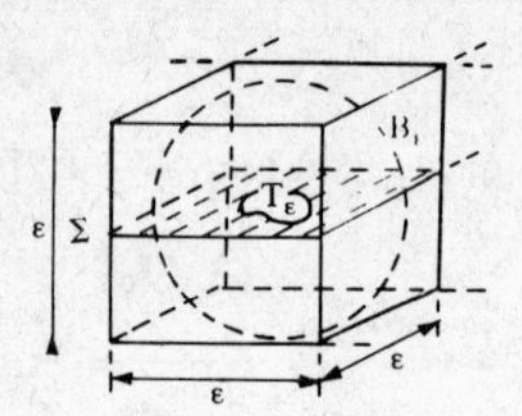

Figure 1.13

Let B_ε be the sphere of diameter ε included in each cell of side ε (in $\mathbb{R}^3$!). Take w_ε the capacity potential of T_ε in B_ε:

$$\begin{cases} \Delta w_\varepsilon = 0 \text{ in } B_\varepsilon \setminus T_\varepsilon \\ w_\varepsilon = 1 \text{ on } T_\varepsilon \\ w_\varepsilon = 0 \text{ on } \partial B_\varepsilon. \end{cases}$$

w_ε is then extended by periodicity to the layer of size ε around Σ and then to the whole of $\mathbb{R}^3$ by zero. Because of the symmetry of the problem with respect to Σ,

$$\frac{\partial w_\varepsilon}{\partial n} = 0 \text{ on } \Sigma \setminus T_\varepsilon.$$

We then easily verify that

$$w_\varepsilon \longrightarrow 0 \text{ weakly in } H^1(\Omega) \text{ and}$$

$$\int_{\Omega_1 \cup \Omega_2} |\text{grad} w_\varepsilon|^2 dx \longrightarrow \begin{cases} 0 \text{ if } r_\varepsilon \ll \varepsilon^2 \\ \text{CapT if } r_\varepsilon = \varepsilon^2 \\ +\infty \text{ if } r_\varepsilon \gg \varepsilon^2. \end{cases}$$

We can now describe the form for every $u \in X$, of an approximating sequence u_ε realizing $\min \{\lim F^\varepsilon(u_\varepsilon) / u_\varepsilon \rightarrow u\}$. Denoting $u = (u_1, u_2) \in H^1(\Omega_1) \times H^1(\Omega_2)$, take

$$u_\varepsilon = (u_1, u_2) - w_\varepsilon(r_1, r_2)$$

$$= (u_1 - w_\varepsilon r_1, u_2 - w_\varepsilon r_2),$$

where r_1 belongs to $H^1(\Omega_1)$ and satisfies $r_{1|\Sigma} = \frac{1}{2}[u]$,

r_2 belongs to $H^1(\Omega_2)$ and satisfies $r_{2|\Sigma} = -\frac{1}{2}[u]$.

Clearly, $u_\varepsilon \longrightarrow u$ in w-X,

$$[u_\varepsilon] = [u] - (r_{1|\Sigma} - r_{2|\Sigma}) = 0 \text{ on } T_\varepsilon.$$

Let us compute

$$F^\varepsilon(u_\varepsilon) = \int_{\Omega_\varepsilon} |\text{grad}u_\varepsilon|^2 dx$$

$$= \sum_{i=1,2} \int_{\Omega_\varepsilon^i} |\text{grad}u_i - \text{grad}w_\varepsilon . r_i - w_\varepsilon . \text{grad}r_i|^2 \, dx$$

$$\sim \sum_{i=1,2} \int_{\Omega_\varepsilon^i} |\text{grad}u_i|^2 + \sum_{i=1,2} \int_{\Omega_\varepsilon^i} |\text{grad}w_\varepsilon|^2 \, r_i^2 \, dx$$

$$\underset{(\varepsilon \to 0)}{\longrightarrow} \int_{\Omega_1 \cup \Omega_2} |\text{grad}u|^2 dx + \frac{C}{4} \int_\Sigma [u]^2 d\sigma.$$

(Take first u regular, then use a density and diagonalization argument).

In order to complete the proof of the epi-convergence result we have to show that, for every $v_\varepsilon \longrightarrow u$ in w-X, $\liminf_{\varepsilon \to 0} F^\varepsilon(v_\varepsilon) \geq F(u)$. Once more, without restriction, one can assume $u = (u_1, u_2)$ with u_i regular. (Otherwise one approaches u by $\phi = (\phi_1, \phi_2)$ with ϕ_i regular). Then, comparing v_ε to u_ε constructed as above,

$$\int_{\Omega_i} |\text{grad}v_\varepsilon|^2 dx \geq 2 \int_{\Omega_i} \text{grad}v_\varepsilon . \text{grad}u_\varepsilon - \int_{\Omega_i} |\text{grad}u_\varepsilon|^2 dx$$

with quite similar arguments as in the proof of Theorem 1.27, we obtain the conclusion. □

1.4 GEOMETRIC INTERPRETATION OF EPIGRAPH-CONVERGENCE

In this section, we show that notions of epi-convergence for sequences of extended real valued functions are associated with set-limit notions for corresponding epigraphs: Epi-convergence is nothing but set-convergence (in

Kuratowski sense [1]) of corresponding epigraphs. Given a function F from X into $\bar{\mathbb{R}}$, its epigraph, denoted epi F, is the subset of $X \times \mathbb{R}$ equal to

$$\text{epi } F = \{(x,\lambda) \in X \times \mathbb{R} \,/\, \lambda \geq F(x)\}.$$

1.4.1 Set-convergence

DEFINITION 1.31 Let $(C^n)_{n\in\mathbb{N}}$ be a sequence of subsets of a topological space (X,τ). *The* τ*-lower limit of the sequence* $(C^n)_{n\in\mathbb{N}}$, *denoted* $\tau\text{-Li}C^n$ (or briefly $\text{Li}C^n$, when there is no ambiguity in the topology τ) is the subset of X defined by:

$$\tau\text{-Li}C^n = c\ell_\tau \bigcup_{n\in\mathbb{N}} \bigcap_{k\geq n} c\ell_\tau C^k \tag{1.110}$$

where $c\ell_\tau$ denotes the τ-closure operation.

The τ*-upper limit of the sequence* $(C^n)_{n\in\mathbb{N}}$, *denoted* $\tau\text{-Ls}C^n$ (or $\text{Ls}C^n$ for short) is the subset of X defined by:

$$\tau\text{-Ls}C^n = \bigcap_{n\in\mathbb{N}} c\ell_\tau \Big(\bigcup_{k\geq n} C^k \Big) \tag{1.111}$$

The sequence $(C^n)_{n\in\mathbb{N}}$ is said to be τ-convergent if the following equality holds:

$$\tau\text{-Li}C^n = \tau\text{-Ls}C^n.$$

Its limit, denoted $C = \tau\text{-Lm}C^n$, is the subset of X equal to this common value

$$C = \tau\text{-lm}C^n = \tau\text{-Li}C^n = \tau\text{-Ls}C^n.$$

We shall sometimes refer to this notion of set-convergence under the name "Kuratowski convergence".

The following technical lemma makes these limits easy to handle:

PROPOSITION 1.32 Let $(C^n)_{n\in\mathbb{N}}$ be a sequence of subsets of a topological space (X,τ). Then,

$$\tau\text{-Li}C^n = \bigcap_{H\in\ddot{\mathbb{N}}} c\ell_\tau \Big(\bigcup_{k\in H} C^k \Big) \tag{1.112}$$

$$\tau\text{-}LsC^n = \bigcap_{H\in\ddot{\mathcal{N}}} cl_\tau(\bigcup_{k\in H} C^k) \tag{1.113}$$

where $\mathcal{N}$ is the Frechet filter on $\mathbb{N}$: $H \in \mathcal{N}$, iff there exists some $n \in \mathbb{N}$ such that $H \supset \{k \in \mathbb{N}/k > n\}$, and $\ddot{\mathcal{N}}$ is the grill of $\mathcal{N}$, that is to say, the set family formed by the subsets of $\mathbb{N}$ which meet all the elements of $\mathcal{N}$, in other words all the (infinite) subsequences of $\mathbb{N}$.

Proof of Proposition 1.32 Formula (1.113) follows directly from definition (1.111) of $\tau\text{-}LsC^n$ and the fact that the set family $(H_n)_{n\in\mathbb{N}}$, $H_n = \{k\in\mathbb{N}/k>n\}$, is a base for the Frechet filter $\mathcal{N}$.

Let us prove (1.112)

(a) $\tau\text{-}LiC^n = cl_\tau \bigcup_{n\in\mathbb{N}} \bigcap_{k>n} cl_\tau C^k \subset \bigcap_{H\in\ddot{\mathcal{N}}} cl_\tau (\bigcup_{k\in H} C^k)$.

Since the second member is closed, it is equivalent to prove that:

$$\forall n \in \mathbb{N}, \forall H \in \ddot{\mathcal{N}} \quad \bigcap_{k>n} cl_\tau C^k \subset cl_\tau (\bigcup_{k\in H} C^k)$$

By definition of $\ddot{\mathcal{N}}$, since $H \in \ddot{\mathcal{N}}$, for any $n \in \mathbb{N}$

$$\{k \in \mathbb{N}/k > n\} \cap H \neq \emptyset.$$

Therefore, $\bigcap_{k>n} cl_\tau C^k \subset \bigcup_{k\in H} cl_\tau C^k \subset cl_\tau \bigcup_{k\in H} cl_\tau C^k$.

We get the desired inclusion noticing that, for any collection of sets $(C^i)_{i\in I}$ in a topological space (X,τ);

$$cl_\tau(\bigcup_{i\in I} C^i) = cl_\tau(\bigcup_{i\in I} cl_\tau C^i) \tag{1.114}$$

(b) Let us now prove the opposite inclusion.

$$cl_\tau \bigcup_{n\in\mathbb{N}} \bigcap_{k>n} cl_\tau C^k \supset \bigcap_{H\in\ddot{\mathcal{N}}} cl_\tau(\bigcup_{k\in H} C^k).$$

From (1.114) it is equivalent to prove that for any sequence of closed sets $(C^n)_{n\in\mathbb{N}}$

$$B = cl_\tau(\bigcup_{n\in\mathbb{N}} \bigcap_{k>n} C^k) \supset \bigcap_{H\in\ddot{\mathcal{N}}} cl_\tau(\bigcup_{k\in H} C^k) = A$$

Let us take $x \notin B$ and prove that $x \notin A$. Since $x \notin B$, there exists an open set $\mathbb{O}$, for the topology τ, which contains x and which satisfies:

$$\forall n \in \mathbb{N} \quad \exists k(n) > n \text{ such that } \mathbb{O} \cap C^{k(n)} = \emptyset. \tag{1.115}$$

Let us define $H = \{k(n)/n \in \mathbb{N}\}$. From (1.115), H belongs to $\ddot{\mathbb{N}}$ and

$$\forall k \in H \quad \mathbb{O} \cap C^k = \emptyset; \text{ equivalently } \mathbb{O} \cap (\bigcup_{k \in H} C^k) = \emptyset.$$

So, we have found an H belonging to $\ddot{\mathbb{N}}$ such that: $x \notin c\ell_\tau(\bigcup_{k \in H} C^k)$, which is equivalent to say that $x \notin A$.

PROPOSITION 1.33 Let $(C^n)_{n \in \mathbb{N}}$ be a sequence of subsets of a topological space (X,τ). Then $\tau\text{-Li}C^n$ and $\tau\text{-Ls}C^n$ are two closed subsets of (X,τ) and

$$\tau\text{-Li}C^n \subset \tau\text{-Ls}C^n.$$

Proof of Proposition 1.33 Since an intersection of closed sets is still closed, from (1.112) and (1.113) it follows that $\tau\text{-Li}C^n$ and $\tau\text{-Ls}C^n$ are closed for the topology τ. The inclusion $\tau\text{-Li}C^n \subset \tau\text{-Ls}C^n$ follows from (1.112) and (1.113) and from the inclusion $\mathcal{N} \subset \ddot{\mathbb{N}}$.

The fact that a filter is included in its grill relies on the following general property of a filter $\mathcal{F}$:

$$\forall A,B \in \mathcal{F} \quad A \cap B \neq \emptyset. \quad \square$$

Let us now give, for the case when the topological space (X,τ) is metrizable, or more generally first countable, the sequential formulations of $\tau\text{-Li}C^n$ and $\tau\text{-Ls}C^n$.

PROPOSITION 1.34 Let (X,τ) be a first countable topological space. For any sequence $(C^n)_{n \in \mathbb{N}}$ of subsets of X, the following sequential formulations hold:

$$\tau\text{-Li}C^n = \{x \in X/\exists (x_n)_{n \in \mathbb{N}}, \forall n \in \mathbb{N} \ x_n \in C^n \text{ and } x_n \xrightarrow[n \to +\infty]{\tau} x\}. \tag{1.116}$$

$$\tau\text{-Ls}C^n = \{x \in X/\exists (n_k)_{k \in \mathbb{N}} \ \exists (x_k)_{k \in \mathbb{N}}, \forall k \in \mathbb{N} \ x_k \in C^{n_k} \text{ and } x_k \xrightarrow[k \to +\infty]{\tau} x\}. \tag{1.117}$$

Proof of Proposition 1.34 Let us first prove (1.116); the inclusion $\tau\text{-Li}C^n \supset \{x = \lim_{n\to+\infty} x_n / x_n \in C^n\}$ is true in a general topological space.

Let $(x_n)_{n\in\mathbb{N}}$ such that $\forall n \in \mathbb{N}\ x_n \in C^n$ and $x_n \xrightarrow[n\to+\infty]{\tau} x$; then, for any τ-neighbourhood $\mathcal{O}$ of x, there exists $H_{\mathcal{O}} \in \mathcal{N}$ such that: $\bigcup_{k\in H_{\mathcal{O}}} x_k \subset \mathcal{O}$. For any $H \in \ddot{\mathcal{N}}$, since $H \cap H_{\mathcal{O}} \neq \emptyset$, it follows from the preceding inclusion that $\mathcal{O} \cap (\bigcup_{k\in H} C^k) \neq \emptyset$. This being true for any τ-neighbourhood $\mathcal{O}$ of x and any $H \in \ddot{\mathcal{N}}$, it follows

$$x \in \bigcap_{H\in\ddot{\mathcal{N}}} c\ell_\tau(\bigcup_{k\in H} C^k) = \tau\text{-Li}C^n, \text{ from (1.112).}$$

Let us prove the opposite inclusion and use now the fact that (X,τ) is first countable. Let $x \in \tau\text{-Li}C^n$ and $(\mathcal{O}_m)_{m\in\mathbb{N}}$ be a countable base of open neighbourhoods of x, which is decreasing and such that $\bigcap_{m\in\mathbb{N}} \mathcal{O}_m = \{x\}$.

By definition (1.110) of $\tau\text{-Li}C^n$, for every $m \in \mathbb{N}$, there exists an integer $N(m)$ such that:

$$\forall n \geq N(m) \quad \mathcal{O}_m \cap C^n \neq \emptyset. \tag{1.118}$$

Without restriction, we may assume the sequence $m \to N(m)$ to be strictly increasing. Let us define a sequence $(x_n)_{n\in\mathbb{N}}$ in the following way: For each n such that $N(m) \leq n < N(m+1)$, take $x_n \in \mathcal{O}_m \cap C^n$, which is non void by (1.118). The so defined sequence $(x_n)_{n\in\mathbb{N}}$ satisfies:

$$\forall n \in \mathbb{N} \quad x_n \in C^n$$

and

$$\forall m \in \mathbb{N} \quad x_n \in \mathcal{O}_m \quad \text{for all } n \geq N(m)$$

(since $\mathcal{O}_m$ is decreasing with m). This is exactly the definition of the convergence of the sequence $(x_n)_{n\in\mathbb{N}}$ to x, for the topology τ.

The proof of (1.117) can be obtained in a similar way. □

From Proposition 1.34, we now derive the sequential formulation of Kuratowski set-convergence:

COROLLARY 1.35 Let (X,τ) be a first countable topological space and let $(C^n)_{n\in\mathbb{N}}$ be a sequence of subsets of X. The following are equivalent:

(i) $C = \tau\text{-Lim } C^n$ (in the Kuratowski sense)

(ii) $\begin{cases} \forall x \in C, \ \exists (x_n)_{n \in \mathbb{N}} \text{ such that: } \forall n \in \mathbb{N} \ x_n \in C^n \text{ and } x_n \xrightarrow[n \to +\infty]{\tau} x. \\ \forall (n_k)_{k \in \mathbb{N}} \ \forall (x_k)_{k \in \mathbb{N}} \text{ such that } \forall k \in \mathbb{N} \ x_k \in C^{n_k}, \ (x_k \xrightarrow[k \to +\infty]{\tau} x) \Rightarrow (x \in C). \end{cases}$

Comments From the sequential formulation of $\tau\text{-Li}C^n$ (Proposition 1.34) and the fact that this limit set is closed (Proposition 1.33), we can derive another proof of the diagonalization result (Corollary 1.18) and extend this result to the case of a first countable space (X,τ).

Let us consider the set $A = \{x_{\nu,\mu} / \nu \in \mathbb{N}, \mu \in \mathbb{N}\}$ and the following partition of A:

$$A = \bigcup_{\nu \in \mathbb{N}} A_\nu \text{ with } A_\nu = \{x_{\nu,\mu} / \mu \in \mathbb{N}\}.$$

Since $x_{\nu,\mu} \xrightarrow[\nu \to +\infty]{} x_\mu$, for every $\mu \in \mathbb{N}$, $x_\mu \in \tau\text{-Li}A_\nu$. Since $x_\mu \xrightarrow[\mu \to +\infty]{\tau} x$ and $\tau\text{-Li}A_\nu$ is closed, $x \in \tau\text{-Li}A_\nu$. (X,τ) being first countable, from Proposition 1.34, there exists a map $\nu \to \mu(\nu)$ such that $x_{\nu,\mu(\nu)} \xrightarrow[\nu \to +\infty]{\tau} x$. □

We are now able to formulate and prove the geometric interpretation of the epi-convergence of a sequence of functions. The following theorem was first proved by Mosco (1969) [2] in the case of convex functions, then extended by De Giorgi (1975) [2] to the general situation.

1.4.2 Set convergence interpretation of epi-convergence

THEOREM 1.36 Let $(F^n)_{n \in \mathbb{N}}$ a sequence of functions from (X,τ), a general topological space, into $\bar{\mathbb{R}}$. The limit sets $\tau\text{-Li(epi } F^n)$ and $\tau\text{-Ls(epi } F^n)$ are still epigraphs. They are equal respectively to the epigraphs of $\tau\text{-ls}_e F^n$ and $\tau\text{-li}_e F^n$:

$$\tau\text{-Li(epi } F^n) = \text{epi}(\tau\text{-ls}_e F^n) \tag{1.119}$$

$$\tau\text{-Ls(epi } F^n) = \text{epi}(\tau\text{-li}_e F^n) \tag{1.120}$$

Proof of Theorem 1.36 Let us first prove (1.119). By definition (1.110) of $\tau\text{-Li}C^n$, $(x,\lambda) \in \tau\text{-Li(epi } F^n)$ if and only if:

$$\forall V \in N_\tau(x), \ \forall \varepsilon > 0, \ \exists n \in \mathbb{N} \text{ s.t. } \forall k \geq n \quad V \times]\lambda-\varepsilon, \lambda+\varepsilon[\cap \text{epi } F^k \neq \emptyset.$$

From the geometrical properties of epigraphs, this is equivalent to

$$\forall V \in N_\tau(x) \quad \forall \varepsilon > 0, \exists n \in \mathbb{N} \text{ s.t. } \forall k \geq n \quad \exists x_k \in V$$

satisfying

$$\lambda + \varepsilon > F^k(x_k).$$

This can be reformulated in the following way:

$$\lambda \geq \sup_{V \in N_\tau(x)} \inf_{n \in \mathbb{N}} \sup_{k \geq n} \inf_{u \in V} F^k(u),$$

that is,

$$\lambda \geq \sup_{V \in N_\tau(x)} \limsup_{n \to +\infty} \inf_{u \in V} F^n(u) = (\tau\text{-ls}_e F^n)(x)$$

which means $(x,\lambda) \in \text{epi}\,(\tau\text{-ls}_e F^n)$.

Let us now prove (1.120). By definition (1.111) of τ-LsC^n, $(x,\lambda) \in \tau$-Ls(epi F^n) if and only if

$$\forall n \in \mathbb{N} \ \forall V \in N_\tau(x) \quad \forall \varepsilon > 0 \quad \exists k \geq n \text{ such that } V \times]\lambda-\varepsilon,\lambda+\varepsilon[\cap \text{ epi } F^k \neq \emptyset.$$

Because the sets are epigraphs, this is equivalent to:

$$\forall n \in \mathbb{N} \ \forall V \in N_\tau(x) \quad \forall \varepsilon > 0 \ \exists k \geq n \ \exists x_k \in V \text{ such that } \lambda + \varepsilon > F^k(x_k).$$

This can be reformulated in the following way

$$\lambda \geq \sup_{V \in N_\tau(x)} \sup_{n \in \mathbb{N}} \inf_{k \geq n} \inf_{u \in V} F^k(u)$$

that is,

$$\lambda \geq \sup_{V \in N_\tau(x)} \liminf_{n \to +\infty} \inf_{u \in V} F^n(u) = (\tau\text{-li}_e F^n)(x),$$

$$(x,\lambda) \in \text{epi}\,(\tau\text{-li}_e F^n). \quad \square$$

REMARK 1.37 It is possible to give a joint proof of (1.119) and (1.120) with the help of formulae (1.112) and (1.113): since we have proved that

$$\tau\text{-Ls(epi } F^n) = \text{epi}(\tau\text{-li}_e F^n),$$

replacing the set family $\mathbb{N}$ (Frechet filter) by its grill $\ddot{\mathbb{N}}$, and using (1.112), we get:

$$\tau\text{-Li}(\text{epi } F^n) = \text{epi } F$$

with

$$F(x) = \sup_{V \in N_\tau(x)} \sup_{H \in \ddot{\mathbb{N}}} \inf_{n \in H} \inf_{u \in V} F^n(u)$$

$$= \tau\text{-ls}_e F^n.$$

This last equality follows from the following elementary lemma:

LEMMA 1.38 Let $\{a_\nu ; \nu \in \mathbb{N}\}$ be a filtered family in $\bar{\mathbb{R}}$. Then,

$$\liminf_\nu a_\nu = \sup_{H \in \mathbb{N}} \inf_{\nu \in H} a_\nu = \inf_{H \in \ddot{\mathbb{N}}} \sup_{\nu \in H} a_\nu$$

$$\limsup_\nu a_\nu = \inf_{H \in \mathbb{N}} \sup_{\nu \in H} a_\nu = \sup_{H \in \ddot{\mathbb{N}}} \inf_{\nu \in H} a_\nu .$$

We are now able to state the main result of this paragraph and establish the equivalence between epi-convergence of a sequence of functions and the Kuratowski convergence of their epigraphs. It is a direct consequence of Definition 1.9 and of Theorem 1.36.

THEOREM 1.39 Let $(F^n)_{n \in \mathbb{N}}$ a sequence of functions from a general topological space (X, τ) into $\bar{\mathbb{R}}$. The sequence $(F^n)_{n \in \mathbb{N}}$ is τ-epi-convergent if and only if the sequences of sets $(\text{epi}F^n)_{n \in \mathbb{N}}$ in the topological space $X \times \mathbb{R}$ (equipped with the product topology) is convergent (in the Kuratowski sense). In that case, the following equality holds:

$$\text{epi } (\tau\text{-lm}_e F^n) = \tau\text{-Lm } (\text{epi } F^n). \quad \square$$

Theorem 1.39 allows us to interpret (by consideration of the epigraphs) the epi-convergence of a sequence a functions in terms of set-convergence. The opposite step is also important and can be realized with the help of the correspondence

$$C \longmapsto I_C$$

which associates with a subset C of an abstract space X the indicator function of C, that is, the function equal to 0 on C and $+\infty$ on $X \setminus C$.

PROPOSITION 1.40 Let $(C^n)_{n\in\mathbb{N}}$ be a sequence of subsets of a topological space (X,τ); then the following equalities hold:

$$\begin{aligned} &\tau\text{-}li_e\, I_{C^n} = I_{\tau\text{-}LsC^n} \\ &\tau\text{-}ls_e\, I_{C^n} = I_{\tau\text{-}LiC^n}. \end{aligned} \tag{1.122}$$

Consequently, the sequence $(C^n)_{n\in\mathbb{N}}$ converges in the Kuratowski sense in (X,τ) iff the sequence $(I_{C^n})_{n\in\mathbb{N}}$ τ-epi converges in (X,τ); in that case the following equality holds:

$$\tau\text{-}lm_e\, I_{C^n} = I_{\tau\text{-}LmC^n}.$$

Proof of Proposition 1.40 Let us prove, for example, the first equality $\tau\text{-}li_e\, I_{C^n} = I_{\tau\text{-}lsC^n}$; the other one is obtained in a similar way. Let

$$x \in \tau\text{-}LsC^n = \bigcap_{H\in\mathcal{N}} c\ell_\tau\Big(\bigcup_{k\in H} C^k\Big);$$

then

$$\forall H \in \mathcal{N} \quad \forall \mathcal{O} \in \mathcal{O}_\tau(x) \quad \exists k \in H \quad \text{s.t. } \mathcal{O} \cap C^k \neq \emptyset;$$

equivalently

$$\forall H \in \mathcal{N} \quad \forall \mathcal{O} \in \mathcal{O}_\tau(x) \quad \exists k \in H \quad \exists u \in \mathcal{O} \text{ s.t. } I_{C^k}(u) = 0;$$

thus

$$\sup_{\mathcal{O}\in\mathcal{O}_\tau(x)} \sup_{H\in\mathcal{N}} \inf_{k\in H} \inf_{u\in\mathcal{O}} I_{C^k}(u) = 0,$$

that is,

$$\tau\text{-}li_e(I_{C^n})(x) = 0.$$

Let $x \notin \tau\text{-}LsC^n$; then

$$\exists H_0 \in \mathcal{N} \quad \exists \mathcal{O}_0 \in \mathcal{O}_\tau(x) \text{ s.t. } \forall k \in H_0 \ \forall u \in \mathcal{O}_0 \ I_{C^k}(u) = +\infty;$$

this implies

$$\sup_{\mathcal{O}\in\mathcal{O}_\tau(x)} \sup_{H\in\mathcal{N}} \inf_{k\in H} \inf_{u\in\mathcal{O}} I_{C^k}(u) \geq \inf_{k\in H_o} \inf_{u\in\mathcal{O}_o} I_{C^k}(u) = +\infty$$

that is,

$$\tau\text{-li}_e(I_{C^n})(x) = +\infty;$$

finally

$$\tau\text{-li}_e(I_{C^n})(x) = \begin{cases} 0 & \text{if } x \in \tau\text{-Ls}C^n \\ +\infty & \text{if } x \notin \tau\text{-Ls}C^n \end{cases} . \quad \square$$

1.4.3 Hypo-convergence

Theorem 1.39, which states that epi-convergence is nothing but Kuratowski convergence of the epigraphs, justifies the terminology we have adopted. For each type of variational problem a certain type of convergence is applied. For example, for maximization problems

$$\sup_{x\in X} F^n(x)$$

one has to develop a convergence theory which is associated with the set convergence of the hypographs of the functions F^n, and which is called *hypo-convergence*. For easy reference, we record the hypograph version of the preceding results. There is no need to rewrite the whole theory, but just to translate the epi-convergence results, noticing that

$$\sup_{x\in X} F^n(x) = -\inf_{x\in X} (-F^n)(x).$$

So the hypo-limits are obtained by taking the opposite of the epi-limits of the $(-F^n)$:

DEFINITION 1.41 Let (X,τ) be a topological space and $(F^n)_{n\in\mathbb{N}}$ a sequence of functions from X into $\bar{\mathbb{R}}$. The τ-hypo-limit inferior and τ-hypo-limit superior of the sequence $(F^n)_{n\in\mathbb{N}}$, denoted respectively $\tau\text{-li}_h F^n$ and $\tau\text{-ls}_h F^n$, are defined by:

$$\tau\text{-li}_h F^n(x) = \inf_{V\in N_\tau(x)} \liminf_{n\to+\infty} \sup_{u\in V} F^n(u) \tag{1.123}$$

$$\tau\text{-}ls_h F^n(x) = \inf_{V \in N_\tau(x)} \limsup_{n \to +\infty} \sup_{u \in V} F^n(u). \tag{1.124}$$

The sequence $(F^n)_{n \in \mathbb{N}}$ is said to be τ-hypo convergent if and only if

$$\forall x \in X \quad \tau\text{-}li_h F^n(x) = \tau\text{-}ls_h F^n(x). \quad \square$$

PROPOSITION 1.42 Let (X,τ) be a first countable topological space and $(F^n)_{n \in \mathbb{N}}$ a sequence of functions from X into $\bar{\mathbb{R}}$. Then,

$$\tau\text{-}li_h F^n(x) = \max_{\{x_n \xrightarrow{\tau} x\}} \liminf_{n \to +\infty} F^n(x_n)$$

$$\tau\text{-}ls_h F^n(x) = \max_{\{x_n \xrightarrow{\tau} x\}} \limsup_{n \to +\infty} F^n(x_n).$$

The sequence $(F^n)_{n \in \mathbb{N}}$ τ-hypo converges to F iff:

$$\begin{cases} \forall x_n \xrightarrow{\tau} x \quad F(x) \geq \limsup_n F^n(x_n) \\ \forall x \quad \exists x_n \xrightarrow{\tau} x \text{ such that } F(x) \leq \liminf_n F^n(x_n). \quad \square \end{cases}$$

PROPOSITION 1.43 Let (X,τ) a general topological space and $(F^n)_{n \in \mathbb{N}}$ a sequence of functions from X into $\bar{\mathbb{R}}$; then,

$$\text{hypo } (\tau\text{-}li_h F^n) = \text{Lim inf}_n \text{ (hypo } F^n) = \bigcap_{H \in \ddot{\mathbb{N}}} c\ell_\tau \left(\bigcup_{n \in H} \text{hypo } F^n \right)$$

$$\text{hypo } (\tau\text{-}ls_h F^n) = \text{Lim sup}_n \text{ (hypo } F^n) = \bigcap_{H \in \mathbb{N}} c\ell_\tau \left(\bigcup_{n \in H} \text{hypo } F^n \right).$$

The hypo-version of Theorem 1.10 is the following:

THEOREM 1.44 Let (X,τ) be a topological space and $(F^n)_{n \in \mathbb{N}}$ a sequence of functions which is τ-hypo convergent:

$$F = \tau\text{-}lm_h F^n.$$

Let $(x_n)_{n \in \mathbb{N}}$ be a sequence of points in X satisfying:

$$F^n(x_n) \geq \sup_{x \in X} F^n(x) - \varepsilon_n, \text{ with } (\varepsilon_n)_{n \in \mathbb{N}} \text{ decreasing to zero.}$$

Then, for every τ-converging subsequence, $x_{n_k} \xrightarrow[k \to +\infty]{\tau} \bar{x}$, we have

$$F(\bar{x}) = \max_{x \in X} F(x) \text{ and } F(\bar{x}) = \lim_{k \to +\infty} F^{n_k}(x_{n_k}). \quad \square$$

1.5 EPI-CONVERGENCE: A PARTICULAR CASE OF Γ-CONVERGENCE

1.5.1 Definitions of De Giorgi's Γ-limits

In order to study convergence of minimizing problems, $\inf_{u \in X} F^n(u)$, we introduced the two limit functions $\tau\text{-li}_e F^n$ and $\tau\text{-ls}_e F^n$ (relatively to the topology τ on X). Considering $F^n(x)$ as a function of n and x and introducing $\mathcal{N}$, the Frechet filter on $\mathbb{N}$, they can be written

$$(\tau\text{-li}_e F^n)(x) = \sup_{V \in N_\tau(x)} \sup_{H \in \mathcal{N}} \inf_{n \in H} \inf_{u \in V} F(n,u)$$

$$(\tau\text{-ls}_e F^n)(x) = \sup_{V \in N_\tau(x)} \inf_{H \in \mathcal{N}} \sup_{n \in H} \inf_{u \in V} F(n,u).$$

Noticing that $\mathcal{N}$ is nothing but the family of neighbourhoods of $+\infty$ for the classical topology of $\bar{\mathbb{N}}$, we can naturally extend the above formulae:

DEFINITION 1.45 Let (X_1,τ_1), (X_2,τ_2) be two topological spaces and

$$F: X_1 \times X_2 \to \bar{\mathbb{R}}$$

$$(u_1,u_2) \to F(u_1,u_2).$$

Then, we define

$$[\Gamma(\tau_1^-,\tau_2^-) \lim_{\substack{u_1 \to x_1 \\ u_2 \to x_2}} F](x_1,x_2) = \sup_{V_2 \in N_{\tau_2}(x_2)} \sup_{V_1 \in N_{\tau_1}(x_1)} \inf_{u_1 \in V_1} \inf_{u_2 \in V_2} F(u_1,u_2)$$

$$[\Gamma(\tau_1^+,\tau_2^-) \lim_{\substack{u_1 \to x_1 \\ u_2 \to x_2}} F](x_1,x_2) = \sup_{V_2 \in N_{\tau_2}(x_2)} \inf_{V_1 \in N_{\tau_1}(x_1)} \sup_{u_1 \in V_1} \inf_{u_2 \in V_2} F(u_1,u_2)$$

$$[\Gamma(\tau_1^-,\tau_2^+) \lim_{\substack{u_1 \to x_1 \\ u_2 \to x_2}} F](x_1,x_2) = \inf_{V_2 \in N_{\tau_2}(x_2)} \sup_{V_1 \in N_{\tau_1}(x_1)} \inf_{u_1 \in V_1} \sup_{u_2 \in V_2} F(u_1,u_2)$$

$$[\Gamma(\tau_1^+,\tau_2^+) \lim_{\substack{u_1 \to x_1 \\ u_2 \to x_2}} F](x_1,x_2) = \inf_{V_2 \in N_{\tau_2}(x_2)} \inf_{V_1 \in N_{\tau_1}(x_1)} \sup_{u_1 \in V_1} \sup_{u_2 V_2} F(u_1,u_2).$$

REMARKS 1.46

(a) These definitions and notations have been introduced by De Giorgi [2], [3]. The notation Γ is justified by the following remark:

When F depends on only one variable, the above formulae reduce to

$$[\Gamma(\tau^-)\lim_{u\to x} F](x) = \sup_{V\in N_\tau(x)} \inf_{u\in V} F(u)$$

$$[\Gamma(\tau^+)\lim_{u\to x} F](x) = \inf_{V\in N_\tau(x)} \sup_{u\in V} F(u)$$

which are respectively the lower semicontinuous and upper semicontinuous regularization of F: the notation Γ for such operations is fairly standard.

(b) With the notations of epi-convergence

$$\Gamma(\tau_1^-,\tau_2^-) \lim_{\substack{u_1\to x_1\\ u_2\to x_2}} F(x_1,x_2) = \underset{(u_1 \xrightarrow{\tau} x_1)}{[\tau_2\text{-}li_e F(u_1,\cdot)](x_2)}$$

$$\Gamma(\tau_1^+,\tau_2^-) \lim_{\substack{u_1\to x_1\\ u_2\to x_2}} F(x_1,x_2) = \underset{(u_1 \xrightarrow{\tau} x_1)}{[\tau_2\text{-}ls_e F(u_1,\cdot)](x_2)}$$

the two last limits are in fact hypo-limits.

The consideration of such limits justifies the study of the epi-convergence of filtered family of functions.

(c) The extension of the above limit concept to three spaces and more generally n spaces is straightforward: let us describe the algorithm:

DEFINITION 1.47 Let (X_1,τ_1) ... (X_N,τ_N) be N topological spaces and

$$F: X_1 \times X_2 \times \ldots \times X_N \to \bar{\mathbb{R}}$$

$$(u_1,u_2,\ldots,u_N) \to F(u_1,u_2,\ldots,u_N).$$

Let us define

$$\Gamma(\tau_1^{\pm},\dots,\tau_{N-1}^{\pm},\tau_N^{+}) \lim_{\begin{cases}u_1 \to x_1 \\ \vdots \\ u_N \to x_N\end{cases}} F(x_1,\dots,x_N) \tag{1.125}$$

$$= \inf_{V\in N_{\tau_N}(x_N)} \Gamma(\tau_1^{\pm},\dots,\tau_{N-1}^{\pm}) \lim_{\begin{cases}u_1 \to x_1 \\ \vdots \\ u_{N-1}\to x_{N-1}\end{cases}} \sup_{u_N\in V} F(u_1,\dots,u_N)$$

and analogously

$$\Gamma(\tau_1^{\pm},\dots,\tau_{N-1}^{\pm},\tau_N^{-}) \lim_{\begin{cases}u_1 \to x_1 \\ \vdots \\ u_N \to x_N\end{cases}} F(x_1,\dots,x_N) \tag{1.126}$$

$$= \sup_{V\in N_{\tau_N}(x_N)} \Gamma(\tau_1^{+},\dots,\tau_{N-1}^{\pm}) \lim_{\begin{cases}u_1 \to x_1 \\ \vdots \\ u_{N-1}\to x_{N-1}\end{cases}} \inf_{u_N\in V} F(u_1,\dots,u_N).$$

When, in the above formulae, the two limits $\Gamma(\dots,\tau_i^{+},\dots)$ and $\Gamma(\dots,\tau_i^{-},\dots)$ are equal, then the sign indexing τ_i is omitted. For example, if

$$\Gamma(\tau_1^{+},\tau_2^{-}) \lim_{\begin{cases}u_1 \to x_1 \\ u_2 \to x_2\end{cases}} F(x_1,x_2) = \Gamma(\tau_1^{-},\tau_2^{-}) \lim_{\begin{cases}u_1 \to x_1 \\ u_2 \to x_2\end{cases}} F(x_1,x_2)$$

then, this common value is denoted by

$$\Gamma(\tau_1,\tau_2^{-}) \lim_{\begin{cases}u_1 \to x_1 \\ u_2 \to x_2\end{cases}} F(x_1,x_2).$$

Moreover, in order to simplify the notation, when the first space (X_1,τ_1) is a parameter space ($\bar{\mathbb{N}}$ or $\bar{\mathbb{R}}$ ...) equipped with the usual topology and if the Γ-limit is independent of the sign indexing τ_1, then we can omit the symbol τ_1. For example, in the above example if $X_1 = \bar{\mathbb{N}}$, we write

$$\Gamma(\tau_2^{-}) \lim_{\substack{n\to+\infty \\ u\to x}} F^n(x) \tag{1.127}$$

instead of

$$\Gamma(\tau_1,\tau_2^-) \lim_{\{\substack{n\to+\infty \\ u\to x}} F(n,x).$$

This is why $\Gamma(\tau^-)$ convergence is an alternative terminology to epi-convergence. Because of its geometrical significant, we prefer the first terminology. However, when working with Γ-limits in three (or more) dimensions the Γ-notations are very useful.

We have already noticed that Γ-convergence in two dimensions is an epi or hypo convergence; thus, it is related to convergence problems for minimization or maximization of sequences of functions.

1.5.2 Epi/hypo-convergence for saddle value problems

When studying convergence notions well adapted to the convergence of *saddle value problems*, the following Γ-limits naturally occur: For any sequence $\{F^n; n \in \mathbb{N}\}$ $F^n:(X,\tau) \times (Y,\sigma) \to \bar{\mathbb{R}}$ of (bivariate) functions, let us introduce

$$h_\sigma/e_\tau\text{-li}F^n(x,y) = \Gamma(N^-,\tau^-,\sigma^+) \lim_{\substack{n\to+\infty \\ u\to x \\ v\to y}} F^n(x,y) \tag{1.128}$$

$$e_\tau/h_\sigma\text{-ls}F^n(x,y) = \Gamma(N^+,\sigma^+,\tau^-) \lim_{\substack{n\to+\infty \\ u\to x \\ v\to y}} F^n(x,y). \tag{1.129}$$

The above Γ-limit functions are called (following Attouch & Wets [2], [3]) respectively, the hypo-epi limit inferior and epi-hypo limit superior of the sequence $\{F^n; n \in \mathbb{N}\}$.

A (bivariate) function F is said to be an *epi/hypo limit* of the sequence $\{F^n;\ n = 1,2,\ldots\}$ if

$$e_\tau/h_\sigma\text{-ls}F^n \leqslant F \leqslant h_\sigma/e_\tau\text{-li}F^n. \tag{1.130}$$

When functions F^n do not depend on y, then the definition of epi/hypo convergence specializes to that of epi-convergence. On the other hand, if the F^n do not depend on x, epi/hypo convergence is simply hypo-convergence. Thus, the theory contains both epi and hypo-convergence. Since the geometric interpretation of this limit involves convergence of both the epigraphs and hypographs of the functions, we obtain a justification of the terminology.

The variational property of the epi/hypo convergence that motivated its introduction is summarized in the following theorem:

THEOREM 1.48 Suppose (X,τ) and (Y,σ) are two topological spaces and $\{F^n;\ n = 1,2,...\}$ a sequence of (bivariate) functions, defined on $X \times Y$ and with values in $\bar{\mathbb{R}}$, that $\text{epi}_\tau/\text{hypo}_\sigma$ converges to a function F. Suppose that $\{(x_n,y_n);\ n = 1,2,...\}$ are saddle points of the functions $\{F^n;\ n = 1,2,...\}$:

$$F^n(x_n,y) \leq F^n(x_n,y_n) \leq F^n(x,y_n) \text{ for all } n \in \mathbb{N},\ x \in X,\ y \in Y$$

and that for some subsequence $\{n_k;\ k = 1,2,...\}$, $\bar{x} = \tau\text{-}\lim_{k\to+\infty} x_{n_k}$, $\bar{y} = \sigma\text{-}\lim_{k\to+\infty} y_{n_k}$.

Then $(\bar{x},\bar{y})$ is a saddle point of F and

$$F(\bar{x},\bar{y}) = \lim_{k\to+\infty} F^{n_k}(x_{n_k},y_{n_k}).$$

The above epi/hypo convergence concept is in some sense minimal in order to get the above variational property. A close concept has been introduced by Cavazzuti [1].

1.6 FURTHER EXAMPLES

In this section, we describe some other limit analysis results where epi-convergence provides a flexible approach.

1.6.1 Homogenization of the elastoplastic torsion of a cylindrical bar

The study of the torsion of an elastoplastic cylindrical homogeneous bar leads to the following variational inequality (refer to Lanchon [1])

$$\min \left\{ \int_\Omega |\text{grad}u|^2 dx - \int_\Omega fu\, dx \;/\; \begin{array}{l} |\text{grad}u(x)| \leq a \text{ on } \Omega \\ u = 0 \text{ on } \partial\Omega \end{array} \right\}$$

involving a constraint on the gradient: the constant a gives the maximum deformation of the material consistent with retaining its elasticity:

$|\text{grad}u(x)| < a$ is the elastic zone

$|\text{grad}u(x)| = a$ is the plastic zone.

When the cylindrical bar is composed of fibres of different natures, then, $a(\cdot)$ is a function of x. When the structure of the material is periodic with period εY, the situation can be modelled just as in Example 1.1.1. Introducing a Y-periodic function, $a(\cdot): \mathbb{R}^N \to \mathbb{R}^+$ the above variational inequality becomes

$$\min \{\int_\Omega |\text{grad}u|^2 dx - \int_\Omega fu\, dx / u \in H_0^1(\Omega),\ |\text{grad}u(x)| \leq a(\frac{x}{\varepsilon}) \text{ on } \Omega\}. \quad (1.131)$$

The limit analysis problem (which corresponds to ε small with respect to $|\Omega|$) treats the limit behaviour, as ε goes to zero, of the sequence of variational inequalities (1.131). One looks for the limit, the "homogenized" problem (Fig. 1.14).

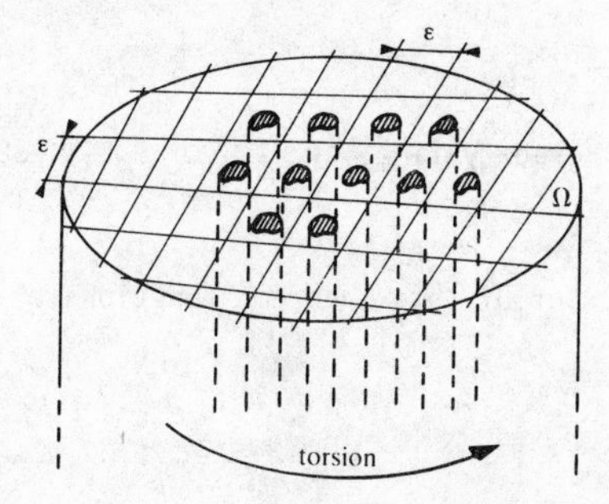

Figure 1.14

The difficulty of this limit analysis problem comes from the varying constrains imposed on the gradient of u. As in Example 1.1.1, the limit problem ($\varepsilon \to 0$) is not obtained by taking the weak limit of the sequence

$$a_\varepsilon(x) = a(\frac{x}{\varepsilon}) \xrightarrow[(\varepsilon \to 0)]{} \mathfrak{m}_Y(a) \text{ (weakly in } \sigma(L^\infty, L^1)).$$

The solutions $\{u_\varepsilon;\ \varepsilon \to 0\}$ of the variational inequalities (1.131) being clearly bounded in $H_0^1(\Omega)$, and hence relatively compact with respect to the strong topology τ of $L^2(\Omega)$, the limit analysis problem is solved in terms of epi-convergence by the following theorem (cf. Attouch [6], Carbone [4], Carbone & Salerno [1]).

THEOREM 1.49 Let $a(\cdot):\mathbb{R}^N \to \mathbb{R}^+$ be a Y-periodic function satisfying

$$0 < \alpha \leq a(y) \quad \text{a.e. } y \in Y.$$

For every $\varepsilon > 0$, let $K^\varepsilon \subset H_o^1(\Omega)$ be the closed convex set of constraints:

$$K^\varepsilon = \{u \in H_o^1(\Omega) \ / \ |\text{grad}u(x)| \leq a(\tfrac{x}{\varepsilon}) \quad \text{a.e. } x \in \Omega\}.$$

Then, the sequence of functionals $\{F^\varepsilon : H_o^1(\Omega) \to \bar{\mathbb{R}}^+ / \ \varepsilon \to 0\}$

$$F^\varepsilon(u) = \int_\Omega |\text{grad}u|^2 dx + I_{K^\varepsilon}(u)$$

τ-epi converges to the functional $F^{hom} : H_o^1(\Omega) \to \bar{\mathbb{R}}^+$ given by

$$F^{hom}(u) = \int_\Omega j^{hom}(\text{grad}u)dx$$

and

$$j^{hom}(z) = \min_{\begin{cases} w \text{ Y-periodic} \\ |\text{grad}w(y)+z| \leq a(y) \end{cases}} \int_Y |\text{grad}w(y) + z|^2 dy \tag{1.132}$$

Proof of Theorem 1.49 Let us sketch the proof of this theorem: Functionals F^ε can be written

$$F^\varepsilon(u) = \int_\Omega j(\tfrac{x}{\varepsilon}, \text{grad}u(x))dx$$

where $j: \mathbb{R}^N \times \mathbb{R}^N \to \bar{\mathbb{R}}^+$ is equal to

$$j(y,z) = \begin{cases} |z|^2 & \text{if } |z| \leq a(y) \\ + \infty & \text{otherwise.} \end{cases}$$

j is Y-periodic in y and convex lower semicontinuous with respect to z. The difference from model example 1.3.2 lies in the fact that j takes $+ \infty$ values. This remark and the form of the limit problem (1.132) suggest that we approach the $\{F^\varepsilon; \varepsilon \to 0\}$ by continuous functionals $\{F_\lambda^\varepsilon; \varepsilon \to 0, \lambda \to 0\}$ in order to reduce to the model situation 1.3.2 already studied. (Theorem 1.20).

This can be done, for example, by penalization of the constraints "$u \in K_\varepsilon$": For every $\varepsilon > 0$ and $\lambda > 0$, let us define

$$\begin{aligned} F_\lambda^\varepsilon(u) &= \int_\Omega |\text{grad}u|^2 dx + \frac{1}{\lambda}\int_\Omega (|\text{grad}u(x)| - a(\tfrac{x}{\varepsilon}))^+ dx \\ &= \int_\Omega j_\lambda(\tfrac{x}{\varepsilon}, \text{grad}u(x))dx \end{aligned} \tag{1.133}$$

where

$$j_\lambda : \mathbb{R}^N \times \mathbb{R}^N \to \mathbb{R}^+$$

$$(y,z) \to j_\lambda(y,z) = |z|^2 + \frac{1}{\lambda}\,(|z| - a(y))^+$$

is Y-periodic in y, convex continuous in z and satisfies

$$|z|^2 \leq j_\lambda(y,z) \leq |z|^2 + \frac{1}{\lambda}\,|z| \leq \Lambda_0(1 + |z|^2)$$

for some $\Lambda_0 > 0$.

We can now apply the conclusion of Theorem 1.20 to the sequence $\{F^\varepsilon_\lambda;\ \varepsilon \to 0\}$:

For every $\lambda > 0$, for every $u \in H^1_0(\Omega)$

$$(\tau\text{-}\lim_e F^\varepsilon_\lambda)(u) = F^{hom}_\lambda(u) \text{ exists,} \tag{1.134}$$

where

$$F^{hom}_\lambda(u) = \int_\Omega j^{hom}_\lambda(\text{grad}u(x))dx$$

with

$$j^{hom}_\lambda(z) = \min_{\{w\ Y\text{-periodic}\}} \int_Y j_\lambda(y, \text{grad}w(y) + z)dy.$$

One can then easily verify that the sequences $\{F^\varepsilon_\lambda;\ \lambda \to 0\}$ and $\{F^{hom}_\lambda;\ \lambda \to 0\}$ do approach respectively, in the epi-convergence sense, the functionals F^ε and F^{hom}. This can be viewed directly or as a consequence of the general result (refer to Section 2) which states that monotone convergence does imply epi-convergence. The preceding results are summarized in Fig. 1.15.

$$\begin{array}{ccc} F^\varepsilon_\lambda & \xrightarrow[\varepsilon \to 0]{} & F^{hom}_\lambda \\ \downarrow{\scriptstyle \lambda \to 0} & & \downarrow{\scriptstyle \lambda \to 0} \\ F^\varepsilon & & F^{hom} \end{array}$$

Figure 1.15

The convergences are taken in the τ-epi convergence sense. The problem is:

"can one close this diagram" or equivalently can one permute the two limits

$$\lim_{\varepsilon} \lim_{\lambda} F_{\lambda}^{\varepsilon} \overset{?}{=} \lim_{\lambda} \lim_{\varepsilon} F_{\lambda}^{\varepsilon}.$$

Such a course is in general rather difficult to justify: it is a particular case of the general problem which consists of solving a limit analysis problem with several parameters. In some examples (refer to 1.3.5) the limit does depend on the way ($\lambda = \lambda(\varepsilon)$) the parameters go to zero and the two limits do not commute.

In the present situation, one can overcome the difficulty by taking advantage of the fact that the approximation $F_{\lambda}^{\varepsilon} \to F^{\varepsilon}$ is monotone: from

$$F^{\varepsilon} \geqslant F_{\lambda}^{\varepsilon}$$

it follows that

$$\mathrm{li}_e F^{\varepsilon} \geqslant \mathrm{li}_e F_{\lambda}^{\varepsilon} = F_{\lambda}^{\mathrm{hom}}.$$

This inequality being true for any $\lambda > 0$

$$\mathrm{li}_e F^{\varepsilon} \geqslant \sup_{\lambda>0} F_{\lambda}^{\mathrm{hom}} = F^{\mathrm{hom}} \tag{1.135}$$

which proves "half" of the convergence result, $F^{\mathrm{hom}} = \tau\text{-}\lim_e F^{\varepsilon}$.

In order to complete the proof, we have to show the existence, for every $u \in H_0^1(\Omega)$, of a sequence $\{u_{\varepsilon}; \varepsilon \to 0\}$ strongly converging to u in $L^2(\Omega)$ such that

$$F^{\mathrm{hom}}(u) \geqslant \limsup_{\varepsilon \to 0} F^{\varepsilon}(u_{\varepsilon}). \tag{1.136}$$

To this end, we follow and adapt the step-argument proof of Theorem 1.20.

(a) Let us first notice that F^{hom} is not everywhere defined: let

$$K^{\mathrm{hom}} = \mathrm{dom}\, j^{\mathrm{hom}} = \{z \in \mathbb{R}^N / \exists w \text{ Y-periodic s.t. } |\mathrm{grad} w(y) + z| \leqslant a(y)\}$$

then

$$\mathrm{dom}\, F^{\mathrm{hom}} = \{u \in H_0^1(\Omega) / \mathrm{grad} u(x) \in K^{\mathrm{hom}} \text{ a.e. } x \in \Omega\}.$$

The density argument can be completed, thanks to the assumption

$$a(y) \geqslant \alpha > 0,$$

which guarantees that $\text{dom}F^{\text{hom}}$ has a non void interior for the $W^{1,\infty}$ topology (refer to Attouch [6] for further details).

(b) It is therefore sufficient to prove (1.136), taking u to be a piecewise affine function which satisfies

$$\text{grad}u(x) \in K^{\text{hom}} \quad \text{a.e. } x \in \Omega \text{ (otherwise } F^{\text{hom}}(u) = +\infty!). \tag{1.137}$$

When u is affine, just take

$$u_\varepsilon(x) = u(x) + \varepsilon w_z(\frac{x}{\varepsilon})$$

with w_z a solution of the local problem (1.132) (the existence of which is guaranteed by assumption (1.137)).

When u is a piecewise affine function, we must be careful how we use the approximating sequences u_ε^i (i = 1,2). If one proceeds as in (1.65) and take

$$u_\varepsilon = u_\varepsilon^{\delta(\varepsilon)} = (1 - \phi_{\delta(\varepsilon)})u_\varepsilon^1 + \phi_{\delta(\varepsilon)}\, u_\varepsilon^2,$$

then this approximating sequence fails to satisfy (in general) the constraint: $u_\varepsilon \in K_\varepsilon$!

We overcome this difficulty, noticing that a piecewise affine function can be obtained from affine functions as the result of a finite number of operations inf and sup!

Take, for example, $u = u_1 \vee u_2$ (the argument can be then easily extended). Then, $u_\varepsilon = u_\varepsilon^1 \vee u_\varepsilon^2$ works: it satisfies $u_\varepsilon \xrightarrow{\tau} u$ as ε goes to zero and $|\text{grad}u_\varepsilon(x)| \leqslant a(\frac{x}{\varepsilon})$ for almost every $x \in \Omega$. On the other hand, using the equality

$$\int_\Omega |\text{grad}(u_\varepsilon^1 \vee u_\varepsilon^2)|^2 dx = \int_\Omega |\text{grad}u_\varepsilon^1|^2 dx$$

$$+ \int_\Omega |\text{grad}u_\varepsilon^2|^2 dx - \int_\Omega |\text{grad}(u_\varepsilon^1 \wedge u_\varepsilon^2)|^2 dx$$

and noticing that

$$\liminf_\varepsilon \int_\Omega |\text{grad}(u_\varepsilon^1 \wedge u_\varepsilon^2)|^2 dx \geqslant \int_\Omega j^{\text{hom}}(\text{grad}(u_1 \wedge u_2))dx$$

(the above inequality is a consequence of (1.135), since

$$u_\varepsilon^1 \wedge u_\varepsilon^2 \xrightarrow[(\varepsilon\to 0)]{\tau} u_1 \wedge u_2 \quad \text{and} \quad |\mathrm{grad}(u_\varepsilon^1 \wedge u_\varepsilon^2)| \leq a(\frac{x}{\varepsilon}))$$

we finally obtain

$$\limsup_\varepsilon \int_\Omega |\mathrm{grad}(u_\varepsilon^1 \vee u_\varepsilon^2)|^2 dx$$

$$\leq \int_\Omega \{j^{hom}(\mathrm{grad}u_1) + j^{hom}(\mathrm{grad}u_2) - j^{hom}(\mathrm{grad}u_1 \wedge u_2)\}dx$$

$$\leq \int_\Omega j^{hom}(\mathrm{grad}(u_1 \vee u_2))dx. \quad \square$$

1.6.2 A singular perturbation problem in optimal control theory

Let us describe the approach via epi-convergence to a singular perturbation problem (refer to J.-L. Lions [1], Haraux & Murat [1], [2]).

For every $\varepsilon > 0$, the state equation is given by

$$\begin{cases} -\varepsilon\Delta y + y^3 = v \text{ on } \Omega \\ y = 0 \qquad \text{on } \partial\Omega \end{cases}$$

where v is the control and $y = y_\varepsilon(v)$ the corresponding state. The cost function J^ε is equal to

$$J^\varepsilon(v) = \int_\Omega \{|y_\varepsilon(v) - z_d|^6 + Nv^2\}dx$$

and the minimal cost is m_ε

$$m_\varepsilon = \inf_{v \in U_{ad}} J^\varepsilon(v),$$

where U_{ad} is the convex set of admissible constraints and N a strictly positive constant.

Let us introduce the corresponding quantities obtained by taking $\varepsilon = 0$:

$$y^3 = v \text{ on } \Omega, \text{ i.e. } y(v) = \sqrt[3]{v}$$

$$J^o(v) = \int_\Omega \{|\sqrt[3]{v} - z_d|^6 + Nv^2\} dx$$

$$m_o = \inf_{v \in U_{ad}} J^o(v).$$

The natural question is: Does m_ε converge to m_o as ε goes to zero? The answer, surprisingly involved, is affirmative for a large class of U_{ad}:

$$U_{ad} = L^2(\Omega),\ U_{ad} = \{v \geq 0\},\ U_{ad} = \{v | \int v\xi_i \leq \alpha_i,\ i \text{ finite}\},$$

U_{ad} bounded in $H^1(\Omega)$,... For a general U_{ad}, closed convex in $L^2(\Omega)$, the problem is still open.

Let us first consider the case $\underline{U_{ad} = L^2(\Omega)}$.

We notice that

$$m_\varepsilon = \inf_y \left[\int_\Omega \{|y-z_d|^6 + N(-\varepsilon\Delta y + y^3)^2\}dx\right],$$

the infimum being taken with respect to all y satisfying

$$y \in L^6,\ -\varepsilon\Delta y + y^3 \in L^2,\ y = 0 \text{ on } \partial\Omega;$$

that is, $y \in H^2 \cap H_o^1 \cap L^6$.

On the other hand,

$$m_o = \inf_{y \in L^6} \int_\Omega \{|y-z_d|^6 + Ny^6\}dx.$$

Introducing on $Y = L^6(\Omega)$, the functionals F^ε, and F^o given by

$$\begin{aligned} F^\varepsilon(y) &= \int_\Omega |y-z_d|^6 + N(-\varepsilon\Delta y + y^3)^2 \\ F^o(y) &= \int_\Omega |y-z_d|^6 + Ny^6 \end{aligned} \tag{1.138}$$

noticing that the corresponding minimizing sequences are weakly relatively compact, the limit analysis problem ($\varepsilon \to 0$) is solved in terms of epi-convergence by the following theorem:

<u>THEOREM 1.50</u> Let $\{F^o, F^\varepsilon : L^6(\Omega) \to \bar{\mathbb{R}}^+ / \varepsilon \to 0\}$ be defined by (1.138).

Denoting w (resp. s) the weak (resp. strong) topology on $L^6(\Omega)$, for every $y \in Y = L^6(\Omega)$, the following epi-convergences hold:

$$w\text{-}lm_e F^\varepsilon(y) = s\text{-}lm_e F^\varepsilon(y) = F^o(y), \text{ as } \varepsilon \to 0.$$

Consequently, $\lim_{\varepsilon \to 0} m_\varepsilon = m_o$. Moreover, the corresponding optimal states

$\{y_\varepsilon ; \varepsilon \to 0\}$ and optimal controls $\{v_\varepsilon ; \varepsilon \to 0\}$ converge respectively strongly in $L^6(\Omega)$ and $L^2(\Omega)$ to the optimal state and optimal control of the limit problem.

Proof of Theorem 1.50

(a) Let us first verify that, for every $y \in L^6(\Omega)$, there exists a sequence $\{y_\varepsilon ; \varepsilon \to 0\}$ strongly converging in $L^6(\Omega)$ to y and which satisfies

$$\lim_{\varepsilon \to 0} F^\varepsilon(y_\varepsilon) = F^0(y).$$

Take y_ε the solution of

$$\begin{cases} -\varepsilon \Delta y_\varepsilon + y_\varepsilon^3 = y^3 \text{ on } \Omega \\ y_\varepsilon = 0 \text{ on } \partial\Omega. \end{cases} \tag{1.139}$$

Multiplying the above equation by y_ε^3 and noticing that

$$\int_\Omega -\varepsilon \Delta y_\varepsilon . y_\varepsilon^3 \, dx = 3\varepsilon \int_\Omega y_\varepsilon^2 |\text{grad} y_\varepsilon|^2 dx$$

we obtain

$$\int_\Omega y_\varepsilon^6 \, dx \leq \int_\Omega y^3 y_\varepsilon^3 \, dx$$

and by Cauchy-Schwarz inequality

$$\int_\Omega y_\varepsilon^6 \, dx \leq \int_\Omega y^6 \, dx.$$

From (1.139) and the above estimation it follows

$$y_\varepsilon^3 \longrightarrow y^3 \text{ weakly in } L^2(\Omega).$$

Noticing that the above inequality implies

$$\limsup_{\varepsilon \to 0} \|y_\varepsilon^3\|_{L^2(\Omega)} \leq \|y^3\|_{L^2(\Omega)},$$

it follows that the sequence $\{y_\varepsilon^3 ; \varepsilon \to 0\}$ converges strongly in $L^2(\Omega)$ to y^3 and hence the sequence $\{y_\varepsilon ; \varepsilon \to 0\}$ converges strongly in $L^6(\Omega)$ to y. From

$$F^{\varepsilon}(y_{\varepsilon}) = \int_{\Omega} |y_{\varepsilon}-z_d|^6 + Ny^6,$$

it follows

$$\lim_{\varepsilon\to 0} F^{\varepsilon}(y_{\varepsilon}) = F^{o}(y).$$

(b) Let us now verify that, for any sequence $\{y_{\varepsilon}; \varepsilon \to 0\}$ which weakly converges in $L^6(\Omega)$, $y_{\varepsilon} \longrightarrow y$, the following inequality holds:

$$\lim_{\varepsilon}\inf F^{\varepsilon}(y_{\varepsilon}) \geqslant F^{o}(y).$$

Let us develop

$$F^{\varepsilon}(y_{\varepsilon}) = \int_{\Omega} |y_{\varepsilon}-z_d|^6 + N\varepsilon^2(\Delta y_{\varepsilon})^2 + Ny_{\varepsilon}^6 - 2N\varepsilon\, \Delta y_{\varepsilon}.y_{\varepsilon}^3.$$

The same computation as above (integration by parts) yields

$$-2N\varepsilon \int_{\Omega} \Delta y_{\varepsilon}.y_{\varepsilon}^3\, dx = 6N\varepsilon \int y_{\varepsilon}^2\, |\text{grad} y_{\varepsilon}|^2 dx \geqslant 0.$$

Therefore

$$F^{\varepsilon}(y_{\varepsilon}) \geqslant \int_{\Omega} \{|y_{\varepsilon}-z_d|^6 + Ny_{\varepsilon}^6\}dx$$

and, from semicontinuity for the weak topology on $L^6(\Omega)$ of the convex continuous functional

$$y \to \int_{\Omega} |y-z_d|^6 + Ny^6,$$

it follows that

$$\lim_{\varepsilon}\inf F^{\varepsilon}(y_{\varepsilon}) \geqslant F^{o}(y).$$

(c) We therefore have

$$w\text{-}lm_e F^{\varepsilon} = s\text{-}lm_e F^{\varepsilon} = F^{o}.$$

From the variational property of epi-convergence (Theorem 1.10), noticing that the sequence of optimal states $\{\bar{y}_{\varepsilon}; \varepsilon \to 0\}$ remains bounded in $L^6(\Omega)$ it follows that

$$\lim_{\varepsilon\to 0} m_\varepsilon = m_o$$

$$\bar{y}_\varepsilon \longrightarrow \bar{y}_o \text{ weakly in } L^6(\Omega)$$

where $\bar{y}_o$ is the optimal state corresponding to the limit problem. From $\lim_{\varepsilon\to 0} m_\varepsilon = m_o$, using the same minorization as in (b), we obtain

$$\limsup_{\varepsilon \to 0} \int_\Omega \{|\bar{y}_\varepsilon - z_d|^6 + N\bar{y}_\varepsilon^6\}dx \leq \int_\Omega \{|\bar{y}_o - z_d|^6 + N\,\bar{y}_o^6\}dx$$

which clearly implies the strong convergence in $L^6(\Omega)$ of $\bar{y}_\varepsilon$ to $\bar{y}_o$. Similarly we can derive strong convergence in $L^2(\Omega)$ of the optimal controls

$$\bar{v}_\varepsilon = -\varepsilon\Delta\bar{y}_\varepsilon + \bar{y}_\varepsilon^3 \text{ to } \bar{v}_o = \bar{y}_o^3. \quad \square$$

REMARKS 1.51

1. The sequence $\{F^\varepsilon;\ \varepsilon \to 0\}$ epi-converges to F^o *both* for the strong and weak topology of $X = L^6(\Omega)$. This type of convergence, called Mosco-convergence, enjoys nice stability properties (refer to Chapter 3) and guarantees in general strong convergence in X (as we verified in the above example) of the solutions of the corresponding minimization problems.

It is interesting to compare with the preceding homogenization examples, where the epi-limits for the weak and strong topology of $X = H_o^1(\Omega)$ are different, and the solutions of the corresponding minimization problems do converge only weakly in X.

2. When the control v is required to belong to some closed convex set $U_{ad} \subset L^2(\Omega)$, the sequence of functionals $\{F^\varepsilon, F^o;\ \varepsilon \to 0\}$ we have now to consider is

$$F^\varepsilon(y) = \int_\Omega \{|y-z_d|^6 + N(-\varepsilon\Delta y + y^3)^2\}dx + I_{K^\varepsilon}(y)$$

$$F^o(y) = \int_\Omega \{|y-z_d|^6 + Ny^6\}dx + I_{K^o}(y)$$

where we denote

$$K^\varepsilon = \{y \in L^6(\Omega)/-\varepsilon\Delta y + y^3 \in U_{ad}, y \in H_o^1(\Omega)\}$$

$$K^o = \{y \in L^6(\Omega)/y^3 \in U_{ad}\}.$$

Part (a) of the proof is unchanged. Part (b) turns to be more difficult since we have now to prove that

$$\left.\begin{array}{l} y_\varepsilon \longrightarrow y \text{ weakly in } L^6(\Omega) \\ -\varepsilon\Delta y_\varepsilon + y_\varepsilon^3 \text{ bounded in } L^2(\Omega) \\ -\varepsilon\Delta y_\varepsilon + y_\varepsilon^3 \in U_{ad}. \end{array}\right\} \Rightarrow y^3 \in U_{ad}$$

When $U_{ad} = \{v \in L^2(\Omega);\ v \geq 0\}$ this clearly follows from the maximum principle since

$$\begin{cases} -\varepsilon\Delta y_\varepsilon + y_\varepsilon^3 \geq 0 \text{ on } \Omega \\ y_\varepsilon = 0 \text{ on } \partial\Omega \end{cases}$$

implies $y_\varepsilon \geq 0$. Thus y, as a weak limit in $L^6(\Omega)$ of y_ε, is still positive and $y^3 \in U_{ad}$.

When U_{ad} is bounded in $H^1(\Omega)$, and hence relatively compact in $L^2(\Omega)$, the same argument as in step (a) yields strong convergence in $L^2(\Omega)$ of y_ε^3 to y^3, and hence $y^3 \in U_{ad}$.

To know if $\lim_{\varepsilon\to 0} m_\varepsilon = m_o$ for a general U_{ad} is as we mentioned, an open problem. In Section 3.6 we shall describe a similar problem where for some U_{ad}, the sequence $\{m_\varepsilon;\ \varepsilon \to 0\}$ fails to be even convergent!

3. Just as for the homogenization problem (Theorem 1.20), it is quite natural to try to prove a convergence result which works for a fairly large class of problems.

Let us consider the unconstrained case. The preceding situation turns out to be a special case of the following problem. Let $\{F^\varepsilon;\ \varepsilon \to 0\}$ given by

$$F^\varepsilon(u) = \int_\Omega j(x,u(x),\varepsilon\Delta u(x))dx.$$

Does this sequence of functionals epi-converge, and if so, is its limit equal to

$$F^o(u) = \int_\Omega j(x,u(x),0)dx?$$

In [5], Buttazzo & Dal Maso give an affirmative answer to the first question: Take

$$j(x,s,\xi) = |s-z_d(x)|^p + j_1(s,\xi),$$

where $z_d \in L^p(\Omega)$ and j_1 satisfies the following growth and continuity properties:

$$-a(x) - c(|s|^p + |\xi|^r) \leq j_1(s,\xi) \leq a(x) + c(|s|^p + |\xi|^r)$$

with

$$a \in L^1(\Omega),\ c > 0,\ 1 < p < +\infty\ ,\ 1 \leq r \leq p,$$

and

$$|j_1(s_1,\xi_1) - j_1(s_2,\xi_2)| \leq \omega(|s_1-s_2| + |\xi_1-\xi_2|)(1 + |s_1|^p + |s_2|^r)$$

where

$\omega:[0,+\infty[\rightarrow [0,+\infty[$ is continuous and $\omega(0) = 0$.

Then, for every $u \in L^p(\Omega)$, taking τ = weak topology of $L^p(\Omega)$,

$\tau\text{-}\lim_e F^\varepsilon(u) = F(u)$ exists

with

$$F(u) = \int_\Omega \psi(x,u(x))dx$$

and

$$\psi(x,s) = \inf_{\varepsilon>0} \inf_{\begin{cases} u-s\in C_0^\infty(Y) \\ \int_Y u(y)dy=s \end{cases}} \int_Y j(x,u(y),\varepsilon\Delta u(y))dy.$$

The second question can be translated into: is $\psi(x,s)$ equal to $j(x,s,0)$? The answer is yes if we take $j_1(s,\xi) = (-\xi+s^3)^2$, $p = 6$, as shown in Theorem 1.50. If $j_1(s,\xi) = \lambda(\xi+s^3)^2$, the answer is known (and is yes) only in some particular cases (λ small): refer to Bensoussan [1]. □

1.6.3 Epi-convergence and correctors

As we have already mentioned, in problems such as homogenization and singular perturbations, the solutions $\{u_\varepsilon;\ \varepsilon \rightarrow 0\}$ of the minimization problems

$$\min_{u\in X} \{F^\varepsilon(u) - \int_\Omega fu\}$$

do converge only *weakly* in the energy space: $u_\varepsilon \rightharpoonup u$ weakly in X, where u is a solution of the limit problem

$$\min_{u \in X} \{F(u) - \int fu\}.$$

In this section, we would like to stress that, since one has obtained this limit analysis result by the epi-convergence method, one can fairly easily derive explicit correctors: Using on the one hand convergence of the energies

$$F^\varepsilon(u_\varepsilon) \to F(u)$$

and existence (explicitly constructed) of a sequence $v_\varepsilon \rightharpoonup u$ weakly in X, such that

$$F^\varepsilon(v_\varepsilon) \to F(u)$$

one can usually derive that: $u_\varepsilon - v_\varepsilon \to 0$ strongly in X, as $\varepsilon \to 0$, and hence derive explicit correctors $\{r_\varepsilon ; \varepsilon \to 0\}$ such that $u_\varepsilon - u - r_\varepsilon \to 0$ strongly in X.

Correctors are important from a numerical point of view: they correct rapid oscillations of the gradient of $(u_\varepsilon - u)$. So, $u + r_\varepsilon$ provides, for ε small, a good approximation of u_ε. Let us illustrate these considerations on model examples 1.1.

1. <u>Homogenization of elliptic operators</u>

Take

$$F^\varepsilon(v) = \int_\Omega \sum_{i,j} a_{ij}\left(\frac{x}{\varepsilon}\right) \frac{\partial v}{\partial x_i} \frac{\partial v}{\partial x_j} \, dx, \quad a_{ij} = a_{ji} \quad 1 < i, j < N.$$

$$F^{hom}(v) = \int_\Omega \sum_{i,j} a_{ij}^{hom} \frac{\partial v}{\partial x_i} \frac{\partial v}{\partial x_j} \, dx.$$

On one hand, by convergence of the energies, $F^\varepsilon(u_\varepsilon) \to F^{hom}(u)$ as $\varepsilon \to 0$. On the other hand, for every $u \in H^1(\Omega)$ we have proved existence of a sequence $\{v_\varepsilon ; \varepsilon \to 0\}$ weakly converging to u such that $F^\varepsilon(v_\varepsilon) \to F^{hom}(u)$. Taking u regular (for simplicity), one can take v_ε given by (cf. (1.64))

$$v_\varepsilon(x) = u(x) + \varepsilon w_{\text{grad}u(x)}\left(\frac{x}{\varepsilon}\right)$$

where w_z is a solution of the local problem (1.62).

Let us verify that

$$u_\varepsilon - v_\varepsilon \to 0 \text{ strongly in } H^1_{loc}(\Omega) \text{ as } \varepsilon \to 0$$

i.e.

$$u_\varepsilon - u - \varepsilon w_{\text{grad}u(x)}\left(\frac{x}{\varepsilon}\right) \to 0 \text{ strongly in } H^1_{loc}(\Omega), \text{ as } \varepsilon \to 0. \quad (1.140)$$

Because of the uniform coercivity of the functionals F^ε, this will result from

$$F^\varepsilon(u_\varepsilon - v_\varepsilon) \to 0 \text{ as } \varepsilon \to 0.$$

Using the quadratic character of F^ε

$$F^\varepsilon(u_\varepsilon - v_\varepsilon) = F^\varepsilon(u_\varepsilon) + F^\varepsilon(v_\varepsilon) - 2\langle A^\varepsilon u_\varepsilon, v_\varepsilon\rangle$$

(in fact one has to use cut-off functions m_ε and take $m_\varepsilon v_\varepsilon$ in order to make $v_\varepsilon = 0$ on $\partial\Omega$ and justify the above integration by parts).

From $A^\varepsilon u_\varepsilon = f$, we derive

$$\lim_{\varepsilon\to 0} F^\varepsilon(u_\varepsilon - v_\varepsilon) = F^{hom}(u) + F^{hom}(u) - 2\langle f,u\rangle$$

and since $A^{hom}u = f$, it follows

$$\langle f,u\rangle = F^{hom}(u) \text{ and } \lim_{\varepsilon\to 0} F^\varepsilon(u_\varepsilon - v_\varepsilon) = 0.$$

2. Problems with holes and Dirichlet boundary conditions

With the notation of Theorem 1.27, let us assume we are in the critical case $r_\varepsilon \sim \varepsilon^3$. Then

$$\int_\Omega |\text{grad}u_\varepsilon|^2 dx \xrightarrow[\varepsilon \to 0]{} \int_\Omega |\text{grad}u|^2 dx + C\int_\Omega u^2.$$

Assuming u to be regular (for simplicity), we know that the sequence

$$v_\varepsilon = u.w_\varepsilon \text{ weakly converges to } u$$

and

$$\int_\Omega |\mathrm{grad} v_\varepsilon|^2 dx \xrightarrow[\varepsilon \to 0]{} \int_\Omega |\mathrm{grad} u|^2 + c \int_\Omega u^2 dx.$$

Let us compute

$$\int_\Omega |\mathrm{grad}(u_\varepsilon - v_\varepsilon)|^2 dx = \int_\Omega |\mathrm{grad} u_\varepsilon|^2 dx + \int_\Omega |\mathrm{grad} v_\varepsilon|^2 dx - 2\int_\Omega \mathrm{grad} u_\varepsilon . \mathrm{grad} v_\varepsilon dx$$

$$\int_\Omega \mathrm{grad} u_\varepsilon . \mathrm{grad} v_\varepsilon dx = \int_\Omega \mathrm{grad} u_\varepsilon . \mathrm{grad} u . w_\varepsilon + \int_\Omega \mathrm{grad} u_\varepsilon . \mathrm{grad} w_\varepsilon . u$$

$$\sim \int_\Omega |\mathrm{grad} u|^2 dx - \int_\Omega u_\varepsilon . \Delta w_\varepsilon . u$$

$$\xrightarrow[\varepsilon \to 0]{} \int_\Omega |\mathrm{grad} u|^2 dx + C \int_\Omega u^2 dx.$$

As in the proof of Theorem 1.27, we have used that $u_\varepsilon = 0$ on the small holes T_ε and hence

$$- \int u_\varepsilon \Delta w_\varepsilon u = \int u_\varepsilon (-\Delta w_\varepsilon)^+ u.$$

Since $u_\varepsilon \longrightarrow u$ weakly in $H_0^1(\Omega)$ and $(-\Delta w_\varepsilon)^+$ converges strongly in $H^{-1}(\Omega)$ to C, one can go to the limit as $\varepsilon \to 0$.

One can summarize the above computation by

$$u_\varepsilon - u - w_\varepsilon u \to 0 \text{ strongly in } H_0^1(\Omega) \text{ as } \varepsilon \to 0. \tag{1.141}$$

The above computation requires u to be regular. In full generality, one can only prove (cf. Cioranescu & Murat [1]) that the above convergence holds in $W_0^{1,1}(\Omega)$ strongly.

3. <u>The Neumann strainer</u>

Similar considerations yield (cf. Picard [1])

$$u_\varepsilon - u + w_\varepsilon r^u \to 0 \text{ strongly in } W^{1,1}(\Omega_1) \times W^{1,1}(\Omega_2) \tag{1.142}$$

where $r^u = (r_1^u, r_2^u)$ is taken such that $r_1^u|_\Sigma = \frac{1}{2}[u]$ and $r_2^u|_\Sigma = -\frac{1}{2}[u]$, and w_ε are the capacitary potentials introduced in the proof of Theorem 1.30. □

2 Properties of epi-convergence

In Chapter 1, we introduced epi-convergence and showed how, by direct application of its definition, many different limit analysis problems can be solved. We now investigate properties of epi-convergence in a general topological setting. Besides variational and lower semicontinuity properties, which are two main features of epi-convergence, we stress the importance of the compactness method.

Because functions are considered to have values in $\bar{\mathbb{R}}$, the compactness theorem states that, from every sequence of functions $\{F^n : X \to \bar{\mathbb{R}}; n \in \mathbb{N}\}$, where X is a metrizable space, one can extract an epi-convergent subsequence. Combining this result and the fact that epi-convergence is in most cases attached to a topology, we establish compactness results for (large) classes of functionals in the calculus of variations. For example, take $\mathcal{F}$ the class of all functionals that are the sum of a fixed energy functional and of an obstacle constraint. Determination of $\bar{\mathcal{F}}$, the compact closure of $\mathcal{F}$, amounts to describing the more general forms of limits of varying obstacle problems.

This approach which allows us to understand in a unified way many a priori distinct phenomena, also provides the right compactness notion for attacking existence problems in optimal control theory (optimal design) and convergence problems in stochastic optimization (homogenization).

We pay particular attention, too, to relations between epi-convergence and pointwise convergence, studying the case of monotone convergence where these two notions coincide, and introducing Moreau-Yosida approximation, which allows us to express epi-convergence in terms of pointwise convergence of the approximation. □

2.1 LOWER SEMICONTINUITY OF EPI-LIMITS

THEOREM 2.1 Let $(F^n)_{n \in \mathbb{N}}$ be a sequence of functions from (X,τ), a topological space, into $\bar{\mathbb{R}}$. Then,

$$\tau\text{-li}_e F^n \text{ and } \tau\text{-ls}_e F^n \text{ are } \tau\text{-lower semicontinuous.} \tag{2.1}$$

Consequently, if the sequence $(F^n)_{n\in\mathbb{N}}$ is τ-epi convergent, its limit $F = \tau\text{-lm}_e F^n$ is τ-lower semicontinuous.

Proof of Theorem 2.1 Because of the importance of this lower semicontinuity property, we give two distinct proofs of this result, each relying on a lemma of independent interest.

The more direct and geometric proof follows from Theorem 1.36, which states that

$$\text{epi}(\tau\text{-ls}_e F^n) = \tau\text{-Li}(\text{epi } F^n)$$

$$\text{epi}(\tau\text{-li}_e F^n) = \tau\text{-Ls}(\text{epi } F^n).$$

It follows from the definitions of the lower and upper limits of a sequence of subsets of a topological space (Definition 1.31), that $\tau\text{-li}_e F^n$ and $\tau\text{-ls}_e F^n$ have a closed epigraph in $X \times \mathbb{R}$ (Proposition 1.33), and hence are τ-lower-semicontinuous.

Let us take this opportunity to remind ourselves of the definition and the various characterizations of the τ-lower semicontinuity of a function F from (X,τ) into $\bar{\mathbb{R}}$.

DEFINITION-PROPOSITION 2.2 Let F be a function from (X,τ), a general topological space, into $\bar{\mathbb{R}}$. Then, the closure of the epigraph of F in $X \times \mathbb{R}$ (equipped with the product topology) is still an epigraph.

(a) By definition, the function whose epigraph is equal to the closure of the epigraph of F is called the τ-closure of F and is denoted by $\tau\text{-c}\ell F$ (or $c\ell_\tau F$):

$$c\ell_\tau(\text{epi } F) = \text{epi}(c\ell_\tau F) \tag{2.2}$$

(b) When the function F has a closed epigraph, or, equivalently, is equal to its closure, $F = c\ell_\tau F$, the function F is said to be closed for the topology τ, or τ-*lower semicontinuous*.

$$\forall x \in X \quad c\ell_\tau F(x) = \sup_{V\in N_\tau(x)} \inf_{u\in V} F(u) \tag{2.3}$$

Proof of Proposition 2.2 All the above properties can be obtained as consequences of the geometric Theorem 1.36, when interpreting the epi-convergence of the stationary sequence $F^n = F \quad \forall n \in \mathbb{N}$.

From Definition (1.110) and (1.111) of the set-limit operations

$$\tau\text{-Li}(\text{epi}F^n) = \tau\text{-Ls}(\text{epi}F^n) = c\ell_\tau(\text{epi}F)$$

(where, for simplicity, we still denote by τ the product topology on $X \times \mathbb{R}$). From Theorem 1.36, epi $[\tau\text{-li}_e F^n] = \text{epi}[\tau\text{-ls}_e F^n] = c\ell_\tau(\text{epi}F)$. This tells us that $c\ell_\tau(\text{epi}F)$ is still an epigraph and, from definition of $c\ell_\tau F$:

$$\tau\text{-li}_e F^n = \tau\text{-ls}_e F^n = c\ell_\tau F.$$

Thus, the stationary sequence, $F^n = F \quad \forall n \in \mathbb{N}$, τ-epi-converges to $c\ell_\tau F$. From Definition 1.8 of $\tau\text{-li}_e F^n$ and $\tau\text{-ls}_e F^n$, it follows that

$$\forall x \in X \quad c\ell_\tau F(x) = \sup_{V \in N_\tau(x)} \inf_{u \in V} F(u). \quad \square$$

We can extract from the above argument the following result of independent interest:

COROLLARY 2.3 Let (X,τ) be a topological space and F a function from X into $\bar{\mathbb{R}}$; then, the stationary sequence $(F^n)_{n \in \mathbb{N}}$ defined by $\forall n \in \mathbb{N} \quad F^n = F$, τ-epi-converges to $c\ell_\tau F$, the τ-closure of F.

A significant consequence of the above result is that epi-convergence theory contains, as a particular case, the important and quite difficult problem of the determination of the lower semicontinuous regularization of a given functional F (cf. for example the minimal surface energy functional, in calculus of variations, with Dirichlet boundary conditions, cf. Miranda [3], Ekeland & Teman [1]).

Another way to derive lower semicontinuity properties of the epi-limits is to use formulae (1.32) and (1.33)

$$\tau\text{-li}_e F^n(x) = \sup_{V \in N_\tau(x)} [\liminf_n \inf_{u \in V} F^n(u)]$$

$$\tau\text{-ls}_e F^n(x) = \sup_{V \in N_\tau(x)} [\limsup_n \inf_{u \in V} F^n(u)]$$

and the followin lemma which is quite useful in a more general context when studying the semicontinuity properties of Γ-limit functions (Section 1.5).

LEMMA 2.4 Suppose (X,τ) is a topological space and q is an extended real valued function defined on the subsets of X. Then, the function $a:X \to \bar{\mathbb{R}}$ defined by

$$\forall x \in X \quad a(x) = \sup_{V \in N_\tau(x)} q(V) \tag{2.4}$$

is τ-lower semicontinuous.

Proof of Lemma 2.4 Let us fix some $V_o \in N_\tau(x)$ and take V_1 equal to the τ-interior of V_o; then V_1 still belongs to $N_\tau(x)$. Moreover, $\forall u \in V_1$, $V_o \in N_\tau(u)$; thus, from the definition of $a(u)$

$$\forall u \in V_1 \quad a(u) \geq q(V_o),$$

that is,

$$\inf_{u \in V_1} a(u) \geq q(V_o);$$

since V_1 belongs to $N_\tau(x)$

$$\sup_{V \in N_\tau(x)} \inf_{u \in V} a(u) \geq q(V_o).$$

This inequality being true for every V_o belonging to $N_\tau(x)$, we obtain

$$\sup_{V \in N_\tau(x)} \inf_{u \in V} a(u) \geq \sup_{V \in N_\tau(x)} q(V) = a(x),$$

i.e. (cf. Proposition 2.2)

$$c\ell_\tau a \geq a.$$

The opposite inequality $a \geq c\ell_\tau a$ being always satisfied, we get $a = c\ell_\tau a$ and a is τ-lower semicontinuous. □

Thus, by the Γ-limit process, we can only define lower semicontinuous or upper semicontinuous functions with respect to the variable x_N (depending on the sign τ_N^+ or τ_N^-). Let us now summarize the various characterizations of the lower semicontinuity.

PROPOSITION 2.5 Let F be a function from (X,τ) a general topological space into $\bar{\mathbb{R}}$; the following statements are equivalent:

(i) F is τ-lower semicontinuous

(ii) $\forall x \in X \quad F(x) = c\ell_\tau F(x) = \sup_{V \in N_\tau(x)} \inf_{u \in V} F(u)$

(iii) $\forall x \in X \quad F(x) \leq c\ell_\tau F(x) = \sup_{V \in N_\tau(x)} \inf_{u \in V} F(u)$

(iv) $\forall \lambda \in \mathbb{R} \quad \{x \in X/F(x) > \lambda\}$ is τ-open

(v) $\forall \lambda \in \mathbb{R} \quad \{x \in X/F(x) \leq \lambda\}$ is τ-closed.

The following result explains why, by the epi-convergence process, one cannot distinguish two sequences $(F^n)_{n\in\mathbb{N}}$ and $(G^n)_{n\in\mathbb{N}}$ such that $c\ell_\tau F^n = c\ell_\tau G^n$:

PROPOSITION 2.6 Let (X,τ) a general topological space, $F:X \to \bar{\mathbb{R}}$ an extended real valued function defined on X and $\mathcal{O}$ a τ-open subset of X. Then,

$$\inf_{u\in\mathcal{O}} F(u) = \inf_{u\in\mathcal{O}} c\ell_\tau F(u). \tag{2.5}$$

Proof of Proposition 2.6 Since $F \geq c\ell_\tau F$ is always true, the right hand side of (2.5) is less than or equal to the left one. Let us prove the opposite inequality: take $y \in \mathcal{O}$; then since $\mathcal{O} \in N_\tau(y)$, from equality (2.3),

$$c\ell_\tau F(y) \geq \inf_{u\in\mathcal{O}} F(u).$$

Since this inequality is true for every $y \in \mathcal{O}$,

$$\inf_{y\in\mathcal{O}} c\ell_\tau F(y) \geq \inf_{u\in\mathcal{O}} F(u)$$

and the equality (2.5) follows. □

COROLLARY 2.7 Let (X,τ) be a topological space and $(F^n)_{n\in\mathbb{N}}$, $(G^n)_{n\in\mathbb{N}}$ two sequences of functions from X into $\bar{\mathbb{R}}$ satisfying:

$$\forall n \in \mathbb{N} \quad c\ell_\tau F^n = c\ell_\tau G^n . \tag{2.6}$$

Then, the two sequences $(F^n)_{n\in\mathbb{N}}$ and $(G^n)_{n\in\mathbb{N}}$ have the same τ-epi limits:

$$\tau\text{-}li_e F^n = \tau\text{-}li_e G^n \text{ and } \tau\text{-}ls_e F^n = \tau\text{-}ls_e G^n.$$

Consequently, $\tau\text{-}lm_e F^n$ exists iff $\tau\text{-}lm_e G^n$ exists, the limits being equal in that case.

The following situation is a particular case of (2.6):

$$\forall n \in \mathbb{N} \quad \tau\text{-}c\ell F^n \leqslant G^n \leqslant F^n. \tag{2.7}$$

<u>Proof of Corollary 2.7</u> By definition of the τ-epi limit of a sequence $(F^n)_{n\in\mathbb{N}}$

$$\tau\text{-}li_e F^n(x) = \sup_{V\in N_\tau(x)} \liminf_n \, [\inf_{u\in V} F^n(u)]$$

and its equivalent formulation with open neighbourhoods (Remark 1.11)

$$= \sup_{\mathcal{O}\in\mathcal{O}_\tau(x)} \liminf_n \, [\inf_{u\in\mathcal{O}} F^n(u)]$$

it follows from Proposition 2.6

$$= \sup_{\mathcal{O}\in\mathcal{O}_\tau(x)} \liminf_n \, \inf_{u\in\mathcal{O}} [c\ell_\tau F^n(u)]$$

i.e., the τ-epi limits do depend only on $(c\ell_\tau F^n)_{n\in\mathbb{N}}$. We could have adopted this point of view from the beginning, working with these equivalence classes. Let us complete these considerations by considering their hypo-version:

<u>COROLLARY 2.8</u> Let $(F^n)_{n\in\mathbb{N}}$ a sequence of functions from (X,τ) a topological space into $\bar{\mathbb{R}}$; then,

$$\tau\text{-}li_h F^n \text{ and } \tau\text{-}ls_h F^n \text{ are } \tau\text{-upper semicontinuous}$$

and so is $\tau\text{-}lm_h F^n$, the τ-hypo limit of the sequence, when it exists.

2.2 VARIATIONAL PROPERTIES OF EPI-CONVERGENCE

In this section, we study the *relations between the epi-convergence of a sequence of functions and the convergence of the solutions of the corresponding minimization problems* (we make precise the results of Theorem 1.10, which motivated the introduction of the notion of epi-convergence).

The notion of a minimizer of a function F on a space X

$$x \in \text{Arg min } F \iff F(x) = \min_{u \in X} F(u)$$

is too restrictive for a proper understanding of these relations. In particular, for a given function F on X, the set Argmin F may be empty. A better notion, in our present context, is the notion of an ε-minimizer: for $\varepsilon > 0$,

$$x \in \varepsilon\text{-Argmin } F \iff F(x) \leq \sup \{-\frac{1}{\varepsilon}; \inf_{u \in X} F(u) + \varepsilon\}. \tag{2.8}$$

This definition is justified because it is possible for the quantity $\inf_{u \in X} F(u)$ to be equal to $-\infty$.

From a numerical point of view, this notion is quite natural; moreover, for every $\varepsilon > 0$, the set ε-Argmin F is non void. We notice that

$$\forall \varepsilon > 0 \quad \text{Argmin } F \subset \varepsilon\text{-Argmin } F,$$

and more precisely

$$\text{Argmin } F = \bigcap_{\varepsilon > 0} \varepsilon\text{-Argmin } F. \tag{2.9}$$

Let us examine the relations between epi-convergence and convergence of infimum and minimizing solutions; without additional assumptions, the more general results are:

PROPOSITION 2.9 Let (X,τ) be a topological space and $(F^n)_{n \in \mathbb{N}}$, a sequence of functions from X into $\bar{\mathbb{R}}$, which is τ-epi convergent: $F = \tau\text{-lm}_e F^n$. Then, the following relations hold:

$$\inf_{u \in X} F(u) \geq \limsup_{n \to +\infty} (\inf_{u \in X} F^n(u)) \tag{2.10}$$

and, for every sequence $(\varepsilon_n)_{n \in \mathbb{N}}$, $\varepsilon_n > 0$ converging to zero,

$$\tau\text{-Lim sup}_{n \to +\infty} (\varepsilon_n\text{-Argmin } F^n) \subset \text{Argmin } F. \tag{2.11}$$

Proof of Proposition 2.9

(a) For every $x \in X$, for every $V \in N_\tau(x)$, and every $n \in \mathbb{N}$

$$\inf_{u \in X} F^n(u) \leq \inf_{u \in V} F^n(u).$$

Taking the upper limit with respect to $n \in \mathbb{N}$,

$$\limsup_n (\inf_{u\in X} F^n(u)) \leq \limsup_n (\inf_{u\in V} F^n(u))$$

$$\leq \sup_{V\in N_\tau(x)} \limsup_n \inf_{u\in V} F^n(u);$$

that is, by definition of $\tau\text{-ls}_e F^n(x)$,

$$\limsup_n (\inf_{u\in X} F^n(u)) \leq \tau\text{-ls}_e F^n(x).$$

The sequence $(F^n)_{n\in\mathbb{N}}$ being τ-epi convergent, $\tau\text{-ls}_e F^n = \tau\text{-li}_e F^n = \tau\text{-lm}_e F^n = F$; the above inequality being true for any $x \in X$, we finally get (2.10):

$$\limsup_n (\inf_{u\in X} F^n(u)) \leq \inf_{u\in X} F(u).$$

(b) The set $\tau\text{-Ls}\ (\varepsilon_n\text{-Argmin } F^n)$ may be empty; in that case, the inclusion (2.11) is clear; so, let us take

$$x \in \tau\text{-}\underset{(n\to+\infty)}{\text{Ls}}\ (\varepsilon_n\text{-Argmin } F^n) = \bigcap_{H\in N} c\ell_\tau (\bigcup_{k\in H} \varepsilon_k\text{-Argmin } F^k);$$

this last equality follows from formulation (1.113) (Proposition 1.32) of the τ-limit superior of a sequence of sets in a general topological space (N is the Frechet filter on $\mathbb{N}$); equivalently,

$$\forall H \in N \quad \forall V \in N_\tau(x), \quad \exists k = k_{H,V} \in H, \quad \exists x_{H,V} \in (\varepsilon_k\text{-Argmin } F^k) \cap V.$$

This last inclusion implies

$$\inf_{u\in V} F^k(u) \leq F^k(x_{H,V}) \leq \sup \{-\frac{1}{\varepsilon_k}; \inf_{u\in X} F^k(u) + \varepsilon_k\}.$$

For every $H \in N$, for every $V \in N_\tau(x)$, $k_{H,V} \in H$; so, for every $V \in N_\tau(x)$, the set $H'_V = \{k_{H,V}; H \in N\}$ belongs to $\ddot{N}$, the grill of N. Denoting

$$\alpha_n = \inf_{u\in V} F^n(u) \text{ and } \beta_n = \sup \{-\frac{1}{\varepsilon_n}; \inf_{u\in X} F^n(u) + \varepsilon_n\},\ n \in \mathbb{N},$$

we have:

$$\forall k \in H'_V \in \ddot{N} \qquad \alpha_k \leq \beta_k.$$

This means that there is a subsequence $(k(n))_{n\in\mathbb{N}}$ such that

$$\forall n \in \mathbb{N} \quad \alpha_{k(n)} \leq \beta_{k(n)};$$

This implies

$$\liminf_n \alpha_n \leq \liminf_n \alpha_{k(n)} \leq \limsup_n \beta_{k(n)} \leq \limsup_n \beta_n;$$

so, for every $V \in N_\tau(x)$

$$\liminf_n \inf_{u\in V} F^n(u) \leq \limsup_n \sup\{-\frac{1}{\varepsilon_n}; \inf_{u\in X} F^n(u) + \varepsilon_n\}.$$

This being true for every $V \in N_\tau(x)$,

$$\sup_{V\in N_\tau(x)} \liminf_n \inf_{u\in V} F^n(u) \leq \limsup_n \sup\{-\frac{1}{\varepsilon_n}; \inf_{u\in X} F^n(u) + \varepsilon_n\}. \quad (2.12)$$

From the definition of $\tau\text{-li}_e F^n$ and τ-epi convergence of the sequence F^n, $F = \tau\text{-lm}_e F^n = \tau\text{-li}_e F^n$, and (2.12), we get:

$$F(x) \leq \limsup_n \sup\{-\frac{1}{\varepsilon_n}; \inf_{u\in X} F^n(u) + \varepsilon_n\}. \quad (2.13)$$

We now notice that for any sequence $(\alpha_n)_{n\in\mathbb{N}}$ in $\bar{\mathbb{R}}$

$$\limsup_n \alpha_n = \limsup_n \sup\{-\frac{1}{\varepsilon_n}; \alpha_n + \varepsilon_n\}: \quad (2.14)$$

clearly, the right member is greater than or equal to the left one; on the other hand, if for some subsequence n_k, $\sup\{-1/\varepsilon_{n_k}; \alpha_{n_k} + \varepsilon_{n_k}\}$ converges to $\alpha > -\infty$, for k sufficiently large, $\sup\{-1/\varepsilon_{n_k}; \alpha_{n_k} + \varepsilon_{n_k}\} = \alpha_{n_k} + \varepsilon_{n_k}$, and α_{n_k} converges to α. Hence $\limsup_n \alpha_n \geq \alpha$ and the equality (2.14) follows. From (2.13) and (2.14) we derive

$$F(x) \leq \limsup_n (\inf_{u\in X} F^n(u))$$

which, combined with inequality (2.10), finally implies

$$F(x) \leq \inf_{u\in X} F(u) \text{ i.e. } x \in \text{Argmin } F$$

and

$$\min_{u \in X} F(u) = \limsup_{n} \inf_{u \in X} F^n(u). \quad \square$$

<u>COROLLARY 2.10</u> Let $F = \tau\text{-lm}_e F^n$ and assume that $\tau\text{-Ls}\,(\varepsilon_n\text{-Argmin } F^n) \neq \emptyset$. Then,

$$\min_{u \in X} F(u) = \limsup_{n \in \mathbb{N}} \inf_{u \in X} F^n(u).$$

<u>Comments:</u> As already noticed, without additional assumptions the inequality (2.10) and inclusion (2.11) are strict. This is illustrated by the following example: Let $X = \mathbb{R}$ and $F^n : X \to \mathbb{R}$ be defined for every $n \in \mathbb{N}$ and $x \in \mathbb{R}$ by:

$$F^n(x) = \begin{cases} x & \text{if } x \geq 0 \\ {}^1/_n x & \text{if } 0 \geq x \geq -n \\ -1 & \text{if } -n \geq x \end{cases}$$

Then, the sequence $(F^n)_{n \in \mathbb{N}}$ epi-converges (for the usual topology on $\mathbb{R}$) to the function F equal to

$$F(x) = \begin{cases} x & \text{if } x \geq 0 \\ 0 & \text{if } x \leq 0. \end{cases}$$

The fact that the sequence epi-converges can be viewed as a consequence of the monotone epi-convergence theorem (cf. Theorem 2.25, Section 2.5), or from the fact that the sequence converges uniformly on bounded subsets of $\mathbb{R}$ (from the Dini theorem) and hence epi-converges. But

$$\inf_{u \in \mathbb{R}} F(u) = \ 0 > \limsup_{n \to +\infty} (\inf_{u \in \mathbb{R}} F^n(u)) = -1.$$

Moreover, for every sequence $\varepsilon_n \to 0$, $\varepsilon_n > 0$,

$$\varepsilon_n\text{-Argmin } F^n =]-\infty, -n(1-\varepsilon_n)] \text{ and } \operatorname{Lim\,sup}_n (\varepsilon_n\text{-Argmin } F^n) = \emptyset.$$

So, the inclusion (2.11) Argmin $F = \mathbb{R}^- \supset \operatorname{Lim\,sup}\,(\varepsilon_n\text{-Argmin } F^n) = \emptyset$ is also strict. $\square$

Let us now state the fundamental result of this section: the minimal assumption one has to make, in order to derive, from τ-epi convergence of a sequence $(F^n)_{n\in\mathbb{N}}$, the convergence of the infimum $(\inf F^n)_{n\in\mathbb{N}}$ and the existence of a minimum for the limit problem is an *inf-compactness property:* one has to verify that for some $\varepsilon_n \to 0$, an ε_n-minimizing sequence $x_n \in \varepsilon_n$-Argmin F^n, is τ-relatively compact. The following theorem, expressed as a necessary and sufficient condition, extends previous results (Attouch & Wets [1], Dolecki [2]).

<u>THEOREM 2.11</u> Let (X,τ) be a first countable topological space and $(F^n)_{n\in\mathbb{N}}$ a sequence of functions from X into $\bar{\mathbb{R}}$ which is assumed τ-epi convergent: $F = \tau\text{-}\mathrm{lm}_e F^n$.

The following statements are equivalent:

(i) $\lim_{n\to\infty} (\inf_{u\in X} F^n(u)) = \inf_{u\in X} F(u)$ and Argmin $F \neq \emptyset$.

(ii) there exists a sequence $(\varepsilon_n)_{n\in\mathbb{N}}$, $\varepsilon_n \xrightarrow[n\to+\infty]{} 0$ and a τ-relatively compact sequence $(x_n)_{n\in\mathbb{N}}$ in X such that: $\forall n \in \mathbb{N}$ $x_n \in \varepsilon_n$-Argmin F^n.

(iii) there exists a sequence $(\varepsilon_n)_{n\in\mathbb{N}}$, $\varepsilon_n \xrightarrow[n\to+\infty]{} 0$ and K a non-void relatively compact set in (X,τ) such that:

$$\forall n \in \mathbb{N} \quad \inf_{u\in K} F^n(u) \leq \sup \{-1/\varepsilon_n; \inf_{u\in X} F^n(u) + \varepsilon_n\}.$$

<u>Proof of Theorem 2.11</u> We first notice that (ii) and (iii) are equivalent:

(ii) $\Rightarrow$ (iii) take $K = \{x_n\}_{n\in\mathbb{N}}$; by assumption, K is τ-relatively compact and, for every $n \in \mathbb{N}$

$$\inf_{u\in K} F^n(u) \leq F^n(x_n) \leq \sup \{-1/\varepsilon_n; \inf_{u\in X} F^n(u) + \varepsilon_n\},$$

the second inequality expressing that $x_n \in \varepsilon_n$-Argmin F^n.

(iii) $\Rightarrow$ (ii) For every $n \in \mathbb{N}$, take $x_n \in K$ satisfying:

$$\text{if } \inf_{u\in K} F^n(u) = -\infty, \quad F^n(x_n) \leq -1/\varepsilon_n$$

$$\text{if } \inf_{u\in K} F^n(u) > -\infty, \quad F^n(x_n) \leq \inf_{u\in K} F^n(u) + \varepsilon_n.$$

By definition, and from assumption (iii), the sequence $(x_n)_{n\in\mathbb{N}}$ satisfies:

$(x_n)_{n\in\mathbb{N}}$ is τ-relatively compact (since $x_n \in K$ for every $n \in \mathbb{N}$)

$$F^n(x_n) \leq \sup\{-1/\varepsilon_n; \inf_{u\in K} F^n(u) + \varepsilon_n\}$$

$$\leq \sup\{-1/\varepsilon_n; \sup\{-1/\varepsilon_n; \inf_{u\in X} F^n(u) + \varepsilon_n\} + \varepsilon_n\}$$

$$\leq \sup\{-1/\varepsilon_n + \varepsilon_n; \inf_{u\in X} F^n(u) + 2\varepsilon_n\}$$

$$\leq \sup\{-1/2\varepsilon_n \; ; \inf_{u\in X} F^n(u) + 2\varepsilon_n\} \quad (\text{for } 0 < \varepsilon_n^2 < \tfrac{1}{2}).$$

i.e. $x_n \in 2\varepsilon_n\text{-Argmin } F^n$.

(ii) $\Rightarrow$ (i) In the proof of Proposition 2.9 we have already obtained that if a sequence $(x_n)_{n\in\mathbb{N}}$ satisfies

$$\forall n \in \mathbb{N} \quad x_n \in \varepsilon_n\text{-Argmin } F^n \quad (\varepsilon_n \to 0),$$

then every τ-cluster point $\bar{x}$ of the sequence $(x_n)_{n\in\mathbb{N}}$ (there exists at least one) minimizes $F = \tau\text{-lm}_e F^n$. So, Argmin $F \ni \bar{x}$ is not empty; let us now prove that

$$\lim_n (\inf_{u\in X} F^n(u)) = \min_{u\in X} F(u).$$

We already know that $\inf_{u\in X} F(u) \geq \limsup_n (\inf_{u\in X} F^n(u))$ (cf. Proposition 2.9). So, the only thing we have to prove is that

$$\inf_{u\in X} F(u) \leq \liminf_n (\inf_{u\in X} F^n(u)). \tag{2.15}$$

By definition of $x_n \in \varepsilon_n\text{-Argmin } F^n$

$$F^n(x_n) \leq \sup\{-1/\varepsilon_n; \inf_{u\in X} F^n(u) + \varepsilon_n\}.$$

Noticing that for any sequence $(\alpha_n)_{n\in\mathbb{N}}$ in $\bar{\mathbb{R}}$,

$$\liminf_n \alpha_n = \liminf_n \sup\{-1/\varepsilon_n; \alpha_n + \varepsilon_n\},$$

we derive

$$\liminf_n F^n(x_n) \leqslant \liminf_n (\inf_{u\in X} F^n(u)) \tag{2.16}$$

By definition of the lim inf operation, there exists a subsequence $(n_k)_{k\in\mathbb{N}}$ such that

$$\liminf_n F^n(x_n) = \lim_k F^{n_k}(x_{n_k}).$$

Using the τ-relative compactness of the sequence $(x_n)_{n\in\mathbb{N}}$, we can extract another subsequence $(x_{n_{k_\ell}})$ such that $\bar{x} = \tau\text{-}\lim x_{n_{k_\ell}}$ exists. Writing n_ℓ instead of n_{k_ℓ}, we get from (2.16)

$$\lim_\ell \inf F^{n_\ell}(x_{n_\ell}) \leqslant \liminf_n (\inf_{u\in X} F^n(u)).$$

Since

$$F(\bar{x}) = \tau\text{-}\mathrm{lm}_e F^n(\bar{x}) = \tau\text{-}\mathrm{li}_e F^n(\bar{x}) \leqslant \liminf_\ell F^{n_\ell}(x_{n_\ell}),$$

(cf. Remark 1.19), we get

$$\inf_{u\in X} F(u) \leqslant F(\bar{x}) \leqslant \liminf_n (\inf_{u\in X} F^n(u));$$

that is, (2.15).

(i) $\Rightarrow$ (ii) By assumption, Argmin F is non-void, so let us take $x_o \in$ Argmin F. Since (X,τ) is first countable (that is the only place where we use this assumption) there exists a sequence $(x_n)_{n\in\mathbb{N}}$ in X such that: $x_n \xrightarrow{\tau} x_o$ and $F(x_o) = \lim_n F^n(x_n)$. Let us prove that:

$$\forall\varepsilon > 0 \quad \exists n(\varepsilon) \in \mathbb{N} \text{ such that: } \forall n \geqslant n(\varepsilon) \quad x_n \in \varepsilon\text{-Argmin } F^n. \tag{2.17}$$

If (2.17) is not true, there exists $\varepsilon_o > 0$ and an increasing mapping $k \to n_k$ from $\mathbb{N}$ into $\mathbb{N}$ such that

$$\forall k \in \mathbb{N} \quad x_{n_k} \notin \varepsilon_o\text{-Argmin } F^{n_k}.$$

This means that

$$F^{n_k}(x_{n_k}) > \sup \{-1/\varepsilon_o; \inf_{u\in X} F^{n_k}(u) + \varepsilon_o\} \text{ for every } k \in \mathbb{N}. \tag{2.18}$$

If $\inf_{u\in X} F(u)$ is equal to $-\infty$, we derive from assumption (i) and (2.18) that

$$F^{n_k}(x_{n_k}) > -1/\varepsilon_o$$

and

$$F(x_o) = \lim_k F^{n_k}(x_{n_k}) > -1/\varepsilon_o;$$

since $F(x_o) = \inf_{u\in X} F(u)$, we get a contradiction.

If $\inf_{u\in X} F(u)$ is finite, we derive from assumption (i) and (2.18) that

$$F^{n_k}(x_{n_k}) > \inf_{u\in X} F^{n_k}(u) + \varepsilon_o;$$

therefore,

$$\operatorname{Inf}_{u\in X} F(u) = F(x_o) = \lim_k F^{n_k}(x_{n_k}) \geq \lim_k \inf_{u\in X} F^{n_k}(u) + \varepsilon_o = \inf_{u\in X} F(u) + \varepsilon_o,$$

which is a clear contradiction.

So (2.17) is proved. It can be expressed, equivalently, in the following form:

$$\forall\varepsilon > 0 \quad \exists(x_n^\varepsilon)_{n\in\mathbb{N}} \text{ s.t. } \forall n \in \mathbb{N} \quad x_n^\varepsilon \in \varepsilon\text{-Argmin } F^n \text{ and } x_n^\varepsilon \xrightarrow[n]{\tau} x_o. \tag{2.19}$$

Note that, in order to obtain $(x_n^\varepsilon)_{n\in\mathbb{N}}$ by using (2.17), one has just to modify the sequence $(x_n)_{n\in\mathbb{N}}$ for a finite number of indexes, $1 \leq n \leq n(\varepsilon)$; for these indexes $(1 \leq n \leq n(\varepsilon))$ one can take for x_n^ε an arbitrary element of ε-Argmin F^n. From Corollary 1.18, by a diagonalization argument, we can find a map $n \to \varepsilon(n)$, decreasing to zero, such that

$$x_o = \tau\text{-}\lim x_n^{\varepsilon(n)}.$$

Taking $\varepsilon_n = \varepsilon(n)$ and $x_n = x_n^{\varepsilon(n)}$, we have realized the conditions of Theorem 2.11, (ii): the sequence $(x_n)_{n\in\mathbb{N}}$ is τ-converging, and hence τ-compact and, for every $n \in \mathbb{N}$, from (2.19) $x_n = x_n^{\varepsilon(n)} \in \varepsilon_n$-Argmin F^n. □

From the preceding argument, we can extract the following result concerning the convergence of the ε-minimizer sets:

THEOREM 2.12 Let $(F^n)_{n\in\mathbb{N}}$ a sequence of functions from (X,τ), a first countable topological space, into $\bar{\mathbb{R}}$. We assume that

$$F = \tau\text{-}\mathrm{lm}_e F^n \text{ exists and } \inf_{u\in X} F(u) = \lim_{n\to+\infty} (\inf_{u\in X} F^n(u)).$$

Then,

$$\bigcap_{\varepsilon>0} \tau\text{-Li}(\varepsilon\text{-Argmin } F^n) = \text{Argmin } F = \bigcap_{\varepsilon>0} \tau\text{-Ls}(\varepsilon\text{-Argmin } F^n). \tag{2.20}$$

Proof of Theorem 2.12 Let us first prove that, for every $\varepsilon > 0$

$$\varepsilon\text{-Argmin } F \supset \tau\text{-Ls } (\varepsilon\text{-Argmin } F^n). \tag{2.21}$$

Let $(n_k)_{k\in\mathbb{N}}$ be a subsequence of $\mathbb{N}$ and $x_k \xrightarrow[k\to+\infty]{\tau} x$ converging in (X,τ) such that:

$$\forall k \in \mathbb{N} \quad x_k \in \varepsilon\text{-Argmin } F^{n_k}, \text{ i.e. } F^{n_k}(x_k) \leq \sup \{-1/\varepsilon; \inf_{u\in X} F^{n_k}(u) + \varepsilon\}.$$

Taking the lower-limit with respect to $k \in \mathbb{N}$,

$$\liminf_{k\to+\infty} F^{n_k}(x_k) \leq \liminf_{k\to+\infty} \sup \{-1/\varepsilon; \inf_{u\in X} F^{n_k}(u) + \varepsilon\}.$$

Since $F = \tau\text{-}\mathrm{lm}_e F^n$, $x_k \xrightarrow[n\to+\infty]{\tau} x$ and $\inf F^n \xrightarrow[n\to+\infty]{} \inf F$,

we get

$$F(x) \leq \sup \{-1/\varepsilon; \inf_{u\in X} F(u) + \varepsilon\} \text{ i.e. } x \in \varepsilon\text{-Argmin } F.$$

So, (2.21) is proved. Noticing that Argmin $F = \bigcap_{\varepsilon>0} \varepsilon$-Argmin F, we obtain:

$$\text{Argmin } F \supset \bigcap_{\varepsilon>0} \tau\text{-Ls}(\varepsilon\text{-Argmin } F^n). \tag{2.22}$$

Let us now prove that

$$\bigcap_{\varepsilon>0} \tau\text{-Li}(\varepsilon\text{-Argmin } F^n) \supset \text{Argmin } F. \tag{2.23}$$

If Argmin $F = \emptyset$, (2.23) is clear. Otherwise, let $x \in$ Argmin F. From (2.19), there exists a sequence $(x_n)_{n\in\mathbb{N}}$ in X satisfying

$$x_n \xrightarrow[n\to+\infty]{\tau} x \text{ and } \forall\varepsilon > 0 \quad \exists n(\varepsilon) \in \mathbb{N} \text{ such that } x_n \in \varepsilon\text{-Argmin } F^n$$

for every $n \geq n(\varepsilon)$. This implies that

$\forall\varepsilon > 0 \quad x \in \tau\text{-Li}(\varepsilon\text{-Argmin } F^n)$ and (2.23) follows.

Combining (2.22) and (2.23), since $\tau\text{-Ls} \supset \tau\text{-Li}$, we obtain the equality (2.20). □

COROLLARY 2.13 Let (X,τ) be a first countable topological space and $(F^n)_{n\in\mathbb{N}}$ a sequence of functions from X into $\bar{\mathbb{R}}$ which is τ-epi convergent, $F = \tau\text{-lm}_e F^n$ and which satisfies:

$$\inf_{u\in X} F(u) = \lim_{n\to+\infty} \inf_{u\in X} F^n(u)$$

Then,

$$\begin{cases} \forall x \in \text{Argmin } F, \ \exists(\varepsilon_n)_{n\in\mathbb{N}} \text{ decreasing to zero}, \ \exists(x_n)_{n\in\mathbb{N}}: \\ \qquad\qquad \forall n \in \mathbb{N} \quad x_n \in \varepsilon_n\text{-Argmin } F^n, \ x_n \xrightarrow[n\to+\infty]{\tau} x \qquad (2.24) \\ \forall(n_k)_{k\in\mathbb{N}}, \ \forall\varepsilon_k \xrightarrow[k\to+\infty]{} 0, \ \forall x_k \in \varepsilon_k\text{-Argmin } F^{n_k}, \ x_k \xrightarrow{\tau} x \Rightarrow x \in \text{Argmin } F. \end{cases}$$

Proof of Corollary 2.13 The first part of (2.24) follows from (2.20) and the diagonalization argument developed in the proof of Theorem 2.11 ((i) ⇒ (ii)).

The second part is a consequence of the general inclusion

$$\tau\text{-Ls}(\varepsilon_n\text{-Argmin } F^n) \subset \text{Argmin } F,$$

proved in Proposition 2.9.

REMARK 2.14 In (2.24), the sequence $(\varepsilon_n)_{n\in\mathbb{N}}$ depends on the family $(F^n)_{n\in\mathbb{N}}$ and the point x picked up in Argmin F: take for example

$$F^n(x) = \begin{cases} 1/n|x| & \text{if } x \neq 0 \\ +\infty & \text{if } x = 0 \end{cases}$$

then, $0 \equiv F = \tau\text{-lm}_e F^n$ (τ = usual topology on $\mathbb{R}$); so, Argmin $F = \mathbb{R}$ and

$\forall\varepsilon > 0$ ε-Argmin $F^n = \mathbb{R}\setminus[-1/n\varepsilon, +1/n\varepsilon]$; for a given sequence $\varepsilon(n) \xrightarrow[n\to+\infty]{} 0$, we have

$$\text{dist}(0,\varepsilon(n)\text{-Argmin } F^n) = 1/n\varepsilon(n) \xrightarrow[(n\to+\infty)]{} 0 \text{ iff } n\varepsilon(n) \xrightarrow[n\to+\infty]{} +\infty;$$

for example, take $\varepsilon(n) = \dfrac{1}{\sqrt{n}}$.

For further developments refer to Wets [4], Dolecki [2].

2.3 PERTURBATION OF EPI-CONVERGENT SEQUENCES

As we noticed in Section 1.2.3, epi-convergence does imply convergence of the infimum and satisfies stability properties with respect to continuous perturbations. It is the conjunction of these two properties which makes epi-convergence deserve the "variational convergence" label. This last aspect is investigated in the present section. We stress the importance of this stability feature since, in physics and optimization one frequently has to consider perturbed or constrained problems.

We like to know what is the *largest class of admissible perturbations* G (with the terminology of Attouch & Sbordone [1]) with respect to a convergence result $F = \lim_e F^n$, that is to say perturbations G for which $F + G = \lim_e(F^n+G)$ still holds. Indeed, one can introduce epi-convergence, $F = \lim_e F^n$, by considering convergence of the infimum $\inf (F^n + G) \to \inf (F + G)$ for a large enough class of perturbations. That is the idea of the Moreau-Yosida approximation approach to epi-convergence as described in Section 2.7.

2.3.1 Continuous perturbations

THEOREM 2.15 Let $\{F^n:X \to \bar{\mathbb{R}};\ n \in \mathbb{N}\}$ a sequence of functions from (X,τ), a topological space, into $\bar{\mathbb{R}}$. Let us assume that the sequence $\{F^n; n \in \mathbb{N}\}$ is τ-epi-convergent. Then, for every τ-continuous function $G:X \to \mathbb{R}$, the sequence $(F^n + G)_{n\in\mathbb{N}}$ is still τ-epi convergent and

$$\tau\text{-lm}_e(F^n + G) = \tau\text{-lm}_e F^n + G. \tag{2.25}$$

More generally, if G^n is a sequence of τ-continuous functions converging uniformly to G, then

$$\tau\text{-lm}_e(F^n + G^n) = \tau\text{-lm}_e F^n + G.$$

In order to prove the above theorem, let us first state the following general inequality.

PROPOSITION 2.16 Let $(F^n)_{n\in\mathbb{N}}$ and $(G^n)_{n\in\mathbb{N}}$ be two sequences of real extended valued functions on the topological space (X,τ), then

$$\tau\text{-}li_e(F^n + G^n) \geq \tau\text{-}li_e F^n + \tau\text{-}li_e G^n. \tag{2.26}$$

Proof of Proposition 2.16 By definition of the τ-epi limit inferior (1.32)

$$\tau\text{-}li_e(F^n + G^n)(x) = \sup_{V\in N_\tau(x)} \liminf_{n\to+\infty} \inf_{u\in V} (F^n + G^n)(u)$$

$$= \sup_{V\in N_\tau(x)} \sup_{H\in\mathbb{N}} \inf_{n\in H} \inf_{u\in V} (F^n + G^n)(u).$$

By subadditivity of the inf operation (we adopt the convention: $\forall a\in\bar{\mathbb{R}}$, $a-\infty=-\infty$)

$$\tau\text{-}li_e(F^n + G^n)(x) \geq \sup_{V\in N_\tau(x)} \sup_{H\in\mathbb{N}} \{\inf_{\substack{n\in H\\ u\in V}} F^n(u) + \inf_{\substack{n\in H\\ u\in V}} G^n(u)\}$$

Let us use the notation

$$a(H,V) = \inf_{\substack{n\in H\\ u\in V}} F^n(u)$$

$$b(H,V) = \inf_{\substack{n\in H\\ u\in V}} G^n(u).$$

Noticing that a and b are decreasing set functions from $\mathbb{N} \times N_\tau(x)$ into $\bar{\mathbb{R}}$, we can complete the proof of Proposition 2.16 with the help of the following lemma:

LEMMA 2.17 Let $\mathfrak{V}$ be a class of subsets of an abstract space Y which is stable for the finite intersection property. Let $a(\cdot)$ and $b(\cdot)$ be two real extended valued functions

$$a:\mathfrak{V} \to \bar{\mathbb{R}} \quad b:\mathfrak{V} \to \bar{\mathbb{R}}$$

which are decreasing; then

$$\sup_{C\in\mathfrak{V}} \{a(C) + b(C)\} = \sup_{C\in\mathfrak{V}} a(C) + \sup_{C\in\mathfrak{V}} b(C). \tag{2.27}$$

Taking $Y = \mathbb{N} \times N_\tau(x)$, which is stable for the finite intersection property (since $\mathbb{N}$ and $N_\tau(x)$ enjoy this property) we get

$$\tau\text{-}li_e(F^n + G^n)(x) \geqslant \sup_{V\in N_\tau(x)} \sup_{H\in\mathbb{N}} \inf_{n\in H} \inf_{u\in V} F^n(u) + \sup_V \sup_H \inf_{n\in H} \inf_{u\in V} G^n(u);$$

that is,

$$\tau\text{-}li_e(F^n + G^n)(x) \geqslant \tau\text{-}li_e F^n(x) + \tau\text{-}li_e G^n(x). \quad \square$$

Noticing that, when (X,τ) is a first countable space, the sup operation on $V \in N_\tau(x)$ is nothing but a monotone sequential limit, one could achieve the proof of Proposition 2.16 directly without using Lemma 2.17.

<u>Proof of Theorem 2.15</u> From Proposition 2.16,

$$\tau\text{-}li_e(F^n + G) \geqslant \tau\text{-}li_e F^n + \tau\text{-}li_e G,$$

where $\tau\text{-}li_e(G)$ denotes the τ-epi limit inferior of the sequence: $G^n = G$, $\forall n \in \mathbb{N}$. From Corollary 2.3, $\tau\text{-}li_e G = c\ell_\tau(G)$.

By assumption, G is τ-continuous and the sequence $(F^n)_{n\in\mathbb{N}}$ τ-epi convergent. Thus $\tau\text{-}li_e G = c\ell_\tau(G) = G$ and $\tau\text{-}li_e F^n = \tau\text{-}lm_e F^n$; so the above inequality becomes

$$\tau\text{-}li_e(F^n + G) \geqslant \tau\text{-}lm_e F^n + G. \tag{2.28}$$

On the other hand, since G is τ-continuous, it is τ-upper semicontinuous and, for every $\varepsilon > 0$, for every $x \in X$, there exists a neighbourhood $V_\varepsilon(x) \in N_\tau(x)$ such that:

$$G(x) + \varepsilon \geqslant \sup_{u\in V_\varepsilon(x)} G(u).$$

Thus for every $V \in N_\tau(x)$ such that $V \subset V_\varepsilon(x)$ and every $n \in \mathbb{N}$,

$$\inf_{u\in V} (F^n + G)(u) \leqslant \inf_{u\in V} F^n(u) + G(x) + \varepsilon.$$

Taking the upper limit with respect to $n \in \mathbb{N}$

$$\limsup_n \inf_{u\in V} (F^n + G)(u) \leq \limsup_n \inf_{u\in V} F^n(u) + G(x) + \varepsilon.$$

The above inequality being true for every $V \subset V_\varepsilon(x)$, noticing that the two members are decreasing set functions of V, we obtain

$$\sup_{V\in N_\tau(x)} \limsup_n \inf_{u\in V} (F^n + G)(u) \leq \sup_{V\in N_\tau(x)} \limsup_n \inf_{u\in V} F^n(u)+G(x)+\varepsilon.$$

This being true for every $\varepsilon > 0$, and $x \in X$,

$$\tau\text{-}ls_e(F^n + G) \leq \tau\text{-}lm_e F^n + G. \tag{2.29}$$

From (2.28) and (2.29), the sequence $(F^n + G)_{n\in\mathbb{N}}$ is τ-epi convergent and

$$\tau\text{-}lm_e(F^n + G) = \tau\text{-}lm_e F^n + G.$$

The above argument is then extended in a straightforward way to the case of uniform convergence $G^n \to G$. □

2.3.2 Examples

In this paragraph, we present various examples illustrating preceding perturbation Theorem 2.15. Take $\{F^\varepsilon; \varepsilon \to 0\}$ as described in model example 1.1.1, homogenization of elliptic equations, then G represents some classical external action or constraint applied to the system.

Let us recall the (unconstrained) convergence Theorem 1.20. Functionals $F^\varepsilon: H^1(\Omega) \to \mathbb{R}$ are given by

$$F^\varepsilon(u) = \int_\Omega j(\frac{x}{\varepsilon}, \text{grad}u(x))dx$$

where $j: \mathbb{R}^N \times \mathbb{R}^N \to \mathbb{R}^+$

$$(y,z) \to y(y,z)$$

is measurable, Y-periodic in y ($Y = [0,1]^N$) and convex continuous in z

$$\lambda_0|z|^2 \leq j(y,z) \leq \Lambda_0(1 + |z|^2) \quad (0 < \lambda_0 \leq \Lambda_0 < +\infty).$$

Then, for every $u \in H^1(\Omega)$

$$L^2(\Omega)\text{-}\mathrm{lm}_e F^\varepsilon(u) = F^{hom}(u) \text{ exists}$$

and

$$F^{hom}(u) = \int_\Omega j_{hom}(\mathrm{grad}u(x))dx$$

with

$$j_{hom}(z) = \min_{w\ Y\text{-periodic}} \int_Y j(y,\mathrm{grad}w(y) + z)dy.$$

EXAMPLE 2.18 Let us first consider a "continuous" perturbation

$$G(u) = \int_\Omega g(x,u(x))dx$$

where

$$g:\mathbb{R}^N \times \mathbb{R} \to \mathbb{R}^+$$

$$(x,\xi) \to g(x,\xi)$$

is measurable in x, continuous in ξ and satisfies:

$$\begin{aligned} &\xi \to g(x,\xi) \text{ is monotone increasing on } [\xi_0, +\infty] \quad (\text{for some } \xi_0 > 0) \\ &\xi \to g(x,\xi) \text{ is monotone decreasing on }]-\infty,\xi_1] \quad (\text{for some } \xi_1 < 0). \end{aligned} \tag{2.30}$$

and

$$\text{for any } r \in \mathbb{R}^+, \quad \sup_{|\xi|<r} g(x,r) = g_r(x) \text{ belongs to } L^1(\Omega). \tag{2.31}$$

A direct attack on the convergence problem-determination of the limit of the infimum,

$$\inf_{u\in H_0^1(\Omega)} \int_\Omega \{j(\frac{x}{\varepsilon}, \mathrm{grad}u(x)) + g(x,u(x))\}dx,$$

is not easy. Let us prove that:

$$\forall u \in H_0^1(\Omega) \quad L^2(\Omega)\text{-}\mathrm{lm}_e(F^\varepsilon+G)(u) = (F^{hom} + G)(u). \tag{2.32}$$

We cannot reach a conclusion directly-via Theorem 2.15 - because G is not continuous for the topology $L^2(\Omega)$. It is only continuous for the $L^\infty(\Omega)$ topology (we make no restriction on the growth of $\xi \to g(x,\xi)$!).

The functional G being (from Fatou's lemma) lower semicontinuous on $L^2(\Omega)$, from Proposition 2.16 (for simplicity, we omit the reference to $\tau = L^2(\Omega)$)

$$li_e(F^\varepsilon + G) \geq li_e F^\varepsilon + G$$
$$\geq F^{hom} + G.$$

The difficult point is to prove that, for every $u \in H_0^1(\Omega)$,

$$(F^{hom} + G)(u) \geq ls_e(F^\varepsilon + G)(u)$$

that is, to find a sequence $(u_\varepsilon)_{\varepsilon>0}$ converging to u in $L^2(\Omega)$ such that

$$(F^{hom} + G)(u) \geq \limsup_{\varepsilon \to 0} (F^\varepsilon + G)(u_\varepsilon).$$

We take this opportunity to stress the fact that, when attacking such a perturbation problem, it is very useful to know the stronger topology τ for which the τ-epi convergence, $F = \tau\text{-lim } F^\varepsilon$, holds. The stronger the topology τ is, the larger is the class of "admissible" constraints G (*by admissible, we mean functionals* G *such that* $\tau\text{-lm}_e(F^\varepsilon + G) = F + G$). In our situation, as we shall see in Section 2.6, topology $\tau = s\text{-}L^2(\Omega)$ is not optimal. We shall prove (Corollary 2.61), that

$$\forall u \in H_0^1(\Omega), \quad \forall p \in [1, +\infty] \quad L^p(\Omega)\text{-lm}_e F^\varepsilon(u) = F^{hom}(u).$$

In particular, taking $p = +\infty$,

$$\forall u \in H_0^1(\Omega) \; \exists u_\varepsilon \to u \text{ in } L^\infty(\Omega) \text{ (i.e. } \|u_\varepsilon - u\|_{L^\infty} \to 0) \text{ such that } F^\varepsilon(u_\varepsilon) \to F^{hom}(u$$

Using the *continuity of* G *for the* $s\text{-}L^\infty(\Omega)$ *topology on* $H_0^1(\Omega) \cap L^\infty$ (this is a direct consequence of the continuity of g, of (2.31) and of the Lebesgue dominated convergence theorem) we obtain:

$$(F^{hom} + G)(u) = \lim_{\varepsilon \to 0} (F^\varepsilon + G)(u_\varepsilon)$$

that is

$$\forall u \in H_0^1 \cap L^\infty \quad (F^{hom} + G)(u) \geq ls_e(F^\varepsilon + G)(u). \tag{2.33}$$

In order to extend this result to any $u \in H_0^1(\Omega)$, we use the following propert; of G (which follows easily from the monotone properties (2.30) of g):

$$\forall u \in H_0^1(\Omega) \quad G(T_k u) \leq G(u) \quad \text{(for k sufficiently large)} \tag{2.34}$$

where T_k is the function

$$T_k(r) = \begin{cases} k & \text{if } r > k \\ r & \text{if } |r| \leq k \\ -k & \text{if } r \leq -k \end{cases} .$$

Applying (2.33) to $u_k = T_k u$,

$$(F^{hom} + G)(u_k) \geq \mathrm{ls}_{e \atop \varepsilon \to 0} (F^{\varepsilon} + G)(u_k).$$

Noticing that both F^{hom} and G satisfy the truncation property (2.34)

$$\begin{aligned} (F^{hom} + G)(u) &\geq (F^{hom} + G)(u_k) \\ &\geq \mathrm{ls}_e (F^{\varepsilon} + G)(u_k) \quad \text{(for k large enough).} \end{aligned}$$

Therefore,

$$(F^{hom} + G)(u) \geq \liminf_{k \to +\infty} [\mathrm{ls}_e(F^{\varepsilon} + G)](u_k).$$

Since $u_k \to u$ in $L^2(\Omega)$ and $L^2(\Omega)\text{-}\mathrm{ls}_e(F^{\varepsilon} + G)$ is $L^2(\Omega)$ lower semicontinuous,

$$\forall u \in H_0^1(\Omega) \ (F^{hom} + G)(u) \geq L^2(\Omega)\text{-}\mathrm{ls}_e(F^{\varepsilon} + G)(u)$$

which achieves the proof of (2.32). □

As a corollary, we obtain convergence of minima and ε-minimizers of the minimization problems

$$\min_{u \in H_0^1(\Omega)} \int_{\Omega} \{j(\tfrac{x}{\varepsilon}, \mathrm{grad}u(x)) + g(x,u(x)) - fu\}dx$$

to the minimum and minimizer of the following problem

$$\min_{u \in H_0^1(\Omega)} \int_{\Omega} \{j_{hom}(\mathrm{grad}u) + g(x,u(x)) - fu\}dx,$$

for every $f \in L^2(\Omega)$ and where g is a Caratheodory function satisfying (2.30) and (2.31). □

The next example is rather surprising, since it demonstrates a functional G which is not continuous for any of the topologies s-$L^p(\Omega)$ $(1 \leq p \leq +\infty)$ and which is *admissible*, i.e.

$$L^2(\Omega)\text{-}\mathrm{lm}_e(F^\varepsilon + G) = F^{hom} + G \tag{2.35}$$

(in the preceding example, the perturbation G was not continuous for the $L^2(\Omega)$ topology, but was continuous for the $L^\infty(\Omega)$ topology).

<u>EXAMPLE 2.19</u> Let us take $G = I_K$ the indicator function of the closed convex set $K = \{u \in H_0^1(\Omega);\ u \geq 0 \text{ a.e. on } \Omega\}$ and $(F^\varepsilon)_{\varepsilon>0}$ as described in the preceding example. Then, we claim that

$$L^2(\Omega)\text{-}\mathrm{lm}_e\ (F^\varepsilon + I_K) = F^{hom} + I_K. \tag{2.36}$$

Noticing that K, the positive cone of $L^2(\Omega)$, is closed for the $L^2(\Omega)$ topology, I_K is lower semicontinuous. From Proposition 2.16

$$\mathrm{li}_e(F^\varepsilon + I_K) \geq F^{hom} + I_K.$$

As before, the difficult point is to prove that

$$F^{hom} + I_K \geq \mathrm{ls}_e(F^\varepsilon + I_K).$$

Equivalently, for every positive function u of $H_0^1(\Omega)$, we look for a sequence $v_\varepsilon \to u$ in $L^2(\Omega)$ such that

$$\begin{cases} v_\varepsilon \geq 0 \\ F^{hom}(u) \geq \limsup_{\varepsilon \to 0} F^\varepsilon(v_\varepsilon). \end{cases}$$

Since $F^{hom} = \mathrm{lm}_e F^\varepsilon$, there exists a sequence $(u_\varepsilon)_{\varepsilon>0}$ converging to u in $L^2(\Omega)$:

$$F^{hom}(u) = \lim_{\varepsilon\to 0} F^\varepsilon(u_\varepsilon). \tag{2.37}$$

Take $v_\varepsilon = u_\varepsilon^+$ (the positive part of u_ε, i.e. $u_\varepsilon^+ = u_\varepsilon \vee 0$). Then $v_\varepsilon \to u^+ = u$ (since $u \geq 0$), in $L^2(\Omega)$. Moreover

$$F^\varepsilon(u_\varepsilon) \geq F^\varepsilon(u_\varepsilon^+) = F^\varepsilon(v_\varepsilon). \tag{2.38}$$

From (2.37) and (2.38) we conclude that

$$F^{hom}(u) \geq \limsup_{\varepsilon \to 0} F^{\varepsilon}(v_{\varepsilon}). \tag{2.39}$$

As a corollary. we obtain the convergence of the minimization problems

$$\min_{\substack{u \in H_o^1(\Omega) \\ u \geq 0}} \int_{\Omega} \{j(\frac{x}{\varepsilon}, \text{grad}u(x)) - fu\}dx$$

to

$$\min_{\substack{u \in H_o^1(\Omega) \\ u \geq 0}} \int_{\Omega} \{j_{hom}(\text{grad}u) - fu\}dx$$

for every $f \in L^2(\Omega)$. (This is a direct consequence of (2.36), of compact injection from $H_o^1(\Omega)$ into $L^2(\Omega)$ and of the variational property of epi-convergence, Theorem 1.10). The preceding result is in fact a particular case of stability results for *general obstacle* constraints $K = \{u \geq g\}$: cf. Boccardo & Marcellini [1], Carbone & Sbordone [2], Attouch [3].

EXAMPLE 2.20 Let us now examine the case of "*non-admissible*" constraints $G = I_K$. Clearly, the smaller the set K is, the less chance it has to be admissible in the sense of (2.35).

The borderline case is when K is reduced to one point. In fact, this turns out to be a specialization of the case where K *is compact for the topology* $H_o^1(\Omega)$: then,

$$s\text{-}L^2(\Omega) - \text{lm}_e(F^{\varepsilon} + I_K) = H + I_K \tag{2.40}$$

where

$$H(u) = \int_{\Omega} \phi(\text{grad}u(x))dx$$

and

$$\phi(z) = \sigma(L^{\infty}, L^1)\lim_{\varepsilon \to 0} j(\frac{\cdot}{\varepsilon}, z) = \mathfrak{m}_Y j(\cdot, z).$$

That is, H is the pointwise limit of the sequence $(F^{\varepsilon})_{\varepsilon > 0}$, which, as we have already noticed, is strictly larger than F^{hom}.

In order to prove (2.40), we first verify that

$$\mathrm{li}_e(F^\varepsilon + I_K)(u) \geq H(u) + I_K(u).$$

Indeed if $u_\varepsilon \in K$ and $u_\varepsilon \to u$ in $L^2(\Omega)$, from the compactness of K in $H_o^1(\Omega)$, u_ε converges strongly to u in $H_o^1(\Omega)$, and

$$\lim_{\varepsilon\to 0} F^\varepsilon(u_\varepsilon) = H(u).$$

This follows from the $\sigma(L^\infty,L^1)$ convergence of $j(\frac{\cdot}{\varepsilon},z)$ for every $z \in \mathbb{R}^N$, the continuity properties of $z \to j(y,z)$ and the strong convergence in $L^2(\Omega)$ of the derivatives of u_ε. Let us make this precise

We first notice that because of the uniform local Lipschitzian property of F^ε on H_o^1, it is enough to prove that

$$\int_\Omega j(\frac{\cdot}{\varepsilon},\mathrm{grad}u) \to \int_\Omega \phi(\mathrm{grad}u) \quad \text{as } \varepsilon \to 0.$$

By the same continuity argument, we can take u regular. Denoting

$$g(x,y) = j(y,\mathrm{grad}u(x))$$

g is regular in x, Y-periodic in y. It is a classical exercise (cf. Suquet [1], Lemma 6, Annexe 2) to prove, then, that

$$g(x,\frac{x}{\varepsilon}) \xrightarrow[\varepsilon \to 0]{} \mathfrak{m}_Y j(\cdot,\mathrm{grad}u(x)) \text{ in } \sigma(L^\infty,L^1)$$
$$\parallel$$
$$\phi(\mathrm{grad}u(x)).$$

On the other hand, we have to verify that

$$H(u) + I_K(u) \geq \mathrm{ls}_e(F^\varepsilon + I_K)(u).$$

Given $u \in K$, it is enough to take $u_\varepsilon = u$, since $F^\varepsilon(u) \to H(u)$ as $\varepsilon \to 0$!

<u>EXAMPLE 2.21</u> An interesting question is to know, given a convex set K, whether it is admissible. At this stage, we can only give an heuristic answer to this problem (we shall discuss it in Chapter 3): roughly speaking we can say that since a constraint involves only u (and not its derivatives: unilateral, bilateral constraint,..., on u) it is admissible. If it involves

some derivatives of u it is no longer admissible (with respect to perturbations of $F^{\varepsilon} \underset{\varepsilon\to 0}{\longrightarrow} F^{hom}$).

Let us describe such an example from the field of elastoplasticity (cf. Section 1.6.1). Take $K = I_{\{|\mathrm{grad}u(x)|\leq 1\}}$ (which is not compact in $H^1(\Omega)$!). Then,

$$\Phi^{hom}(u) = L^2(\Omega)\text{-}\lim_e(F^{\varepsilon}+I_K)(u) \text{ exists for every } u \in H_o^1(\Omega)$$

$$= \int_{\Omega} \theta_{hom}(\mathrm{grad}u(x))dx$$

with

$$\theta_{hom}(z) = \min_{\begin{cases} v \text{ Y-periodic} \\ |\mathrm{grad}v+z|\leq 1 \end{cases}} \int_Y j(y,\mathrm{grad}v(y)+z)dy. \tag{2.41}$$

This result can be obtained in a quite similar way to Theorem 1.49, (cf. also Carbone [1], [2]). One can easily verify that

$$\theta_{hom}(z) = +\infty \text{ if } |z| > 1$$

$$\theta_{hom}(z) > j_{hom}(z) \text{ if } |z| \leq 1 \quad \text{(in general)}.$$

Thus

$$\lim_e(F^{\varepsilon} + I_K) = \Phi^{hom}$$

$$> F^{hom} + I_K! \quad \square$$

We have sketched examples related to the model problems we described at the beginning of Chapter 1. In fact, it is a very general and important problem to know how to go to the limit (in a variational sense) on the sum of two sequences $(F^n)_{n\in\mathbb{N}}$ and $(G^n)_{n\in\mathbb{N}}$ and to know when the following equality holds:

$$\lim_e(F^n + G^n) = \lim_e(F^n) + \lim_e(G^n). \tag{2.42}$$

We shall present other related situations in the following chapters, coming from perturbations (sometimes singular, sometimes not) in the calculus of variations and optimal control. $\square$

2.4 COMPACTNESS RESULTS

2.4.1 Setting of the problem

Up to now, we have been interested with epi-convergence of sequences of functions. In this section, we adopt a more general point of view and study compactness properties of classes of functions with respect to epi-convergence.

(a) The basic abstract theorem used for this type of approach (Theorem 2.22) states that, from each sequence $\{F^n : X \to \bar{\mathbb{R}};\ n \in \mathbb{N}\}$ of real extended valued functions defined on a second countable space (X,τ), one can extract a τ-epi-convergent subsequence $\{F^{n_k};\ k \in \mathbb{N}\}$. Using geometric interpretation of epi-convergence, this theorem can be viewed as an equivalent formulation of the celebrated Kuratowski compactness theorem, which states that every sequence of sets in a second countable topological space contains a set-convergent subsequence. In these theorems, compactness comes respectively from compactness of $\bar{\mathbb{R}}$, and from the fact that the limit set may be empty.

(b) We then describe two compactness results for classes of functionals corresponding to the model examples described in Chapter 1.

The first one states compactness in the epi-convergence sense (for the s-$L^2(\Omega)$ topology) of the class $\mathcal{F}_{\lambda,\Lambda}$ of functionals F of the form

$$F(u) = \int_\Omega j(x, \operatorname{grad} u(x))dx,$$

where j satisfies uniform bounds from below and above:

$$\lambda|z|^2 \leq j(x,z) \leq \Lambda(1 + |z|^2) \qquad 0 < \lambda \leq \Lambda < +\infty,$$

(this corresponds to Example 1.1.1, homogenization of elliptic operators).

The second one describes the compact closure (in the epi-convergence sense) for the s-$L^2(\Omega)$ topology of the class $\mathcal{F}$ of functionals F_T of the form

$$F_T(u) = \int_\Omega |\operatorname{grad} u|^2 dx + I_{K_T}(u)$$

where with each subset $T \subset \Omega$ is associated the "bilateral" constraint convex set $K_T = \{u \in H^1(\Omega);\ u = 0 \text{ on } T\}$ and I_{K_T} its indicator functional.

As described in the "cloud of ice" (Example 1.1.2), it may happen that, for some sequence $\{T_n;\ n \in \mathbb{N}\}$ of subsets of Ω, the limit problem takes a "relaxed form" and is no more associated with a constraint $\{u = 0 \text{ on } T\}$, for some $T \subset \Omega$. The compact closure $\bar{\mathcal{F}}$ of $\mathcal{F}$ is indeed equal to

$$\bar{\mathcal{F}} = \{F:H^1(\Omega) \to \bar{\mathbb{R}}^+/F(u) = \int_\Omega |\mathrm{grad}u|^2 dx + \int_\Omega a(x)\tilde{u}^2(x)^2 d\mu(x)$$

where $\mu \in H^{-1}(\Omega)^+$, $a \geq 0$ μ-measurable$\}$.

When taking the density measure $a d\mu$ equal to $+\infty$ on some subset T of Ω, $+\infty$ elsewhere, we obtain the functional F_T. (So, the above class does contain $\mathcal{F}$ and the relaxed constraint functionals).

More generally, we describe the compact closure of the class $\mathcal{F}$

$$\mathcal{F} = \{F_g/F_g(u) = \int_\Omega |\mathrm{grad}u|^2 dx + I_{K_g}(u);\ g:\Omega \to \bar{\mathbb{R}}\}$$

where

$$K_g = \{u \in H^1(\Omega)/u \geq g \text{ on } \Omega\}$$

(Theorems 2.29 and 2.35).

(c) Following De Giorgi [3], in order to establish such theorems, one considers the above integral functionals as functions not only of u, but also of the set of observations ω:

$$F(u,\omega) = \int_\omega f(x,u(x),\mathrm{grad}u(x))dx.$$

As a function of u, F enjoys some continuity or semicontinuity properties, as a function of ω it is usually a Borel measure. The two variables are linked by the following local property:

$$u|_\omega = v|_\omega \Rightarrow F(u,\omega) = F(v,\omega).$$

So we do an axiomatization of $\mathcal{F}$ and list the properties of this class which are preserved by the epi-limit process. Relying on the abstract compactness theorem, we thus obtain a compact class $\mathcal{F}_1$ containing $\mathcal{F}$. Then we prove an integral representation of such functionals and describe precisely the compact closure $\bar{\mathcal{F}}$ of $\mathcal{F}$ in $\mathcal{F}_1$.

(d) Obtaining such compactness results for (possibly) large classes of functionals is one of the main objectives of epi-convergence theory. In turn, given a sequence of functionals, such a theorem gives us at once all possible forms of the limit problem. Convergence then follows by a classical compactness argument and identification (by using specific properties of the so-studied sequence) of the limit function (cf. Section 1.3.1).

As we shall see, in some optimal control problems, epi-convergence provides the right compactness notion needed to prove existence of optimal control. Typically, in optimal design, a minimizing sequence of domains may tend to be more and more fragmented and tend to minimize a limit homogenized or relaxed functional.

Finally, we shall describe how such compactness results allow us to attack stochastic homogenization of (possibly non-linear!) variational inequalities (Section 2.8). □

2.4.2 The abstract compactness theorem

In this section, we prove the abstract compactness theorem which states that every sequence of functions $\{F^n : X \to \bar{\mathbb{R}}\}$ from (X,τ), a second countable topological space, into $\bar{\mathbb{R}}$ contains a τ-epi-convergent subsequence. Because of its importance we shall give throughout this chapter, different proofs of this theorem, which are of independent interest.

In this section, we give a blitz proof of this theorem relying on the definition of epi-convergence by formulae (1.32) and (1.33). Then, we present the historical proof, relying on the concept of set-convergence, which can be found in Kuratowski [1]. In Section 2.7, we give another proof (and get in fact a slightly sharper result) via Moreau-Yosida proximal approximation. Finally, in Section 2.8, assuming the space (X,τ) to be locally compact, we give a proof relying on the construction of a topology (on the family of τ-lower semicontinuous functions on X) which induces the τ-epi-convergence. The compactness theorem is then obtained by verification of the finite intersection property.

Let us stress the fact that, in the compactness theorem stated in this section, there are no compactness assumptions on (X,τ). One has to consider this theorem as a compactness theorem for $\bar{\mathbb{R}}^X$, for the epi-convergence topology (which is not comparable with the product topology). Clearly, compactness comes from compactness of $\bar{\mathbb{R}}$; consequently, we have to admit $F \equiv +\infty$ (i.e. with an empty epigraph) as a possible limit function.

Let us state the theorem.

THEOREM 2.22 Let (X,τ) be a second countable topological space. Then, each sequence $(F^n)_{n\in\mathbb{N}}$ of functions from X into $\bar{\mathbb{R}}$ contains a subsequence which is τ-epi convergent.

Proof of Theorem 2.22 Let $\{G_k; k = 1,2,\ldots\}$ be a denumerable base of τ-open sets in X ((X,τ) is assumed to be second countable).

By a classical diagonalization process, we can extract a subsequence $\{F^{n_\nu}; \nu \in \mathbb{N}\}$ such that

$$\forall k = 1,2,\ldots \quad \lim_{\nu\to+\infty} [\inf_{u\in G_k} F^{n_\nu}(u)] \text{ exists in } \bar{\mathbb{R}}.$$

(we have only used the compactness of $\bar{\mathbb{R}}$). In other words,

$$\forall k = 1,2,\ldots \liminf_{\nu\to+\infty} \inf_{u\in G_k} F^{n_\nu}(u) = \limsup_{\nu\to+\infty} \inf_{u\in G_k} F^{n_\nu}(u).$$

It follows that, for every $x \in X$,

$$\sup_{G_k\ni x} \liminf_{\nu\to+\infty} \inf_{u\in G_k} F^{n_\nu}(u) = \sup_{G_k\ni x} \limsup_{\nu\to+\infty} \inf_{u\in G_k} F^{n_\nu}(u).$$

The density of the family $\{G_k; k = 1,2,\ldots\}$ yields (cf. Remark 1.11)

$$\sup_{\mathcal{O}\in\mathcal{O}_\tau(x)} \liminf_{\nu\to+\infty} \inf_{u\in\mathcal{O}} F^{n_\nu}(u) = \sup_{\mathcal{O}\in\mathcal{O}_\tau(x)} \limsup_{\nu\to+\infty} \inf_{u\in\mathcal{O}} F^{n_\nu}(u).$$

From Definition 1.8 of the τ-epi limits,

$$\forall x \in X \quad \tau\text{-}\mathrm{li}_e F^{n_\nu}(x) = \tau\text{-}\mathrm{ls}_e F^{n_\nu}(x).$$

This means that the subsequence $\{F^{n_\nu}; \nu = 1,2,\ldots\}$ τ-epi converges. □

Using the geometric interpretation of epi-convergence (Theorems 1.36, 1.39) the above theorem turns to be equivalent to the celebrated compactness theorem for set convergence of Kuratowski [1] (take $Y = X \times \mathbb{R}$ and $S^n = \mathrm{epi} F^n$):

THEOREM 2.23 Let (Y,σ) be a second countable topological space. From each sequence $(S^n)_{n\in\mathbb{N}}$ of subsets of Y, one can extract a subsequence $(S^{n_k})_{k\in\mathbb{N}}$

converging in the Kuratowski sense: $\sigma\text{-Li}S^{n_k} = \sigma\text{-Ls}\, S^{n_k}$.

Proof of Theorem 2.23 We record the historical proof (which enhances the simplicity of the above proof by epi-convergence).

Let $\{G_i ; i = 1,2,\ldots\}$ be a denumerable base of (Y,σ). We construct a doubly indexed sequence $(A_n^i)_{(i,n)\in\mathbb{N}\times\mathbb{N}}$ in the following way.

Define $A_1^i = S^i$ for every $i = 1,2,\ldots$. Then, for $n > 1$, let us define (A_n^i) as a subsequence of $(A_{n-1}^i)_{i=1,2,\ldots}$.

(a) if there exists a subsequence $(k_i)_{i\in\mathbb{N}}$ $k_1 < k_2 < \ldots$ such that

$$G_n \cap (\sigma\text{-Ls}A_{n-1}^{k_i}) = \emptyset \text{ take } A_n^i = A_{n-1}^{k_i}.$$

(b) otherwise, take $A_n^i = A_{n-1}^i$, $i = 1,2,\ldots$.

Let us prove that the sequence $(D_n)_{n\in\mathbb{N}}$, defined by $D_n = A_n^n$, converges in the Kuratowski sense. (Clearly, the sequence $(D_n)_{n\in\mathbb{N}}$ is extracted from $(S^n)_{n\in\mathbb{N}}$). Let us argue by contradiction. If the sequence $(D_n)_{n\in\mathbb{N}}$ does not converge, then

$$\sigma\text{-Ls}D_n \neq \sigma\text{-Li}D_n.$$

Noticing that

$$\sigma\text{-Li}D_n = \bigcap_{\substack{\text{all the subsequences}\\ \{j_n\}_{n\in\mathbb{N}} \subset \mathbb{N}}} \sigma\text{-Ls}D_{j_n},$$

we derive the existence of a subsequence $\{j_n ; n \in \mathbb{N}\}$ such that

$$\sigma\text{-Ls}D_n \neq \sigma\text{-Ls}D_{j_n}.$$

Since $\sigma\text{-Ls}D_{j_n}$ is closed for the topology σ, the above relation implies the existence of some G_m belonging to the base such that

$$\sigma\text{-Ls}D_n \cap G_m \neq \emptyset \text{ and } \sigma\text{-Ls}D_{j_n} \cap G_m = \emptyset.$$

By construction, each sequence $(A_n^i)_{i\in\mathbb{N}}$ is extracted from the preceding one

$(A^i_{n-1})_{i\in\mathbb{N}}$. Therefore, the diagonal sequence $D_n = A^n_n$ is extracted from each of the sequences $(A^i_\ell)_{i\in\mathbb{N}}$ (from the index $n \geq \ell$).

Since the sequence $(D_n)_{n\in\mathbb{N}}$ is extracted from $(A^i_m)_{i\in\mathbb{N}}$ and $(\sigma\text{-}\mathrm{Ls}D_n) \cap G_m \neq \emptyset$, it follows that

$$\sigma\text{-}\mathrm{Ls}(A^i_m) \cap G_m \neq \emptyset. \tag{2.43}$$

On the other hand, since the sequence $(D_{j_n})_{n\in\mathbb{N}}$ is extracted from $(A^i_{m-1})_{i\in\mathbb{N}}$ and $(\sigma\text{-}\mathrm{Ls}D_{j_n}) \cap G_m = \emptyset$, from the construction of the sequence $(A^i_n)_{i\in\mathbb{N}}$, there exists a subsequence $\{\ell_i\} \subset \mathbb{N}$ such that:

$$A^i_m = A^{\ell_i}_{m-1}$$

and $(\sigma\text{-}\mathrm{Ls}\ A^{\ell_i}_{m-1}) \cap G_m = \emptyset$. These two last properties imply

$$(\sigma\text{-}\mathrm{Ls}A^i_m) \cap G_m = \emptyset,$$

which is contradictory with (2.43).

<u>REMARK</u> In preceding proof, we have used the following result:

$$\sigma\text{-}\mathrm{Li}\ D_n = \bigcap_{\{j_n;n=1,2,\ldots\}\subset\mathbb{N}} \sigma\text{-}\mathrm{Ls}\ D_{j_n}. \tag{2.44}$$

Clearly, the left member is included in the right one. In order to prove the opposite inclusion, take $x \notin \sigma\text{-}\mathrm{Li}\ D_n$ and prove that there exists $\{j_n\} \subset \mathbb{N}$ such that $x \notin \sigma\text{-}\mathrm{Ls}\ D_{j_n}$. By characterization (1.112) of $\sigma\text{-}\mathrm{Li}(D_n)$

$$\sigma\text{-}\mathrm{Li}\ D_n = \bigcap_{H\in\tilde{\mathbb{N}}} c\ell_\sigma\Big(\bigcup_{k\in H} D_k\Big).$$

Since $x \notin \sigma\text{-}\mathrm{Li}D_n$, there exists $H \in \tilde{\mathbb{N}}$ (i.e. a subsequence $(j_n)_{n\in\mathbb{N}}$) and a σ-open set G such that:

$$\forall n \in \mathbb{N} \quad G \cap D_{j_n} = \emptyset.$$

This is equivalent to $x \notin \sigma\text{-}\mathrm{Ls}(D_{j_n})$ (cf. Definition (1.113)). □

2.4.3 A compactness result for a class of integral functionals

A. Main theorem

Let us sketch the proof of the following compactness theorem for a class of integral functionals which are supposed to be uniformly bounded and coercive. Then, corresponding limiting cases, explosion or degeneracy of the coefficients, will be explored. We write $\mathbb{O}$ (resp. $\mathbb{O}_b$) the family of open (resp. open bounded) subsets of $\mathbb{R}^N$.

THEOREM 2.24 Given $0 < \lambda \leqslant \Lambda < +\infty$, let us define the following class $\mathcal{F}_{\lambda,\Lambda}$:

$$\mathcal{F}_{\lambda,\Lambda} = \{F : H^1_{loc}(\mathbb{R}^N) \times \mathbb{O}_b \to \mathbb{R}^+ / F(u,\omega) = \int_\omega j(x,\mathrm{grad}u(x))dx$$

where

$$j : \mathbb{R}^N \times \mathbb{R}^N \to \mathbb{R}^+$$

$$(x,z) \to j(x,z)$$

is measurable with respect to x, convex continuous with respect to z and satisfies

$$\lambda|z|^2 \leqslant j(x,z) \leqslant \Lambda(1 + |z|^2)\}.$$

Then, $\mathcal{F}_{\lambda,\Lambda}$ is sequentially compact with respect to epi-convergence. More precisely, from every sequence $\{F^n; n \in \mathbb{N}\}$ of functionals of $\mathcal{F}_{\lambda,\Lambda}$, one can extract a subsequence $\{F^{n_k}; k \in \mathbb{N}\}$ and find some function F still belonging to $\mathcal{F}_{\lambda,\Lambda}$ such that

$$F(u,\omega) = \tau\text{-}\mathrm{lm}_e\, F^{n_k}(u,\omega)$$

holds for every $u \in H^1_{loc}(\mathbb{R}^N)$, for every bounded open set ω with boundary of zero Lebesgue measure, and τ equal to the strong topology of $L^2(\omega)$.

Proof of Theorem 2.24

Step one: Let us extract the properties of the class $\mathcal{F}$ which are preserved by epi-limit, and introduce the abstract class $\mathbb{A}$:

$$\mathbb{A} = \{F : H^1_{loc}(\mathbb{R}^N) \times \mathbb{O}_b \to \mathbb{R}^+ \qquad (2.45)$$

satisfying the following properties (i), (ii), (iii), (iv):

(i) for every $u \in H^1_{loc}(\mathbb{R}^N)$, $\omega \to F(u,\omega)$ is the restriction to $\mathcal{O}_b$ of a Borel measure (still denoted $B \to F(u,B)$, where $\mathcal{B}$ = Borel subsets of $\mathbb{R}^N$);

(ii) for every $\omega \in \mathcal{O}_b$ and every $u,v \in H^1_{loc}(\mathbb{R}^N)$

$$u|_\omega = v|_\omega \Rightarrow F(u,\omega) = F(v,\omega);$$

(iii) for every $\omega \in \mathcal{O}_b$, $u \to F(u,\omega)$ is convex continuous on $H^1(\omega)$;

(iv) for every $\omega \in \mathcal{O}_b$, for every $u \in H^1_{loc}(\mathbb{R}^N)$,

$$0 \leq F(u,\omega) \leq \Lambda \int_\omega (1 + |\text{grad}u(x)|^2)dx.\}$$

Step two: We now prove that the class $\mathbb{A}$ is "compact" with respect to epi-convergence. To that end, let us introduce $\mathbb{D}$ a dense denumerable subset of $\mathcal{O}$, for example the family of all finite unions of N-dimensional open rectangles with vertices in $\mathbb{Q}^N$. By the compactness Theorem 2.22 and by a classical diagonalization argument, from every sequence $\{F^n; n \in \mathbb{N}\}$ of functionals of $\mathbb{A}$, one can extract a subsequence $\{F^{n_k}; k \in \mathbb{N}\}$ which satisfies

$$\text{for every } u \in H^1_{loc}(\mathbb{R}^N), \text{ for every } \omega \in \mathbb{D}, \ \text{lm}_e F^{n_k}(u,\omega) \tag{2.46}$$

exists.

Equivalently, introducing for every $u \in H^1_{loc}(\mathbb{R}^N)$, for every $B \in \mathcal{B}$

$$\begin{aligned} F_i(u,B) &= L^2(B) - \lim\inf_e F^{n_k}(u,B) \\ F_s(u,B) &= L^2(B) - \lim\sup_e F^{n_k}(u,B) \end{aligned} \tag{2.47}$$

we have

$$F_i(u,\omega) = F_s(u,\omega) \text{ for every } u \in H^1_{loc}(\mathbb{R}^N) \text{ and } \omega \in \mathbb{D}. \tag{2.48}$$

As we already noticed (Section 1.3.1), epi-convergence preserves the order: if $F^n \leq G^n$, then $\text{li}_e F^n \leq \text{li}_e G^n$ and $\text{ls}_e F^n \leq \text{ls}_e G^n$. Consequently, for every $u \in H^1_{loc}(\mathbb{R}^N)$, $B \to F_i(u,B)$ and $B \to F_s(u,B)$ are increasing functions of sets. From (2.48) and the density of $\mathbb{D}$, it follows that they have the same "inner regularization":

$$\text{for every } \omega \in \mathcal{O}_b, \ \sup_{\substack{A \in \mathcal{O} \\ A \subset\subset \omega}} F_i(u,A) = \sup_{\substack{A \in \mathcal{O} \\ A \subset\subset \omega}} F_s(u,A) \doteq H(u,\omega) \tag{2.49}$$

where $A \subset\subset \omega$ means that A is strongly included in ω, i.e. $\bar{A} \subset \omega$. We then verify that H *belongs to* $\mathbb{A}$ and is equal to *the epi-limit of the sequence* F^{n_k}. The difficult (and decisive) point in order to establish the above statement is to prove that $H(u,\cdot)$ *is still a measure:* Indeed, one verifies that $F_i(u,\cdot)$ is super-additive while $F_s(u,\cdot)$ satisfies subadditive properties. From (2.49), $H(u,\cdot)$ inherits both properties and is a measure. Let us make this more precise:

Superaddivity of F_i is a general feature: given ω_1, ω_2 two disjoint open subsets of $\mathbb{R}^N$, for every sequence $u_k \to u$ strongly in $L^2(\omega_1 \cup \omega_2)$,

$$\begin{aligned}
\liminf_k F^{n_k}(u_k,\omega_1 \cup \omega_2) &\geq \liminf_k \{F^{n_k}(u_k,\omega_1) + F^{n_k}(u_k,\omega_2)\} \\
&\geq \liminf_k F^{n_k}(u_k,\omega_1) + \liminf_k F^{n_k}(u_k,\omega_2) \\
&\geq F_i(u,\omega_1) + F_i(u,\omega_2)
\end{aligned}$$

since $u_k \to u$ in $L^2(\omega_1)$ and $L^2(\omega_2)$.

Subaddivity of F_s is more delicate to verify in such an abstract setting. Let us assume for simplicity that the functionals F^n are given in an integral form as in Theorem 2.24 (in fact, as we shall see, axiomatization $\mathbb{A}$ does characterize such integral functionals). Then, let us prove that, for every regular bounded open sets A_1, A_2, for every $u \in H^1_{loc}(\mathbb{R}^N)$

$$F_s(u,A_1 \cup A_2) \leq F_s(u,A_1) + F_s(u,A_2). \tag{2.50}$$

Noticing that

$$A_1 \cup A_2 \supset A_1 \cup (A_2 \setminus \bar{A}_1)$$

and that A_1 and $A_2 \setminus \bar{A}_1$ are two regular disjoint open sets with a common boundary of zero Lebesgue measure, (2.50) will follow from the following inequality:

for every $\omega_1, \omega_2 \in \mathcal{O}_b$ such that $\omega_1 \cap \omega_2 = \emptyset$ and every $\omega \in \mathcal{O}_b$ such that

$\omega = \omega_1 \cup \omega_2 \cup \Delta$ with meas $\Delta = 0$,

$$F_s(u,\omega) \leqslant F_s(u,\omega_1) + F_s(u,\omega_2) \tag{2.51}$$

For every $\delta > 0$, let us introduce Δ_δ, (Fig. 2.1) a δ-neighbourhood of Δ,

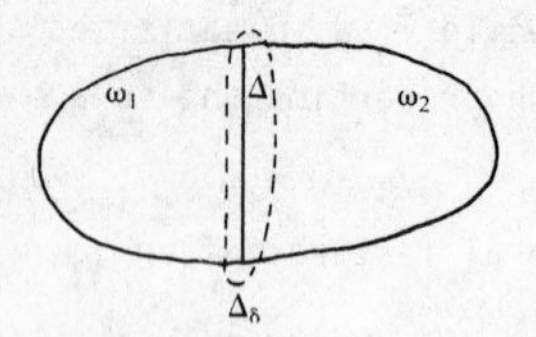

Figure 2.1

$$\Delta_\delta = \{x \in \mathbb{R}^N / \mathrm{dist}(x,\Delta) < \delta\},$$

and $\phi_\delta \in \mathcal{C}^\infty(\mathbb{R}^N)$ satisfying

$$0 \leqslant \phi_\delta \leqslant 1$$

$$\phi_\delta = 1 \text{ on } \Delta_\delta,\ \phi_\delta = 0 \text{ on } \mathbb{R}^N \setminus \Delta_{2\delta}.$$

By definition of $F_s(u,\omega)$,

$$F_s(u,\omega_1) = \limsup_k F^{n_k}(u_k^1,\omega_1) \text{ for some sequence } u_k^1 \to u \text{ strongly in } L^2(\omega_1)$$

$$F_s(u,\omega_2) = \limsup_k F^{n_k}(u_k^2,\omega_2) \text{ for some sequence } u_k^2 \to u \text{ strongly in } L^2(\omega_2). \tag{2.52}$$

Let us define

$$u_k = \begin{cases} (1-\phi_\delta)u_k^1 + \phi_\delta u & \text{on } \omega_1 \\ u & \text{on } \Delta_\delta \\ (1-\phi_\delta)u_k^2 + \phi_\delta u & \text{on } \omega_2. \end{cases}$$

Clearly, u_k belongs to $H^1_{loc}(\mathbb{R}^N)$ and $u_k \to u$ strongly in $L^2(\omega)$. For technical reasons, let us first argue at tu, $0 < t < 1$, then let t go to one. By definition of $F_s(tu,\omega)$

$$F_s(tu,\omega) \leqslant \limsup_k F^{n_k}(tu_k,\omega). \tag{2.53}$$

Let us compute

$$F^{n_k}(tu_k,\omega) = \sum_{p=1,2} \int_{\omega_p} j_{n_k}(x, t(1-\phi_\delta)\mathrm{grad}u_k^p$$

$$+ t\phi_\delta \mathrm{grad}u + (1-t)\frac{t}{1-t}(u-u_k^p)\mathrm{grad}\phi_\delta)dx.$$

Noticing that $t(1-\phi_\delta) + t\phi_\delta + (1-t) = 1$ and $0 < t < 1$, by convexity of j_{n_k},

$$F^{n_k}(tu_k,\omega) \leqslant \sum_{p=1,2} \int_{\omega_p} j_{n_k}(x,\mathrm{grad}u_k^p)dx$$

$$+ \int_{\Delta_{2\delta}} j_{n_k}(x,\mathrm{grad}u)dx + (1-t)\sum_p \int_{\omega_p} j_{n_k}(x,\frac{t}{1-t}(u-u_k^p)\mathrm{grad}\phi_\delta)dx.$$

From the majorization property (iv) in (2.45),

$$F^{n_k}(tu_k,\omega) \leqslant \sum_{p=1,2} \int_{\omega_p} j_{n_k}(x,\mathrm{grad}u_k^p)dx$$

$$+ \Lambda\int_{\Delta_{2\delta}} (1+|\mathrm{grad}u|^2)dx+(1-t)\Lambda. \sum_{p=1,2}\int_{\omega_p}(1+|\frac{t}{1-t}(u-u_k^p)\mathrm{grad}\phi_\delta|^2)dx.$$

When k goes to $+\infty$, by (2.52) and (2.53)

$$F_s(tu,\omega) \leqslant F_s(u,\omega_1) + F_s(u,\omega_2) + \Lambda\int_{\Delta_{2\delta}} (1+|\mathrm{grad}u|^2)dx+(1-t)\Lambda.\mathrm{meas}(\Omega).$$

Letting δ and t go respectively to zero and one, and using the lower semi-continuity of $F_s(\cdot,\omega)$, we get (2.51).

Applying (2.51) with $\omega_1 = A_1$, $\omega_2 = A_2 \setminus \bar{A}_1$, $\omega = A_1 \cup A_2$, we finally obtain

$$F_s(u,A_1 \cup A_2) \leqslant F_s(u,A_1) + F_s(u,A_2\setminus\bar{A}_1)$$

$$\leqslant F_s(u,A_1) + F_s(u,A_2),$$

that is, (2.50). □

Let us summarize the properties satisfied by $H(u,\cdot)$:

$$H(u,\cdot) \text{ is a positive, increasing, inner regular, sub-additive and super-additive set function. By a classical measure theory argument, (De Giorgi \& Letta [1]) it can be extended into a Borel measure on } \mathbb{R}^N. \tag{2.54}$$

It is now fairly easy to verify that equality (2.48)

$$F_i(u,\omega) = F_s(u,\omega)$$

can be extended from the dense set $\mathcal{D}$ to any open regular ω (i.e. with boundary of zero Lebesgue measure), this common value being precisely equal to $H(u,\omega)$.

On one hand, using the increasing property of $F_i(u,\cdot)$,

$$F_i(u,A) \leqslant F_i(u,\omega) \text{ for every } A \subset\subset \omega,$$

by Definition (2.49) of $H(u,\cdot)$, it follows that

$$H(u,\omega) \leqslant F_i(u,\omega). \tag{2.55}$$

On the other hand, given ω an open bounded subset of $\mathbb{R}^N$, for every $\delta > 0$ let us introduce

$$A_{1,\delta} = \{x \in \omega / \mathrm{dist}(x,\partial\omega) > \delta\}$$

$$A_{2,\delta} = \{x \in \mathbb{R}^N / \mathrm{dist}(x,\partial\omega) < 2\delta\}.$$

These two open sets, denoted A_1 and A_2 for brevity, satisfy:

$$\omega \subset A_1 \cup A_2, \quad A_1 \subset\subset \omega, \quad \bigcap_{\delta>0} A_{2,\delta} = \partial\omega.$$

From the above properties and subadditivity of $F_s(u,\cdot)$ (cf. (2.50)) it follows that

$$\begin{aligned} F_s(u,\omega) &\leqslant F_s(u,A_1 \cup A_2) \\ &\leqslant F_s(u,A_1) + F_s(u,A_2) \end{aligned} \tag{2.56}$$

By definition of $H(u,\cdot)$, using the fact that $A_1 \subset\subset \omega$,

$$F_s(u,A_1) \leqslant H(u,\omega).$$

By definition of $F_s(u,\cdot)$, taking $u_k = u$ and using majorization (iv),

$$F_s(u,A_2) \leq \Lambda.\int_{A_2} (1 + |\text{grad}u|^2)dx.$$

Combining (2.56) and the two preceding inequalities, we obtain

$$F_s(u,\omega) \leq H(u,\omega) + \Lambda.\int_{A_{2,\delta}} (1 + |\text{grad}u|^2)dx.$$

The above computation being valid for any $\delta > 0$, by letting δ go to zero and using that $\text{meas } A_{2,\delta} \underset{\delta \to 0}{\to} 0$, we finally get

$$F_s(u,\omega) \leq H(u,\omega). \tag{2.57}$$

From (2.55) and (2.57), it follows that

$$F_s(u,\omega) = H(u,\omega) = F_i(u,\omega),$$

i.e.

$$\text{lm}_e F^{n_k}(u,\omega) = H(u,\omega) \tag{2.58}$$

for every $u \in H^1_{loc}(\mathbb{R}^N)$ and every open set ω with boundary of zero Lebesgue measure. □

Step 3 We have now to give an integral representation of functionals of the "abstract" class $\mathbb{A}$. In the above context, it is rather classical to obtain such representation, thanks to the Radon-Nikodym theorem, to the local property of the functionals and to their local Lipschitz property on $H^1_{loc}(\mathbb{R}^N)$. This last point follows from majorization (iv) (clearly preserved by epi-limit) and from convexity (preserved too by epi-limit, this type of property is systematically studied in Chapter 3).

B. Critical cases: degeneracy of the ellipticity condition

In order to complete the proof of Theorem 2.23, we notice that because of the uniform coerciveness assumption

$$\lambda|z|^2 \leq j(x,z), \tag{2.59}$$

functionals of the class $\mathcal{F}$ are naturally defined on $H^1_{loc}(\mathbb{R}^N)$ with $+\infty$ values on $L^2_{loc}(\mathbb{R}^N) \setminus H^1_{loc}(\mathbb{R}^N)$, and so are their epi-limits. We stress the fact that, as far as one is concerned only with identification of limit functionals on $H^1_{loc}(\mathbb{R}^N)$, the minorization assumption (2.59) does not play any role: indeed, it is only used at the end of the proof in order to guarantee that limit functionals are $+\infty$ outside of $H^1_{loc}(\mathbb{R}^N)$.

When such a uniform coercivity property fails to be satisfied it may happen, as we have already encountered (material with holes and Neumann boundary conditions, transmission condition through thin layers) that corresponding limit functionals do live on a larger space than $H^1_{loc}(\mathbb{R}^N)$: their domains may include functions with jumps over some manifolds. It is quite difficult, and still a largely open question, whether we can weaken assumption (2.59) and completely determine on such possibly enlarged spaces corresponding limit functionals. Let us indicate some results in this direction. Take for simplicity a sequence $\{F^n;\ n \in \mathbb{N}\}$ of integral functionals describing isotropic materials:

$$F^n(u) = \int_\Omega a_n(x)|\text{grad}u|^2 dx.$$

A first type of assumption (cf. Marcellini & Sbordone [5]) is to assume that the sequence $\{\frac{1}{a_n};\ n \in \mathbb{N}\}$ is equintegrable. More precisely, if

$$a_n \geqslant \lambda(x) \text{ with } \frac{1}{\lambda} \in L^s(\Omega) \text{ for some } s > 1, \tag{2.60}$$

one can easily verify, by the Hölder inequality, that the epi-limit of the sequence $\{F^n; n \in \mathbb{N}\}$ is identically $+\infty$ outside the weighted Sobolev space $H^1(\Omega, d\lambda)$. When the sequence $\{\frac{1}{a_n};\ n \in \mathbb{N}\}$ is only uniformly bounded in $L^1(\Omega)$, the limit functional may take finite values (as described in transmission problem 1.1.3) for functions u having a possible jump over an (N-1)-manifold.

In a second type of assumption (cf. L. Tartar [2], Marchenko & Hruslov [2]) the conductivity coefficients $\{a_n; n \in \mathbb{N}\}$ take the value zero on possibly large sets $\{T_n; n \in \mathbb{N}\}$ and are uniformly bounded away from zero on $\Omega_n = \Omega \setminus T_n$:

$$a_n \geqslant \lambda > 0 \text{ on } \Omega_n. \tag{2.61}$$

The above critical situation, where there may be some splitting of Ω with jumps over the interfaces, is then avoided thanks to the uniform connectedness assumption (cf. Section 1.3.3): there exist extension operators $\mathbb{P}^n: H^1(\Omega_n) \to H^1(\Omega)$ such that

$$\sup_{n\in\mathbb{N}} \|\mathbb{P}^n\|_{\mathcal{L}(H^1(\Omega_n),H^1(\Omega))} < +\infty. \tag{2.62}$$

Let us introduce χ_n, the characteristic function of Ω_n (i.e. $\chi_n = 1$ on Ω_n, 0 on T_n). The sequence $\{\chi_n;\ n \in \mathbb{N}\}$ being bounded in $L^2(\Omega)$, let us extract a weakly converging sequence (still denoted χ_n):

$$\chi_n \to \chi \quad \text{weakly in } L^2(\Omega). \tag{2.63}$$

Then, for every convergent sequence $u_n \to u$ such that $\liminf_n F^n(u_n) < +\infty$ from (2.61),

$$\sup_n \|u_n\|_{H^1(\Omega_n)} < +\infty.$$

Thus, from (2.62)

$$\sup_n \|\mathbb{P}_n(u_n)\|_{H^1(\Omega)} < +\infty.$$

By compact embedding $H^1(\Omega) \hookrightarrow L^2(\Omega)$, we can extract some subsequence (still denoted $\mathbb{P}^n(u_n)$)

$$\mathbb{P}^n(u_n) \longrightarrow u^* \text{ strongly in } L^2(\Omega). \tag{2.64}$$

Going to the limit on

$$u_n.\chi_n = \mathbb{P}^n(u_n).\chi_n$$

we obtain

$$u.\chi = u^*.\chi.$$

Assuming for example $\chi > 0$, which contains the homogenization of materials with holes and Neumann boundary conditions (Section 1.3.3), we obtain $u = u^*$, that is $u \in H^1(\Omega)$, i.e. $\text{lm}_e F^n = +\infty$ outside of $H^1(\Omega)$.

C. Critical cases: explosion of the coefficients

This situation is in some sense the "dual" of the preceding one. In the above situation the domain of the limit functional could be larger than $H^1(\Omega)$, now it may happen to be smaller as described in the following example (Pham Huy & E. Sanchez-Palencia [1], Carbone & Sbordone [2]...). As in

Example 1.1.3, let us consider a "thin layer" Σ_ε of thickness ε; take for simplicity (Fig. 2.2)

$$\Sigma = \{(x_1,x_2,x_3) \in \mathbb{R}^3 / x_3 = 0\},$$

$$\Sigma_\varepsilon = \{(x_1,x_2,x_3) \in \mathbb{R}^3 / |x_3| < \varepsilon/2\}.$$

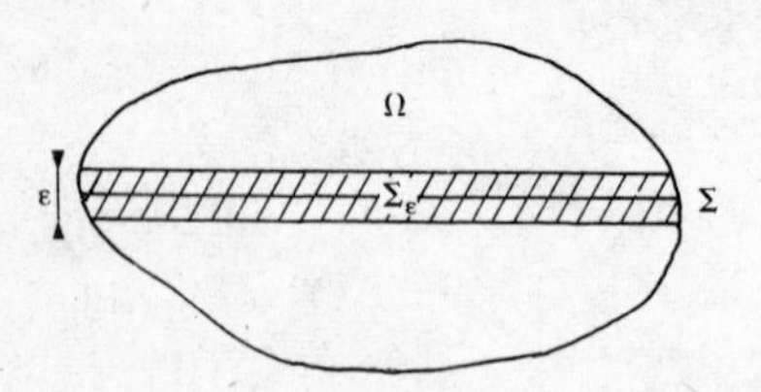

Figure 2.2

The layer Σ_ε is now supposed to have "high conductivity" λ. We model a thin highly conducting layer by considering the limit analysis problem: $\varepsilon \to 0$ and $\lambda \to +\infty$.

Introducing the conductivity coefficient $a_{\varepsilon,\lambda}$,

$$a_{\varepsilon,\lambda}(x) = \begin{cases} \lambda \text{ on } \Sigma_\varepsilon \\ 1 \text{ on } \Omega \setminus \Sigma_\varepsilon, \end{cases}$$

the corresponding energy functional $F^{\varepsilon,\lambda}$, $H^1(\Omega) \to \mathbb{R}^+$ is given by

$$F^{\varepsilon,\lambda}(u) = \int_\Omega a_{\varepsilon,\lambda}(x) |\text{grad} u|^2 dx.$$

<u>THEOREM 2.25</u> The limit behaviour as $\varepsilon \to 0$, $\lambda \to +\infty$ of the sequence $\{F_{\varepsilon,\lambda}\}$ depends on the limit of the quantity $\lim\limits_{\substack{\varepsilon\to 0\\ \lambda\to+\infty}} \varepsilon\lambda = \alpha$:

(a) <u>When $\varepsilon\lambda \to \alpha$, $0 < \alpha < +\infty$</u> (as $\varepsilon \to 0$ and $\lambda \to +\infty$) the sequence $\{F_{\varepsilon,\lambda}\}$ τ-epi converges to F (τ = weak topology of $H^1(\Omega)$)

$$F = \tau\text{-lm}_e F^{\varepsilon,\lambda}$$

where $F:H^1(\Omega) \to \mathbb{R}^+$ is given by

$$F(u) = \int_\Omega |\text{grad}u|^2 dx + \alpha \int_\Sigma (|\frac{\partial u}{\partial x_1}(x_1,x_2,0)|^2 + |\frac{\partial u}{\partial x_2}(x_1,x_2,0)|^2) dx_1 dx_2$$

(its domain is the subspace of $H^1(\Omega)$ obtained by completion of $\mathbb{C}^\infty(\Omega)$ for the semi-norm $\sqrt{F(u)}$).

(b) <u>When $\varepsilon\lambda \to +\infty$,</u> $F = \tau\text{-}\lim_e F^{\varepsilon,\lambda}$ is equal to

$$F(u) = \begin{cases} \int_\Omega |\text{grad}u|^2 dx \text{ if } u \in H^1(\Omega) \text{ is constant on } \Sigma \\ +\infty \text{ elsewhere.} \end{cases}$$

(c) <u>When $\varepsilon\lambda \to 0$,</u> then $F = \tau\text{-}\lim_e F^{\varepsilon,\lambda}$ is equal to

$$F(u) = \int_\Omega |\text{grad}u|^2 dx \text{ on } H^1(\Omega).$$

<u>Proof of Theorem 2.25</u> First take u regular, $u \in \mathbb{C}^1$. Then consider the following approximating sequence $\{u_\varepsilon; \varepsilon \to 0\}$ which is taken independent of x_3 where the conductivity coefficient is high, that is on Σ_ε (compare with Theorem 1.29):

$$u_\varepsilon(x_1,x_2,x_3) = \begin{cases} u(x_1,x_2,x_3) \text{ if } |x_3| > \varepsilon \\ u(x_1,x_2,0) \text{ if } |x_3| < \varepsilon/2 \\ [u(x_1,x_2,\varepsilon)-u(x_1,x_2,0)]\frac{2x_3}{\varepsilon} + 2u(x_1,x_2,0)-u(x_1,x_2,\varepsilon), \\ \qquad \text{if } \varepsilon/2 < x_3 < \varepsilon. \\ [u(x_1,x_2,0)-u(x_1,x_2,-\varepsilon)]\frac{2x_3}{\varepsilon} + 2u(x_1,x_2,0)-u(x_1,x_2,-\varepsilon) \\ \qquad \text{if } -\varepsilon/2 < x_3 < -\varepsilon. \end{cases}$$

Then, taking $\lambda \sim \frac{\alpha}{\varepsilon}$,

$$F^{\varepsilon,\lambda} \sim \int_{\Omega\setminus\Sigma_\varepsilon} |\text{grad}u|^2 dx + \frac{\alpha}{\varepsilon}\int_{\Sigma_\varepsilon} (|\frac{\partial u}{\partial x_1}(x_1,x_2,0)|^2 + |\frac{\partial u}{\partial x_2}(x_1,x_2,0)|^2) dx$$

and

$$\lim_{\substack{\varepsilon\to 0 \\ \lambda = \frac{\alpha}{\varepsilon}}} F^{\varepsilon,\lambda}(u_\varepsilon) = F(u).$$

On the other hand, given $v_\varepsilon \longrightarrow u$ weakly in $H^1(\Omega)$

$$\begin{aligned}\liminf_{\varepsilon\to 0} F^\varepsilon(v_\varepsilon) &\geq \liminf_{\varepsilon\to 0}\int_{\Omega\setminus\Sigma_\varepsilon} |\mathrm{grad} v_\varepsilon|^2 dx \\ &+ \liminf_{\varepsilon\to 0}\frac{\alpha}{\varepsilon}\int_{\Sigma_\varepsilon}\left(\left|\frac{\partial v_\varepsilon}{\partial x_1}\right|^2 + \left|\frac{\partial v_\varepsilon}{\partial x_2}\right|^2\right)dx_1dx_2dx_3 \\ &\geq \int_\Omega |\mathrm{grad} v|^2 dx + \liminf_{\varepsilon}\frac{\alpha}{\varepsilon}\int_{\Sigma_\varepsilon}\left(\left|\frac{\partial v_\varepsilon}{\partial x_1}\right|^2\right. \\ &\left. + \left|\frac{\partial v_\varepsilon}{\partial x_2}\right|^2\right)dx_1dx_2dx_3.\end{aligned} \tag{2.65}$$

Let us introduce, for any $v \in H^1(\Omega)$, the function $\tilde{v}$ defined by

$$\tilde{v}(x_1,x_2) = \frac{1}{\varepsilon}\int_{|x_3|<\varepsilon/2} v(x_1,x_2,t)dt.$$

Then, using Hölder's inequality

$$\begin{aligned}\int_\Sigma |\mathrm{grad}\tilde{v}|^2 dx_1dx_2 &= \int_\Sigma \left(\left|\frac{1}{\varepsilon}\int_{|x_3|<\varepsilon/2}\frac{\partial v}{\partial x_1}(x_1,x_2,t)\right|^2\right. \\ &\left. + \left|\frac{1}{\varepsilon}\int_{|x_3|<\varepsilon/2}\frac{\partial v}{\partial x_2}(x_1,x_2,t)\right|^2\right)dx_1dx_2 \\ &\leq \frac{1}{\varepsilon}\int_\Sigma dt \int_{|x_3|<\varepsilon/2}\left|\frac{\partial v}{\partial x_1}(x_1,x_2,t)\right|^2 \\ &+ \left|\frac{\partial v}{\partial x_2}(x_1,x_2,t)\right|^2 dx_1dx_2.\end{aligned}$$

Returning to (2.65), from the above inequality

$$\liminf_{\varepsilon} F^\varepsilon(v_\varepsilon) \geq \int_\Omega |\mathrm{grad} v|^2 dx + \alpha \liminf_{\varepsilon}\int_\Sigma |\mathrm{grad}\tilde{v}_\varepsilon|^2 dx_1dx_2.$$

Noticing that $\tilde{v}_\varepsilon \longrightarrow v(x_1,x_2,0)$ in $H^1(\Sigma)$ when $\liminf_\varepsilon F^\varepsilon(v_\varepsilon) < +\infty$, we finally obtain

$$\liminf_{\varepsilon} F^\varepsilon(v_\varepsilon) \geq F(v).$$

The proof is then completed by a density argument. □

Let us now examine what in general can be said about the limit functional when the conductivity coefficients may be large, but satisfy a uniform integrability property (which excludes the above situation, Theorem 2.25). The following theorem is from Carbone & Sbordone [2]:

THEOREM 2.26 With the notation of Theorem 2.24, let us assume that a sequence $\{F^n;\ n \in \mathbb{N}\}$ of integral functionals

$$F^n(u,\omega) = \int_\omega j_n(x,\mathrm{grad}u(x))dx$$

satisfies:

$$j_n : \mathbb{R}^N \times \mathbb{R}^N \to \mathbb{R}^+$$

$$(x,z) \to j_n(x,z)$$

is measurable with respect to x, convex continuous with respect to z and

$$0 \leqslant j_n(x,z) \leqslant a_n(x)(1 + |z|^p)$$

for some $p \geqslant 1$ where the sequence $\{a_n;\ n \in \mathbb{N}\}$ is assumed to be weakly relatively compact in $L^1(\Omega)$, take $a_\nu \longrightarrow a\ \sigma(L^1,L^\infty)$.

Then, one can extract a subsequence $\{F^{n_k};\ k \in \mathbb{N}\}$ and find some F

$$F(u) = \int_\Omega j(x,\mathrm{grad}u(x))dx$$

with

$$0 \leqslant j(x,z) \leqslant a(x)(1 + |z|^p)$$

such that

$$\tau\text{-}\mathrm{lm}_e F^{n_k}(u,\omega) = F(u,\omega)$$

holds for any $u \in W^{1,\infty}_{loc}(\mathbb{R}^N)$, for any bounded open set ω and τ equal to the strong topology of $L^r(\Omega)$ for any $1 \leqslant r \leqslant +\infty$.

2.4.4 Compactness of the class of obstacle constraint functionals

A. Examples of varying obstacle limit analysis

Considering "varying obstacles" $g_n : \Omega \to \bar{\mathbb{R}}$, $n = 1,2,\ldots$, we first describe all possible limit forms, as n goes to $+\infty$, of the sequence of variational inequalities

$$\min \{ \int_{\Omega} |\text{grad} u|^2 dx - \int_{\Omega} fu \, dx / u(x) \geq g_n(x) \text{ a.e. on } \Omega \}.$$

We can distinguish two different types of situation:

(a) the limit problem is still an obstacle problem

(b) the limit problem is no longer an obstacle problem.

Let us first notice that, even when starting with inequalities taken in the sense almost everywhere, it may happen that the limit constraint is still of obstacle type, but has to be interpreted in a more precise way than almost everywhere. This is illustrated by the following example:

EXAMPLE 2.27 "Thin obstacles" Let Σ be a smooth (N-1)-dimensional manifold in $\Omega \subset \mathbb{R}^N$ and Σ_n a 1/n-neighbourhood of Σ,

$$\Sigma_n = \{x \in \mathbb{R}^N; \text{dist}(x,\Sigma) < 1/n\}.$$

Take g_n equal to 1 on Σ_n and 0 elsewhere. The corresponding convex of constraints K_{g_n} is equal to

$$K_{g_n} = \{u \in H^1(\Omega); u \geq g_n \text{ a.e. } \Omega\}.$$

Then, for every $u \in H^1(\Omega)$, one can easily verify that

$$s\text{-}L^2(\Omega)\text{-}\lim_e \left[\int_{\Omega} |Du|^2 + I_{K_{g_n}}(u)\right] = \int_{\Omega} |Du|^2 dx + I_{K_g}(u),$$

where g is equal to 1 on Σ and 0 elsewhere, i.e.

$$K_g = \{u \in H^1(\Omega); \ u|_{\Sigma} \geq 1, \ u \geq 0 \text{ a.e. on } \Omega\} .$$

The limit obstacle exists on a set of zero Lebesgue measure which has a physical meaning with respect to Dirichlet energy $\int |\text{grad} u|^2$. To consider functions of $H^1(\Omega)$ as defined almost everywhere is not precise enough. In this example, we have to use the trace of $u \in H^1(\Omega)$ on Σ in order to interpret the limit constraint. Such an obstacle, negligible with respect to Lebesgue measure but not to H^1-energy, is called a *thin obstacle*. Let us notice that the sequence $\{g_n; n \in \mathbb{N}\}$ converges in any $L^p(\Omega)$ to the function identically equal to zero, which is different, indeed, from the correct limit obstacle!

These considerations lead us naturally to introduce the $H^1(\Omega)$ notions, corresponding respectively to Lebesgue measure and convergence in measure, which allow us to interpret in a general way such phenomena. So, let us briefly introduce classical elements of potential theory related to capacity and energy: (cf. Ancona [1], Deny [1]...).

Capacity of an open subset ω of $\mathbb{R}^N$ (relatively to the norm $H^1(\mathbb{R}^N)$) is defined by

$$\text{Cap } \omega = \inf \{\|u\|_{H^1(\mathbb{R}^N)}; u \in H^1(\mathbb{R}^N), u \geq 1 \text{ a.e. on } \omega\}. \tag{2.66}$$

Capacity of an arbitrary set A of $\mathbb{R}^N$ is then defined by

$$\text{Cap } A = \inf \{\text{Cap } \omega / \omega \text{ open}, \omega \supset A\}.$$

Capacity is a positive real valued function of sets which is increasing and denumerably subadditive. Let us notice that, as far as one is only concerned with topological notions attached to capacity theory, one may define in (2.66) the capacity by using an equivalent norm (refer to Attouch & Picard [1]). A set A is said to be of capacity zero (in H^1 sense) if Cap A = 0. A property P(x) is said to hold quasi everywhere on A if

$$\text{Cap } \{x \in A / P(x) \text{ is false}\} = 0.$$

A function u is said to be quasi continuous (resp. quasi upper semicontinuous) if, for each $\varepsilon > 0$, there exists an open set ω of capacity less than ε, such that the restriction of u to the complement of ω_ε is continuous (resp. upper semicontinuous). (2.67)

The notion of quasi continuity plays an important role in the functional representation of space H^1. With each function u of H^1 one can associate the class of its quasi continuous representations $\tilde{u}$ (defined up to the equality quasi everywhere). For every u in H^1, one can prove that the following limit exists for quasi every x

$$\tilde{u}(x) = \lim_{r \to 0} \frac{1}{|B_r(x)|} \int_{B_r(x)} u(y)dy \tag{2.68}$$

and the function $\tilde{u}$ so defined is a quasi-continuous representant of u:

The following result makes connection between convergence in H^1 and convergence quasi everywhere.

Let $\{u_n; n = 1,2,...\}$ be a converging sequence in H^1, $u_n \to u$; there exists a subsequence $\{u_{n_k}; k = 1,2,...\}$ such that

$$\tilde{u}_{n_k} \xrightarrow[k \to +\infty]{} \tilde{u} \quad \text{quasi-uniformly (and hence quasi-everywhere).} \tag{2.69}$$

(By quasi-uniform convergence, we mean that for every $\varepsilon > 0$, there exists an open set ω_ε of capacity less than ε, such that the sequence converges uniformly on the complement of ω_ε). Noticing that H^1 is the completion of $\mathbb{C}^0 \cap H^1$ with respect to norm H^1, the above result explains the quasi continuous representation (2.68) of elements of H^1.

Another notion which plays an important role in the theory of unilateral constraints in calculus of variations is the notion of *finite energy measure:*

A positive Radon measure μ is said to be of finite energy if $\mu \in H^{-1}$. Equivalently, there exists a constant $C > 0$ such that for every continuous function u with compact support,

$$\left|\int_{\mathbb{R}^N} u \, d\mu\right| \leq C \, \|u\|_{H^1} \tag{2.70}$$

One can prove that the quasi continuous representation $u \to \tilde{u}$ is, for every positive μ of finite energy, a continuous embedding from H^1 into $L^1(\mu)$. More precisely,

(i) for each Borel subset B such that $Cap(B) = 0$, we have $\mu(B) = 0$.

(ii) for every $u \in H^1$, $\langle \mu, u \rangle_{(H^{-1}, H^1)} = \int_{\mathbb{R}^N} \tilde{u}(x) d\mu(x)$. □ (2.71)

Let us now examine two model situations where the limit constraint is no longer of obstacle type:

EXAMPLE 2.28: "Fakir's bed of nails"

This is the obstacle version of Theorem 1.3 which has been called the "cloud of ice". With the same notations, let us consider the $\frac{1}{n}$-Y periodic structure, where in each small cell of size 1/n is included an r_n homothetic of a fixed "hole" $T \subset Y =]-\frac{1}{2}, +\frac{1}{2}[^N$ (Fig. 2.3). Take

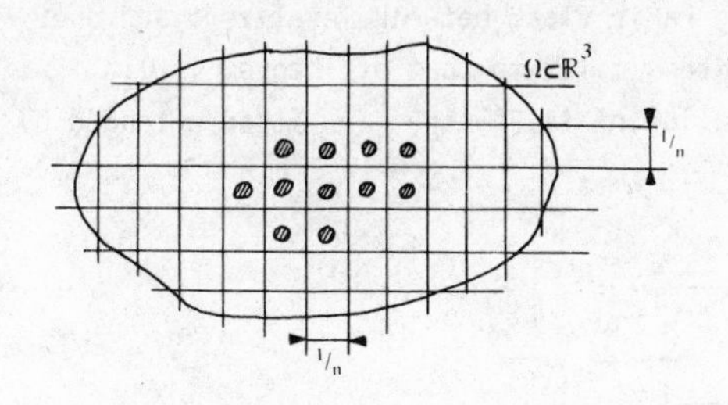

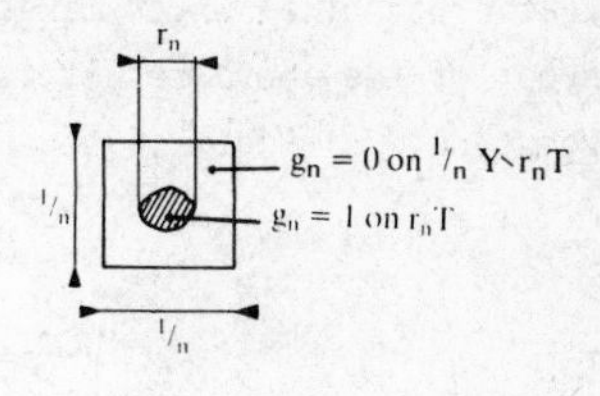

Figure 2.3

$$g_n = \begin{cases} 1 & \text{on } \bigcup_i (r_n T)_i \\ 0 & \text{elsewhere} \end{cases},$$

the corresponding convex set of constraints is

$$K_{g_n} = \{u \in H_0^1(\Omega);\ u \geq g_n \text{ a.e. on } \Omega\} .$$

Let us describe the critical size, that is $r_n \sim \frac{1}{n^3}$ (when N = 3):

For every $u \in H_0^1(\Omega)$,

$$s\text{-}L^2(\Omega)\text{lm}_e[\int_\Omega |\text{grad}u|^2 dx + I_{K_{g_n}}(u)] = \int_\Omega |\text{grad}u|^2 dx$$
$$+ C \int_\Omega [(1-u)^+]^2 dx + I_{\{v \geq 0 \text{ on } \Omega\}}(u) \tag{2.72}$$

where C is equal to the "capacity of T":

$$C = \inf \{\int_{\mathbb{R}^3} |\text{grad}v|^2 dx;\ v \in H^1(\mathbb{R}^3),\ v \geq 1 \text{ on } T\}.$$

When $r_n << \frac{1}{n^3}$, then C = 0, when $r_n >> \frac{1}{n^3}$ then C = + ∞: in these cases the limit constraint is associated respectively with the obstacles $g \equiv 0$ and $g \equiv 1$. This justifies the terminology we adopted: one may imagine that the obstacles

$\{g_n;\ n \in \mathbb{N}\}$ describe a "Fakir's bed of nails". If the size of the nails is too small (that is $r_n \ll \frac{1}{n^3}$), it is as though there were no nails: The Fakir lies at level zero. If the size of the nails is too large (that is $r_n \gg \frac{1}{n^3}$) the Fakir lies at level one. In the critical case, there is a quadratic potential which makes the Fakir float between level zero and one.

Proof of the above result is quite similar to that of Theorem 1.27: Introducing w_n, (Fig. 2.4) the solution of (1.90) on $\frac{1}{n}$ Y (then extended by

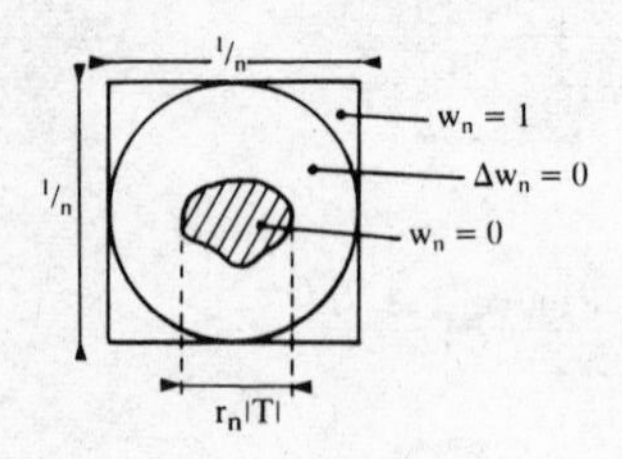

Figure 2.4

$\frac{1}{n}$ Y periodicity) take for every $u \in H_0^1(\Omega)$ the following approximating sequence:

$$u_n = u + (1-w_n)(1-u)^+ \tag{2.73}$$

One easily verifies that

$$u_n \to u \text{ in s-}L^2(\Omega) \text{ (since } w_n \to 1 \text{ is s-}L^2(\Omega))$$

$$u_n = u + (1-u)^+ = u \vee 1 \geqslant 1 \text{ on } \cup(r_n T),$$

hence $u_n \geqslant g_n$. The same computation as in Theorem 1.27 yields that

$$\int_\Omega |\text{grad}u_n|^2 dx \xrightarrow[n \to +\infty]{} \int_\Omega |\text{grad}u|^2 dx + C \int_\Omega (1-u)^{+2} dx$$

and that u_n does realize the infimum of $\lim_n \int_\Omega |\text{grad}v_n|^2 dx$ among all sequences $\{v_n;\ n \in \mathbb{N}\}$ converging to u in $L^2(\Omega)$ and satisfying $v_n \geqslant g_n$ for every $n \in \mathbb{N}$.

In this example the limit constraint functional

$$u \to C \int_\Omega (1-u)^{+2} dx + I_{\{v\geq 0\}}(u)$$

takes a relaxed or fuzzy form: for positive functions u, violating the constraint $u \geq 1$ is penalized only by a positive quantity, $C \int (1-u)^{+2}$. □

EXAMPLE 2.29: "Blowing up" of the obstacles

The following situation is quite classical in potential theory. It is related to the fact that the elementary solution of

$$-\Delta u = \delta$$

where δ is the Dirac measure, does not belong to $H^1(\mathbb{R}^N)$ when $N \geq 2$. (It has been used to provide counterexamples by G. Choquet [1], Ancona [1], Murat [5]). Let us consider $\Omega = B_R$, the open ball of radius R centred at the origin.

For every integer $n \in \mathbb{N}$, let us define $g_n : \Omega \to \mathbb{R}$

$$g_n(x) = \frac{1}{\sqrt{n}} \inf \{n; \frac{1}{|x|^{N-2}} - \frac{1}{R^{N-2}}\}. \tag{2.74}$$

One can easily verify that the sequence $\{g_n; n \in \mathbb{N}\}$ converges to zero for the weak topology of $H_o^1(B_R)$ and that

$$||g_n||_{H_o^1(B_R)} = C$$

where C is a constant depending only on the dimension N of space $\mathbb{R}^N$.

Then, for every $h \in L^2(\Omega)$, for every open set ω in Ω, such that $0 \notin$ boundary of ω,

$$\min \{\int_\Omega |\text{grad}u|^2 dx - \int_\Omega hu \, dx / u \in H_o^1(\Omega),\ u \geq g_n \text{ a.e. on } \omega\}$$

$$\xrightarrow[(n \to +\infty)]{} \min \{\int_\Omega |\text{grad}u|^2 dx + C.\delta(\omega) - \int_\Omega hu \, dx / u \in H_o^1(\Omega),\ u \geq 0 \text{ a.e. } \omega\}$$

where δ denotes the Dirac measure at the origin:

$$C.\delta(\omega) = \begin{cases} C \text{ if } 0 \in \omega \\ 0 \text{ otherwise} \end{cases}.$$

In other words, for every $u \in H_o^1(\Omega)$

$$L^2(\Omega)\text{-}\lim_e\{\int_\Omega |\text{grad}u|^2dx + I_{\{u \geq g_n \text{ a.e. } \omega\}}\}$$

$$= \int_\Omega |\text{grad}u|^2dx + C.\delta(\omega) + I_{\{v \geq 0 \text{ on } \omega\}}(u). \tag{2.75}$$

This phenomenon can still be interpreted as a relaxation phenomenon, since despite the original constraints functionals take only the values 0 and $+\infty$, the limit constraint functional $C\delta(\omega) + I_{\{v \geq 0 \text{ on } \omega\}}(u)$ takes the value C with $0 < C < +\infty$: it is no longer the indicator functional of a closed convex set!

As we shall see, this phenomenon is related to the fact that there does not exist a function $u \in H_o^1(\Omega)$ such that $u \geq g_n$ a.e. on Ω. In a more precise way, a weaker assumption which prevents this phenomenon from occurring is:

There exists a sequence $\{u_n; n = 1,2,...\}$ strongly convergent in $H_o^1(\Omega)$ such that $u_n \geq g_n$ a.e. on Ω. (2.76)

Clearly, this condition is not satisfied here, because

$$\frac{1}{|x|^{N-2}} - \frac{1}{R^{N-2}} \notin H_o^1(B_R)!$$

This justifies the terminology "blowing up of the obstacles" which has been chosen to describe this situation.

Let us now prove (2.75):

<u>We first consider the case $0 \in \omega$.</u>

(a) Let us give $u \in H_o^1(\Omega)$ such that $u \geq 0$ on ω, and take $u_n = u + g_n$; then, $u_n \geq g_n$ on ω and, since $g_n \longrightarrow 0$ in $w\text{-}H_o^1(\Omega)$, we have $u_n \longrightarrow u$ in $w\text{-}H_o^1(\Omega)$ and $s\text{-}L^2(\Omega)$. Let us compute

$$\int_\Omega |\text{grad}u_n|^2dx = \int_\Omega |\text{grad}u|^2dx + \int_\Omega |\text{grad}g_n|^2dx + 2\int_\Omega \text{grad}u.\text{grad}g_n\,dx$$

$$= C + \int_\Omega |\text{grad}u|^2dx + 2\int_\Omega \text{grad}u.\text{grad}g_n\,dx.$$

Since $\text{grad}g_n \longrightarrow 0$ in $w\text{-}L^2(\Omega)$,

$$\lim_{n\to+\infty} \int_\Omega |\mathrm{grad} u_n|^2 dx = C + \int_\Omega |\mathrm{grad} u|^2 dx. \tag{2.77}$$

(b) Let us complete the proof of (2.75) by proving that for every sequence $\{u_n;\ n = 1,2,\ldots\}$ converging in $L^2(\Omega)$, $u_n \xrightarrow[n\to+\infty]{} u$, and satisfying $u_n \geq g_n$ on ω, the following inequality holds:

$$\liminf_{(n\to+\infty)} \int_\Omega |\mathrm{grad} u_n|^2 dx \geq C + \int_\Omega |\mathrm{grad} u|^2 dx. \tag{2.78}$$

Let us decompose u_n in the following way:

$$u_n = u_n - g_n + g_n.$$

Then

$$\int_\Omega |\mathrm{grad} u_n|^2 dx = \int_\Omega |\mathrm{grad}(u_n-g_n)|^2 dx + \int_\Omega |\mathrm{grad} g_n|^2 dx + 2\int_\Omega \mathrm{grad} g_n . \mathrm{grad}(u_n-g_n)$$

$$= C + \int_\Omega |\mathrm{grad}(u_n-g_n)|^2 dx + 2\int_\Omega (-\Delta g_n).(u_n-g_n) dx$$

and

$$\begin{aligned}\liminf_n \int_\Omega |\mathrm{grad} u_n|^2 dx \geq C &+ \liminf_n \int_\Omega |\mathrm{grad}(u_n-g_n)|^2 dx \\ &+ 2 \liminf_n \int_\Omega (-\Delta g_n)(u_n-g_n) dx\end{aligned} \tag{2.79}$$

Since $u_n - g_n \longrightarrow u$ in $w\text{-}H_o^1(\Omega)$,

$$\liminf_n \int_\Omega |\mathrm{grad}(u_n-g_n)|^2 dx \geq \int_\Omega |\mathrm{grad} u|^2 dx \tag{2.80}$$

On the other hand, $-\Delta g_n$ is a positive measure supported by the sphere of radius r_n, where r_n satisfies $\frac{1}{r_n^{N-2}} - \frac{1}{R^{N-2}} = n$. Noticing that $r_n \to 0$ as n goes to $+\infty$ and that $u_n - g_n$ is positive on a fixed open set $\omega \ni 0$, for n large

$$\int_\Omega (-\Delta g_n)(u_n-g_n) dx \geq 0. \tag{2.81}$$

Combining (2.79), (2.80) and (2.81) we get the inequality (2.78).

In the case $0 \notin \bar{\omega}$, noticing that $g_n \to 0$ uniformly on ω, it follows easily that the limit constraint is $\{u \geq 0 \text{ on } \omega\}$:

Let us consider a function $\phi \in \mathcal{C}_o^1(\Omega)$ such that $\phi \geq 0$, $\phi = 1$ on $\bar{\omega}$. Denoting

$$\|g_n\|_{L^\infty(\omega)} = \varepsilon_n \xrightarrow[(n\to+\infty)]{} 0,$$

for every $u \in H_o^1(\Omega)$ such that $u \geq 0$ on ω

$$u_n = u + \varepsilon_n\phi \geq g_n \text{ on } \omega$$

and

$$\|u_n\|^2_{H_o^1(\Omega)} \to \|u\|^2_{H_o^1(\Omega)} \quad \text{as } n \to +\infty.$$

On the other hand, if $u_n \xrightarrow[n\to+\infty]{} u$ in $L^2(\Omega)$ and $u_n \geq g_n$ on ω,

$$\liminf_n \|u_n\|^2_{H_o^1(\Omega)} \geq \|u\|^2_{H_o^1(\Omega)} \quad \text{and } u \geq 0 \text{ on } \omega. \quad \square$$

This example brings to light an interesting feature that appears when studying convergence problems (for unilateral constraints...) and the dependence on the set of observation ω: there is a family of *exceptional sets* for which the conclusion of the general convergence result may fail. Here, the particular and difficult analysis of the case where 0 belongs to the frontier of ω has to be made: it requires sharp tools of potential theory. In the example of the Fakir's bed of nails (as well as homogenization of elliptic operators) one has to drop open sets ω whose boundaries are not of zero Lebesgue measure! □

B. Variational inequalities with varying obstacles: General form of the limit problem

The following theorem (De Giorgi, Dal Maso & Longo [1], Attouch & Picard [4], Dal Maso [4] [5]) explains in a unified way the various phenomena encountered in the preceding examples.

In order to state it we need, besides the notion of potential theory already introduced, the following notion of "rich family":

A subfamily $\mathbf{R}$ of the family $\mathbf{B}$ of bounded Borel subsets of Ω is said to be rich in $\mathbf{B}$ if, for each family

$$\{B_t,\ t \in [0,1] / \forall t \in [0,1]\ B_t \in \mathcal{B},\ \forall s < t\ \ \bar{B}_s \subset \mathring{B}_t\} \tag{2.82}$$

the set $\{t \in [0,1] / B_t \notin \mathcal{R}\}$ is at most countable.

<u>THEOREM 2.30</u> Let Ω be an open (not necessary bounded) subset of $\mathbb{R}^N$.

(a) Let us give Φ a convex continuous quadratic energy functional on $H^1(\Omega)$: Given $\{a_{ij};\ i,j = 1,2,\ldots,N\}$ belonging to $L^\infty(\Omega)$ and satisfying

$$\begin{aligned} &a_{ij} = a_{ji} \quad \text{for every } i,j = 1,2,\ldots,N \\ &\sum_{i,j=1}^{N} a_{ij}(x)\xi_i\xi_j \geq 0 \text{ for every } (\xi_1,\ldots,\xi_N) \in \mathbb{R}^N \end{aligned} \tag{2.83}$$

and given c_o a positive function of $L^\infty(\Omega)$, for every open set ω in Ω and every u belonging to $H^1(\omega)$, we denote:

$$\Phi(u,\omega) = \int_\omega \{ \sum_{i,j=1}^{N} a_{ij}(x) \frac{\partial u}{\partial x_i} \frac{\partial u}{\partial x_j} + c_o(x)u^2\}dx.$$

(b) Let us give $\{g_n : \Omega \to \bar{\mathbb{R}};\ n \in \mathbb{N}\}$ a sequence of obstacles satisfying:

For every $\omega \in \mathcal{O}$, there exists a sequence $\{u_o^n;\ n = 1,2,\ldots\}$, which remains bounded in $H^1(\Omega)$ and such that for every $n = 1,2,\ldots$

$$\tilde{u}_o^n(x) \geq g_n(x) \text{ for quasi every } x \text{ in } \omega. \tag{2.84}$$

Then, there exists a subsequence $\{n_k;\ k = 1,2,\ldots\}$, two positive Radon measures μ and ν, with $\mu \in H^{-1}(\Omega)$, ν singular with respect to μ, and a Borel function $j : \Omega \times \mathbb{R} \to \bar{\mathbb{R}}^+$ with $t \to j(x,t)$ convex, decreasing, lower semicontinuous such that

$$\begin{aligned} &\underset{(k\to+\infty)}{L^2(\Omega)\text{-}\lim_e} [\Phi(u,\Omega) + I_{\{\tilde{v} \geq g_{n_k} \text{ q.e. on } B\}}(u)] \\ &\qquad = \Phi(u,\Omega) + \int_B j(x,\tilde{u}(x))d\mu(x) + \nu(B) \end{aligned} \tag{2.85}$$

holds for every $u \in H^1(\Omega)$ and $B \in \mathcal{R}$, where $\mathcal{R}$ is a rich family of Borelian subsets of $\mathcal{B}$.

Moreover, for every $\omega \in \mathcal{R} \cap \mathcal{O}$, and $u \in H^1(\omega)$

$$L^2(\omega)\text{-}\lim_e [\Phi(u,\omega) + I_{\{\tilde{u} \geq g_{n_k} \text{ q.e. on } \omega\}}(u)]$$
$$(k \to +\infty)$$

$$= \Phi(u,\omega) + \int_\omega j(x,\tilde{u}(x))d\mu(x) + \nu(\omega). \qquad (2.86)$$

The family $\mathbb{R}$ contains the sets B for which there exist a compact K and an open set ω such that $K \subset \mathring{B} \subset B \subset \omega$ and

$$g_n \equiv -\infty \text{ on } \omega \setminus K.$$

Finally, one may replace in (2.85) the topology $L^2(\Omega)$ by the topology $L^2_{loc}(\Omega)$. □

<u>Comments on Theorem 2.30 and interpretation of previous examples</u>

<u>REMARK 2.31</u> We first notice that, in the above statement, no converging assumption is made for the sequence of obstacles $\{g_n; n = 1,2,...\}$. Assumption (2.84) assures us only that the convex sets of constraints

$$K_{g_n} = \{u \in H^1(\Omega)/\tilde{u}(x) \geq g_n(x) \text{ q.e.}\}$$

are non-void, and that the solutions of the corresponding minimization problems remain bounded in $H^1(\Omega)$. So, Theorem 2.30 can be interpreted as a *relative compactness theorem* for the class of obstacle constrained functionals. We shall make this precise later.

<u>REMARK 2.32</u> We then notice that, given a function $g:\Omega \to \bar{\mathbb{R}}$ measurable in $x \in \Omega$, it is always possible to write in an *integral form the indicator functional of the corresponding set of constraints.* This is clear when the constraint is taken in the sense almost everywhere: given $\omega \in \mathcal{O}$, let us denote

$$K_g = \{u \in H^1(\Omega)/u(x) \geq g(x) \text{ almost everywhere on } \omega\}.$$

Then we have

$$I_{K_g}(u) = \int_\omega j(x,u(x))dx$$

where

$$j(x,t) = \begin{cases} 0 & \text{if } t \geqslant g(x) \\ +\infty & \text{otherwise} \end{cases} .$$

This result is more surprising when the constraint is taken in the sense quasi everywhere. It is a corollary of Theorem 2.30. It yields a new characterization of obstacle constraints and can be formulated in the following way: Let $g:\Omega \to \bar{\mathbb{R}}$ be an extended real-valued function such that there exists some $u_o \in H^1(\Omega)$ such that $u_o \geqslant g$ quasi everywhere on Ω. Then, there exists a positive finite energy measure μ ($\mu \in H^{-1}(\Omega)$), such that for all $u \in H^1(\Omega)$ and $\omega \in \mathcal{O}$, $\bar{\omega} \subset \Omega$,

$$\tilde{u} \geqslant g \text{ quasi-everywhere on } \omega \iff \tilde{u} \geqslant \tilde{g} \ \mu\text{-almost everywhere on } \omega, \tag{2.87}$$

where $\tilde{g}$ is the quasi-upper semicontinuous regularization of g (cf. Attouch & Picard [1]). Equivalently denoting

$$K_g = \{u \in H^1(\Omega);\ \tilde{u}(x) \geqslant g(x) \text{ quasi-everywhere on } \omega\},$$

we have

$$I_{K_g}(u) = \int_\omega j(x,\tilde{u}(x))d\mu(x) \tag{2.88}$$

with

$$j(x,t) = \begin{cases} 0 & \text{if } t \geqslant \tilde{g}(x) \text{ on } \omega \\ +\infty & \text{otherwise.} \end{cases}$$

Via the introduction of a regularized function $\tilde{g}$ of the obstacle g, the obstacle constraint which was initially formulated in terms of capacity (quasi everywhere) can be reformulated in terms of measure (μ-almost everywhere) and hence in an integral way. Let us notice that the measure μ very much depends on the obstacle g! For example, if the constraint $u \geqslant g$ is taken in the sense quasi everywhere on a manifold of dimension (N-1) in $\mathbb{R}^N$ (which is of strictly positive capacity), one may take for μ the Hausdorff measure of dimension (N-1) supported by the manifold (Example 2.27, thin obstacle). So, integral representation (2.88) of the limit constraint functional allows us to interpret any type of obstacle constraint.

REMARK 2.33 This integral formulation of the limit constraint functional is very flexible, since it allows us to represent in a unified way *all the possible forms of the limit constraint* - we examined above the case of "pure" limit obstacle functionals. In the "Fakir's bed of nails", Example 2.28, where a *relaxation phenomenon* occurs, the limit constraint functional G can be written

$$G(u,B) = \int_B j(u(x))dx$$

with $\mu = dx$ and

$$j(t) = \begin{cases} C[(1-t)^+]^2 & \text{if } t \geq 0 \\ +\infty & \text{if } t < 0. \end{cases}$$

The measure ν which appears in formula (2.85) allows us to interpret the second type of relaxation phenomenon described in Example 2.29, "blow-up of the obstacles": take $\nu = C\delta$, where δ is the Dirac measure. So, *through the introduction of the integral functionals*

$$G(u,B) = \int_B j(x,\tilde{u}(x))d\mu(x) + \nu(B)$$

as described in Theorem 2.30, we have enlarged the class of the obstacle constraint functionals and built a class which is compact in the epi-convergence sense. Let us stress that in order to interpret relaxation phenoma, one has to consider epi-convergence of the sum $\Phi + G^n$ of the energy and constraint functional. One cannot treat separately the obstacle constraint functionals $\{G^n; n = 1,2,...\}$ and study their convergence. This is made clear when noticing that in the "Fakir's bed of nails", if the energy functional Φ is multiplied by a constant $\lambda > 0$, so is the limit constraint functional G!).

REMARK 2.34 Another important feature of Theorem 2.30 is its *local aspect*. As in Theorem 2.24, it is the preservation by epi-limit of the local character of the functionals which allows us to identify the limit functional. Let us stress the fact that so far we are not able to state a compactness or convergence result which holds for all observation sets $B \in \mathcal{B}$ and corresponding sequences of functionals

$$F^n(u,B) = \Phi(u) + G^n(u,B).$$

The obverse of the simplicity and generality of Theorem 2.30 (which relies on an integral representation of all possible limit constraint functionals) is that we have left aside a family of exceptional sets, $\mathbb{B}\setminus\mathbb{R}$, for which we can reach no conclusion. The fact that the family $\mathbb{R}$ is rich leads to the exceptional character of these sets.

As we already noticed a particular study is required in the "Fakir's bed of nails" for open sets whose boundaries are not of zero Lebesgue measure while in the "blow-up of the obstacles" one has to consider separately open sets whose boundary contains the origin. □

Let us now sketch the proof of Theorem 2.30. Take for simplicity of notation $\Phi(u,\omega) = \int_\omega |\mathrm{grad}\, u|^2 dx$, the Dirichlet energy.

Step one We first list the properties of the constraint functionals G^n preserved by epi-limit. As already pointed out, functionals G^n are viewed as functions of u and ω, where ω is the set of observation. This allows us to formulate local character and measure dependence with respect to ω of the functionals G^n:

$$G^n(u,\omega) = I_{\{v \geq g_n \text{ on } \omega\}}(u).$$

We are led to introduce the following class (G):

DEFINITION 2.35 We denote by $\mathbb{B}$ (resp. $\mathbb{O}$) the family of bounded Borel (resp. bounded open) subsets of Ω, and by (G) the class of functionals

$$G: H^1(\Omega) \times \mathbb{B} \to \bar{\mathbb{R}}^+$$

which satisfy the following five properties $(G_1),\ldots,(G_5)$:

Measure dependence with respect to $\mathbb{B}$

(G_1) For every $u \in H^1(\Omega)$, the function $B \to G(u,B)$ from $\mathbb{B}$ into $\bar{\mathbb{R}}^+$ is increasing on $\mathbb{B}$, and there exists a functional $\tilde{G}: \mathbb{B} \times H^1(\Omega) \to \bar{\mathbb{R}}^+$ which coincides with G on $H^1(\Omega) \times \mathbb{O}$ and such that, for every u in $H^1(\Omega)$,

$$B \to \tilde{G}(u,B) \text{ is a Borel measure on } \mathbb{B}.$$

Dependence on u

(G_2) For every $\omega \in \mathcal{O}$, $u \to G(u,\omega)$ is a closed convex proper function from $H^1(\Omega)$ into $\bar{\mathbb{R}}^+$.

(G_3) For every $\omega \in \mathcal{O}$, $u \to G(u,\omega)$ is decreasing, i.e.

$$u \leq v \text{ a.e. on } \Omega \Rightarrow G(u,\omega) \geq G(v,\omega).$$

Local properties

(G_4) For every $u,v \in H^1(\Omega)$, for every $\omega \in \mathcal{O}$, $u|_\omega = v|_\omega$ implies $G(u,\omega) = G(v,\omega)$.

(G_5) For every $u,v \in H^1(\Omega)$, for every $\omega \in \mathcal{O}$, $G(u \wedge v,\omega) + G(u \vee v,\omega) = G(u,\omega) + G(v,\omega)$.

Step two We now prove a compactness theorem for the abstract class of functionals $\mathcal{F} = \{\Phi+G/G \in (G)\}$.

Let us consider a sequence $\{F^n; n \in \mathbb{N}\}$ of functionals of the class $\mathcal{F}$:

$$F^n = \Phi + G^n.$$

There are two natural ways to localize them:

(a) In the *complete localization method* we localize both Φ and G^n:

$$F^n(u,\omega) = \Phi(u,\omega) + G^n(u,\omega).$$

(b) In the *partial localization method*, taking advantage of the fact that Φ is known, we localize only the varying constraints

$$F^n(u,\omega) = \Phi(u,\Omega) + G^n(u,\omega).$$

A posteriori, referring to Theorem 2.30, formulae (2.85) and (2.86), the two above approaches lead to the same results. But, from a technical point of view some results are easier to prove with one formulation than with the other. In order to include both aspects, for any open set $A \in \mathcal{O}$, Borel set $B \in \mathcal{B}$ such that $B \subset A$ and any $u \in H^1(A)$, let us introduce

$$\Phi(u,A) + G_s(u,A,B) = L^2(A)\text{-}\mathrm{ls}_e[\Phi(u,A) + G^n(u,B)] \tag{2.89}$$

$$\Phi(u,A) + G_i(u,A,B) = L^2(A)\text{-}\mathrm{li}_e[\Phi(u,A) + G^n(u,B)] \tag{2.90}$$

The situation $A = \Omega$, $B \in \mathcal{B}$, $B \subset \Omega$ corresponds to the so called partial localization method, while $A = B = \omega$ corresponds to the complete localization method. By definition of the epi-limits

$$\int_\Omega |Du|^2 dx + G_s(u,\Omega,\omega) = \min[\limsup_n \{\int_\Omega |Du_n|^2 dx + G^n(u_n,\omega)\};\ u_n \longrightarrow u \text{ weakly in } H^1(\Omega)]$$

Writing $u_n = u + z_n$ with $z_n \longrightarrow 0$ weakly in $H^1(\Omega)$, we obtain

$$G_s(u,\Omega,\omega) = \min [\limsup_n \{\int_\Omega |Dz_n|^2 dx + G^n(u+z_n,\omega)\};\ z_n \longrightarrow 0 \text{ weakly in } H^1(\Omega)] \tag{2.91}$$

$$G_i(u,\Omega,\omega) = \min [\liminf_n \{\int_\Omega |Dz_n|^2 dx + G^n(u+z_n,\omega)\};\ z_n \longrightarrow 0 \text{ weakly in } H^1(\Omega)] \tag{2.92}$$

and similar formulae hold for $G_s(u,\omega,\omega)$ and $G_i(u,\omega,\omega)$.

Let us examine the properties of functionals G_i and G_s:

Clearly, from formulations (2.91) and (2.92), G_i and G_s as functions of u, are still *decreasing* and still satisfy the *local properties*.

Let us verify that $u \to G_s(u,\Omega,\omega)$ is *convex:*

Take u_1 and u_2 belonging to $H^1(\Omega)$ and $0 < \lambda < 1$. By (2.91), there exists two sequences $\{z_n^1;\ n \in \mathbb{N}\}$ and $\{z_n^2;\ n \in \mathbb{N}\}$ weakly converging to zero in $H^1(\Omega)$ such that

$$G_s(u_1,\Omega,\omega) = \limsup_n \{\int_\Omega |Dz_n^1|^2 + G^n(u_1 + z_n^1,\omega)\}$$

$$G_s(u_2,\Omega,\omega) = \limsup_n \{\int_\Omega |Dz_n^2|^2 + G^n(u_2 + z_n^2,\omega)\}.$$

Since $\lambda z_n^1 + (1-\lambda) z_n^2$ still weakly converges to zero in $H^1(\Omega)$, from (2.91)

$$\begin{aligned} G_s(\lambda u_1+(1-\lambda)u_2,\Omega,\omega) &\leq \limsup_n \{\int_\Omega |\lambda Dz_n^1+(1-\lambda)Dz_n^2|^2 dx \\ &\quad + G^n(\lambda u_1+(1-\lambda)u_2+\lambda z_n^1 + (1-\lambda)z_n^2,\omega)\} \\ &\leq \limsup_n \{\lambda(\int_\Omega |Dz_n^1|^2+G^n(u_1+z_n^1,\omega)) \\ &\quad + (1-\lambda)(\int_\Omega |Dz_n^2|^2 + G^n(u_2 + z_n^2,\omega))\} \end{aligned}$$

$$\leq \lambda G_s(u_1, \Omega,\omega) + (1-\lambda)G_s(u_2,\Omega,\omega)$$

Once more, the difficult point is to verify the measure properties of G_i and G_s:

Superadditivity is quite easy to verify on G_i when using the complete localization method: Let $u \in H^1(\Omega)$ and $A,B \in \mathcal{O}$ such that $A \cap B = \emptyset$. From (2.92) there exists a sequence $\{z_n; n \in \mathbb{N}\}$ weakly converging to zero in $H^1(A \cup B)$ such that

$$G_i(u,A\cup B, A\cup B) = \liminf_n \{\Phi(z_n,A\cup B) + G^n(u+z_n, A\cup B)\}$$

$$\geq \liminf_n \{\Phi(z_n,A)+G^n(u+z_n,A)\}+\liminf_n\{\Phi(z_n,B) +G^n(u+z_n,B)\}$$

i.e.

$$G_i(u,A\cup B, A\cup B) \geq G_i(u,A,A) + G_i(u,B,B). \tag{2.93}$$

Subadditivity is (on the other hand) easier to verify by the partial localization method: By definition (2.89) of G_s, given two open sets A and B there exists two approximating sequences $\{u_n^A; n \in \mathbb{N}\}$ and $\{u_n^B; n \in \mathbb{N}\}$ of u such that

$$\Phi(u,\Omega) + G_s(u,\Omega,A) = \limsup_n \{\Phi(u_n^A,\Omega) + G^n(u_n^A,A)\}$$

$$\Phi(u,\Omega) + G_s(u,\Omega,B) = \limsup_n \{\Phi(u_n^B,\Omega) + G^n(u_n^B,B)\}.$$

Then take $u_n = u_n^A \vee u_n^B$, which still converges to u weakly in $H^1(\Omega)$. By definition of G_s,

$$\Phi(u,\Omega)+G_s(u,\Omega,A\cup B) \leq \limsup_n \{\Phi(u_n^A \vee u_n^B,\Omega)+G^n(u_n^A\vee u_n^B, A\cup B)\}. \tag{2.94}$$

By the local property of Φ,

$$\Phi(u_n^A \vee u_n^B) + \Phi(u_n^A \wedge u_n^B) = \Phi(u_n^A) + \Phi(u_n^B).$$

By properties (G_1) and (G_3) of G^n

$$G^n(u_n^A \vee u_n^B, A \cup B) \leq G^n(u_n^A \vee u_n^B, A) + G^n(u_n^A \vee u_n^B, B)$$

$$\leq G^n(u_n^A, A) + G^n(u_n^B, B).$$

Returning to (2.94) and using the two above inequalities, we obtain

$$\Phi(u,\Omega) + G_s(u,\Omega,A \cup B) \leq \limsup_n \{\Phi(u_n^A,\Omega) + G^n(u_n^A,A)\}$$

$$+ \limsup_n \{\Phi(u_n^B,\Omega) + G^n(u_n^B,B)\}$$

$$+ \limsup_n \{-\Phi(u_n^A \wedge u_n^B)\}$$

$$\leq \Phi(u,\Omega)+G_s(u,\Omega,A) + \Phi(u,\Omega)+G_s(u,\Omega,B)-\Phi(u,\Omega)$$

that is

$$G_s(u,\Omega,A \cup B) \leq G_s(u,\Omega,A) + G_s(u,\Omega,B). \tag{2.95}$$

In order to obtain both super and subadditivity properties we need to find the *relations between the two localization methods:*

One easily verifies that for every $u \in H^1(\Omega)$ and $\omega \in \mathcal{O}$,

$$\begin{aligned} &G_s(u,\omega,\omega) \leq G_s(u,\Omega,\omega) \\ &G_i(u,\omega,\omega) \leq G_s(u,\Omega,\omega) \end{aligned} \tag{2.96}$$

Conversely (and this is indeed the only technical part of the proof), for every $u \in H^1(\Omega)$, for every $\omega, A \in \mathcal{O}$ such that $\bar{A} \subset \omega$,

$$\begin{aligned} &G_s(u,\Omega,A) = G_s(u,\omega,A) \leq G_s(u,\omega,\omega) \\ &G_i(u,\Omega,A) = G_i(u,\omega,A) \leq G_i(u,\omega,\omega). \end{aligned} \tag{2.97}$$

One can now, as in Theorem 2.24, develop the compactness argument. Let us introduce $\mathcal{D}$, a dense denumerable family of $\mathcal{B}$. By abstract compactness Theorem 2.22 and by the diagonalization argument one can extract a subsequence $\{n_k;\ k \in \mathbb{N}\}$ such that

for every $u \in H^1(\Omega)$ and every $\omega \in \mathcal{D}$, $L^2(\Omega)\text{-}\lim_e[\Phi(u,\Omega)+G^{n_k}(u,\omega)]$ exists.

With notations (2.89) and (2.90), working now on the subsequence $\{F^{n_k} = \Phi + G^{n_k};$ $k \in \mathbb{N}\}$

$$G_s(u,\Omega,\omega) = G_i(u,\Omega,\omega) \text{ for every } u \in H^1(\Omega), \text{ for every } \omega \in \mathbb{D} \quad (2.98)$$

We now rely on a general regularity property of functionals $G(u,B)$

$$G:H^1(\Omega) \times \mathbb{B} \to \bar{\mathbb{R}}^+$$

(the difficulty comes from the fact that functionals G may take $+\infty$ values) which, as functions of u, are lower semicontinuous with respect to the strong topology of $H^1(\Omega)$ and are increasing functions of B. A Borel set $B \in \mathbb{B}$ is said to be regular with respect to G if, for any $u \in H^1(\Omega)$,

$$G(u,B) = \sup \{G(u,B');\ B' \in \mathbb{B} \text{ and } \bar{B}' \subset \mathring{B}\}. \quad (2.99)$$

One can prove that for such functional G, the class of its regular sets forms a rich family of $\mathbb{B}$.

By density of $\mathbb{D}$ in $\mathbb{B}$ and (2.98) it follows that, for every u in $H^1(\Omega)$ and every $B \in \mathbb{B}$

$$\sup_{\bar{\omega} \subset \mathring{B}} G_s(u,\Omega,\omega) = \sup_{\bar{\omega} \subset \mathring{B}} G_i(u,\Omega,\omega) \doteq G(u,B) \quad (2.100)$$

where $G(u,B)$ is precisely defined as the above common value. From (2.99), taking $\mathbb{R}$ the family of Borel sets which are regular both for G_i and G_s, we easily get that $\mathbb{R}$ is still a rich family and

$$G_i(u,\Omega,B) = G_s(u,\Omega,B) = G(u,B) \quad (2.101)$$

for all $u \in H^1(\Omega)$ and $B \in \mathbb{R}$.

Using (2.96) and (2.97) it follows that the limit constraint functionals obtained by the complete or partial localization method have the same "inner regularization" (2.100). Hence $G(u,\cdot)$ inherits both superadditivity and subadditivity properties (2.93) and (2.95), which makes it a measure.

Other properties of G follow easily from decreasing, local and convexity properties of G_s.

<u>Step three</u> We have now to prove an integral representation formula for functionals of the class (G). Indeed, this is a non-trivial result, because

of the non-continuity on $H^1(\Omega)$ of such functionals, which are only lower semicontinuous. In order to overcome this difficulty, a natural idea is to approximate functionals of class (G) by continuous ones for which a classical argument would apply (first apply the Radon-Nikodym theorem to affine functions, then by local property and density pass from piecewise affine functions to any function u of $H^1(\Omega)$). When using such approximation method, in order to preserve properties of class (G) one needs a non-standard approximation scheme (which relies on approximation of convex functionals by convex polyhedral functionals - see Theorem 3.40). Following this, one obtains

$$G(u,B) = \int_B j(x,\tilde{u}(x))d\mu(x) + \nu(B) \text{ for every } u \in H^1(\Omega),\ B \in \mathbb{B}$$

where the integrand j is decreasing with respect to u and may take + ∞ values. Let us notice that this formula makes sense thanks to (2.71). □

<u>The converse problem</u> Theorem 2.30 tells us that any constraint functional G that is the limit of obstacle constraint functionals can be written

$$G(u,\omega) = \int_\omega j(x,\tilde{u}(x))d\mu(x) + \nu(\omega).$$

A natural question is: Can one effectively obtain all these functionals as limits of obstacle problems? The following theorem gives a fairly precise answer to this question:

<u>THEOREM 2.36</u> Given any functional $F:H^1(\Omega) \times \mathbb{B} \to \bar{\mathbb{R}}^+$

$$F(u,B) = \int_\Omega |\text{grad}u|^2 dx + \int_B j(x,\tilde{u}(x))d\mu(x)$$

where μ is a positive finite energy Radon measure ($\mu \in H^{-1}(\Omega)^+$), j(x,t) is Borel measurable in x and, as a function of t, satisfies

$$j \geq 0,\ j' \leq 0 \text{ (decreasing)},\ j'' \geq 0 \text{ (convexity) and } j''' \leq 0, \qquad (2.102)$$

then there exists a sequence of obstacles $\{g_n;\ n \in \mathbb{N}\}$ and a rich family $\mathbb{R}$ of Borel sets such that for every $u \in H^1(\Omega)$, for every $B \in \mathbb{R}$,

$$F(u,B) = \underset{(n\to+\infty)}{s\text{-}L^2(\Omega)\text{lm}_e}\left[\int_\Omega |\text{grad}u|^2 dx + I_{\{v \geq g_n \text{ a.e. } B\}}(u)\right].$$

Proof of Theorem 2.36

(a) In order to obtain any $\mu \in H^{-1}(\Omega)^+$, let us first notice that, when considering the "Fakir's bed of nails" example one obtains $\mu = Cdx$, with constant C equal to the capacity in $\mathbb{R}^3$ of the set T. Thus, one can obtain any constant C belonging to $\mathbb{R}^+$. Thanks to the local character of the constraint functionals $G \in (G)$, one can then obtain any density measure $\mu = hdx$, where h is a positive step function from Ω into $\mathbb{R}^+$. Noticing that such functions form a dense subspace of $H^{-1}(\Omega)^+$, one can then pass by density to any $\mu \in H^{-1}(\Omega)^+$. Note that when developing such an argument one works with a fixed integrand, let us say $j(t)=[(k-t)^+]^2$ where $k \in \mathbb{R}^+$. Moreover, in order to make the density argument rigorous, one has to use a diagonalization process for epi-convergence, the justification of which (it relies on the existence, in the above case, of a metric inducing epi-convergence) can be found in Section 2.8.

(b) Explicit construction of all possible limit integrands j is a difficult problem. Properties $j > 0$, $j' < 0$, $j'' > 0$ are quite natural since they are related to positivity, decreasing and convexity of constraint functionals $G \in (G)$. On the other hand, property $j''' < 0$ is rather surprising and related to the quadratic property of the energy functional! Once more, in order to obtain such j, obstacles g_n are constructed from the Fakir's bed of nails example. But now, we have to superpose them: in order to analyse such limit problem let us first consider the case of a two-level sequence of obstacles $\{g_n; n \in \mathbb{N}\}$, $g_n: \mathbb{R}^3 \to \mathbb{R}$ (Fig. 2.5).

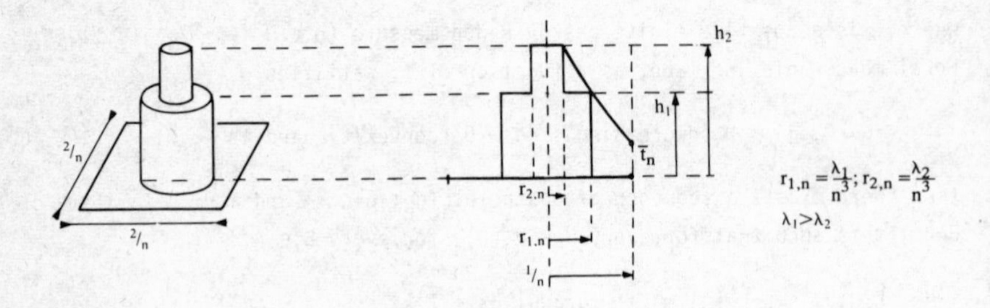

Figure 2.5

We recall that given an obstacle of only one stage, let us say of level h and diameter r_n, the limit constraint functional is equal to

$$G(t,\omega) = \min \{ \lim_{n\to+\infty} \int_\omega |\text{grad} w_n|^2 ; \; w_n \longrightarrow t, \; w_n \geq g_n \text{ on } \omega\},$$

the minimum being achieved for w_n equal in each small cell to w_n^t, the solution of the following capacity problem:

$$\min \{\int_{\frac{1}{n}Y} |\text{grad} w|^2 dx; \; w = h \text{ on } B_{r_n}, \; w = t \text{ on } \frac{1}{n} Y \setminus B_{1/n}\}, \tag{2.103}$$

where B_{r_n} is the ball centred in $\frac{1}{n}Y$ of radius r_n. A direct computation yields

$$w_n^t(\rho) = h - \frac{h-t}{1/r_n - n} (1/r_n - 1/\rho) \tag{2.104}$$

and

$$\int_{\frac{1}{n}Y} |\text{grad} w_n^t|^2 dx = 4\pi . \frac{(h-t)^2}{1/r_n - n} .$$

Thus,

$$G(t,\omega) \sim \frac{\text{meas } \omega}{(2/n)^3} \cdot 4\pi \cdot \frac{(h-t)^2}{1/r_n - n} ,$$

and when taking r_n equal to the critical size $r_n = \frac{\lambda}{n^3}$, one obtains

$$G(t,\omega) = \frac{\pi}{2} \cdot \lambda \cdot (h-t)^2 \text{ meas } \omega \text{ for } 0 \leq t \leq h. \tag{2.105}$$

Let us now return to the case of a two-level obstacle $\{g_n; n \in \mathbb{N}\}$. The first layer has level h_1 and radius $r_{1,n}$, the second has level h_2 and radius $r_{2,n}$. Given some $t > 0$, when computing $G(t,\omega)$, the only obstacle one has to consider is the upper one $(h_2, r_{2,n})$ as long as the corresponding capacity potential remains above the lower obstacle $(h_1, r_{1,n})$.

The critical value $\bar{t}_n$ for which there is "contact" is obtained by using (2.104)

$$h_2 - \frac{h_2 - \bar{t}_n}{1/r_{n,2} - n} (1/r_{n,2} - 1/r_{n,1}) = h_1$$

that is (taking $r_{n,1} = \frac{\lambda_1}{n^3}$, $r_{n,2} = \frac{\lambda_2}{n^3}$),

$$\bar{t}_n \xrightarrow[n \to +\infty]{} \bar{t} = h_2 - \frac{h_2 - h_1}{1 - \lambda_2/\lambda_1} . \tag{2.106}$$

The above construction works as soon as $\bar{t} > 0$, i.e.

$$h_2\lambda_2 < h_1\lambda_1. \tag{2.107}$$

Similar computation to the above yields the following value for $G(t,\omega)$ (Fig. 2.6):

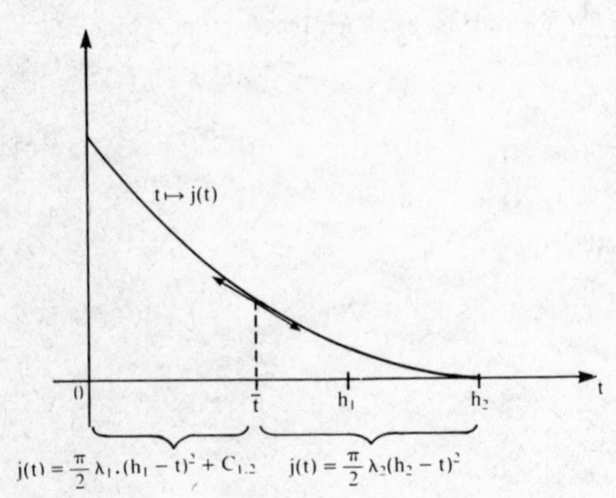

j is a piecewise quadratic function of class C^1.

$$\begin{cases} C_{1,2} = \frac{\pi}{2} \frac{(h_1 - h_2)^2}{1/\lambda_2 - 1/\lambda_1} \\ \bar{t} = h_2 - \frac{h_2 - h_1}{1 - \lambda_2/\lambda_1} \end{cases}$$

Figure 2.6

$$G(t,\omega) = \frac{\pi}{2} \lambda_2(h_2-t)^2 \text{ meas } \omega \text{ for } \bar{t} \leq t \leq h_2.$$

When $0 \leq t < \bar{t}$ and n is large enough, $0 \leq t \leq \bar{t}_n$. The corresponding capacity potential w_n^t has a fixed energy contribution $C_{1,2}$ on the ball $B_{r_{1,n}}$, while on the annulus $B_{1/n} - B_{r_{1,n}}$ it is equal to the capacity potential of the obstacle $(h_1, r_{1,n})$ and has an energy contribution equal to $\frac{\pi}{2} \lambda_1(h_1-t)^2$. Its total energy is equal to the sum of these two quantities. Continuity of j and (2.106) yield the value of $C_{1,2}$.

We are now going to play on the parameters λ_i and h_i and the number of stages $i = 1,2,\ldots,\ell$ in order to obtain any j satisfying (2.102).

We first observe that, by the above construction, function j" is a positive step function decreasing with t. (Fig. 2.7). Then we verify that, by suitable

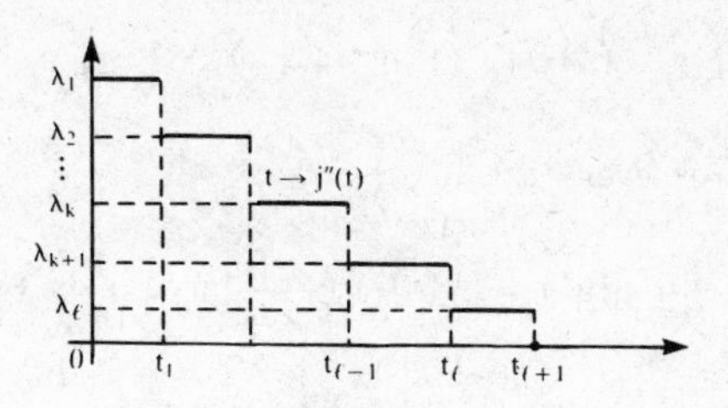

Figure 2.7

choice of λ_i and h_i, we can realize any such step function. First start by the highest stage at level $h_\ell = t_{\ell+1}$ and radius $r_{\ell,n} = \frac{\lambda_\ell}{n^3}$. Then, choose the stage below at level $h_{\ell-1}$ such that $\bar{t}_\ell = t_\ell$. Equivalently, from (2.106),

$$t_\ell = h_\ell - \frac{h_\ell - h_{\ell-1}}{1-\lambda_\ell/\lambda_{\ell-1}},$$

which yields

$$h_{\ell-1} = h_\ell - (h_\ell - t_\ell)(1-\lambda_\ell/\lambda_{\ell-1}) \quad (h_\ell = t_{\ell+1},\ t_\ell,\ \lambda_\ell \text{ and } \lambda_{\ell-1}$$

are prescribed!) and this provides a uniquely determined value of $h_{\ell-1}$. Because $\lambda_\ell < \lambda_{\ell-1}$ we have $h_{\ell-1} < h_\ell$. So we can reiterate this operation. Then, by a density argument, we can pass from functions j such that j" is a positive *decreasing* step function with compact support to any function j satisfying (2.102).

C. Bilateral constraints

In this paragraph, we show how one can reduce the bilateral obstacle problem "$g_n \leqslant u \leqslant h_n$" to the unilateral ones $g_n \leqslant u$ and $u \leqslant h_n$.

This can be done (cf. Attouch & Picard [4]) under the following hypothesis:

There exists a function $V \in H^1(\Omega)$ such that $g_n \leqslant V \leqslant h_n$ for every $n \in \mathbb{N}$. (2.108)

Take for simplicity V = 0 and denote for every $u \in H^1(\Omega)$,

$$\int_\Omega |\mathrm{grad}u|^2dx + G(u) = L^2(\Omega)\text{-}\mathrm{lm}_e \, \{\int_\Omega |\mathrm{grad}u|^2dx + I_{\{v\geq g_n\}}(u)\} \qquad (2.109)$$

$$\int_\Omega |\mathrm{grad}u|^2dx + H(u) = L^2(\Omega)\text{-}\mathrm{lm}_e \, \{\int_\Omega |\mathrm{grad}u|^2dx + I_{\{v\leq h_n\}}(u)\}. \qquad (2.110)$$

Then, let us prove that, for every $u \in H^1(\Omega)$

$$\int_\Omega |\mathrm{grad}u|^2dx + G(u^-) + H(u^+) = L^2(\Omega)\text{-}\mathrm{lm}_e \, \{\int_\Omega |\mathrm{grad}u|^2dx + I_{\{g_n\leq v\leq h_n\}}(u)\},$$

where we write $u = u^+ + u^-$, $u^+ = u \vee 0$, $u^- = u \wedge 0$.

(a) Given any sequence $\{u_n;\ n \in \mathbb{N}\}$, such that $g_n \leq u_n \leq h_n$ for every $n \in \mathbb{N}$, and which converges in $L^2(\Omega)$ to u,

$$\liminf_n \int_\Omega |\mathrm{grad}u_n|^2dx \geq \liminf_n \int_\Omega |\mathrm{grad}u_n^-|^2dx + \liminf_n \int_\Omega |\mathrm{grad}u_n^+|^2dx. \qquad (2.111)$$

From $g_n \leq 0$ and $g_n \leq u_n$ it follows that $u_n^- \geq g_n$. From $u_n^- \to u^-$ in $L^2(\Omega)$ and (2.109),

$$\liminf_n \int_\Omega |\mathrm{grad}u_n^-|^2dx \geq \int_\Omega |\mathrm{grad}u^-|^2dx + G(u^-).$$

From $0 \leq h_n$ and $u_n \leq h_n$, it follows that $u_n^+ \leq h_n$. From $u_n^+ \to u^+$ in $L^2(\Omega)$ and (2.110)

$$\liminf_n \int_\Omega |\mathrm{grad}u_n^+|^2dx \geq \int_\Omega |\mathrm{grad}u^+|^2dx + H(u^+).$$

Returning to (2.111) by using the two above inequalities, we obtain

$$\liminf_n \int_\Omega |\mathrm{grad}u_n|^2dx \geq \int_\Omega |\mathrm{grad}u|^2dx + G(u^-) + H(u^+).$$

(b) Then, given u belonging to $H^1(\Omega)$, by (2.109) and (2.110) there exist two sequences $\{v_n;\ n \in \mathbb{N}\}$ and $\{w_n;\ n \in \mathbb{N}\}$ such that

$$\int_\Omega |\mathrm{grad}u^-|^2dx + G(u^-) \geq \limsup_n \int_\Omega |\mathrm{grad}v_n|^2dx \text{ with } \begin{cases} v_n \to u^- \text{ in } L^2(\Omega) \\ v_n \geq g_n \end{cases}$$

and

$$\int_\Omega |\text{grad}u^+|^2dx + H(u^+) \geq \limsup_n \int_\Omega |\text{grad}w_n|^2dx \text{ with } \begin{cases} w_n \to u^+ \text{ in } L^2(\Omega). \\ w_n \leq h_n \end{cases}$$

Noticing that

$$\int_\Omega |\text{grad}w_n|^2 \, dx \geq \int_\Omega |\text{grad}w_n^+|^2dx \text{ and } \begin{cases} w_n^+ \to u^+ \text{ in } L^2(\Omega) \\ w_n^+ \leq h_n \quad (\text{since } h_n \geq 0) \end{cases}$$

and

$$\int_\Omega |\text{grad}v_n|^2dx \geq \int_\Omega |\text{grad}v_n^-|^2dx \text{ and } \begin{cases} v_n^- \to u^- \text{ in } L^2(\Omega) \\ v_n^- \geq g_n \quad (\text{since } g_n \leq 0), \end{cases}$$

we can assume without restriction that w_n is positive and v_n negative.

Then, take

$$u_n = (w_n \wedge u^+) + (v_n \vee u^-).$$

Clearly

$$u_n \to u^+ + u^- = u \text{ in } L^2(\Omega) \text{ and } g_n \leq u_n \leq h_n.$$

$w_n \wedge u^+$ and $v_n \vee u^-$ having disjoint supports,

$$\begin{aligned}\int_\Omega |\text{grad}u_n|^2dx &= \int_\Omega |\text{grad}(w_n \wedge u^+)|^2dx + \int_\Omega |\text{grad}(v_n \vee u^-)|^2dx \\ &= \int_\Omega |\text{grad}w_n|^2dx + \int_\Omega |\text{grad}u^+|^2dx - \int_\Omega |\text{grad}(w_n \vee u^+)|^2dx \\ &+ \int_\Omega |\text{grad}v_n|^2dx + \int_\Omega |\text{grad}u^-|^2dx - \int_\Omega |\text{grad}(v_n \wedge u^-)|^2dx.\end{aligned}$$

It follows from definition of w_n and v_n and lower semicontinuity of the energy that

$$\begin{aligned}\limsup_n \int_\Omega |\text{grad}u_n|^2dx \leq &\int_\Omega |\text{grad}u^+|^2dx + H(u^+) + \int_\Omega |\text{grad}u^-|^2dx + G(u^-) \\ &+ \int_\Omega |\text{grad}u^+|^2dx + \int_\Omega |\text{grad}u^-|^2dx \\ &- \int_\Omega |\text{grad}u^+|^2dx - \int_\Omega |\text{grad}u^-|^2dx\end{aligned}$$

$$\leq \int_\Omega |\text{grad}u|^2 dx + H(u^+) + G(u^-). \quad \square$$

From Theorem 2.30 and the above argument, we derive the following compactness theorem for bilateral obstacle constraints:

<u>THEOREM 2.37</u> Let $\{g_n; n \in \mathbb{N}\}$ and $\{h_n; n \in \mathbb{N}\}$ be two sequences of functions from Ω into $\bar{\mathbb{R}}$ such that there exists a sequence $\{V_n; n \in \mathbb{N}\}$ strongly convergent in $H^1(\Omega)$ satisfying, for every $n \in \mathbb{N}$,

$$g_n \leq V_n \leq h_n \text{ on } \Omega. \tag{2.112}$$

Then, there exists a subsequence $\{n_k; k \in \mathbb{N}\}$, and a rich subset $\mathbb{R}$ of $\mathbb{B}$ such that, for every $u \in H^1(\Omega)$ and $B \in \mathbb{R}$,

$$L^2(\Omega)\text{-}\lim_e\left[\int_\Omega |\text{grad}u|^2 dx + I_{\{g_{n_k} \leq v \leq h_{n_k} \text{ on } B\}}(u)\right]$$

$$= \int_\Omega |\text{grad}u|^2 dx + \int_B j_1(x,\tilde{u}(x))d\mu_1(x)$$

$$+ \int_B j_2(x,\tilde{u}(x))d\mu_2(x),$$

where μ_1 and μ_2 belong to $H^{-1}(\Omega)^+$, j_1 and j_2 are closed convex functions respectively decreasing and increasing with respect to u.

<u>Comments</u>: Indeed, assumption (2.112), which is a straight generalization of (2.108), can be dropped (cf. Dal-Maso [3]) but, in that case, there may be interaction between the upper and lower obstacles and the bilateral problem cannot be reduced to the unilateral one.

If we take

$$\begin{cases} g_n = 0 \text{ on } T_n, & -\infty \text{ elsewhere} \\ h_n = 0 \text{ on } T_n, & +\infty \text{ elsewhere} \end{cases}$$

which corresponds to imposing "$u = 0$ on T_n", the above theorem specializes into

<u>THEOREM 2.38</u> Let $\{T_n; n \in \mathbb{N}\}$ be a sequence of "holes" in Ω. Then, there

exists a subsequence $\{n_k;\ k \in \mathbb{N}\}$ and a rich family of Borel sets $\mathcal{R}$ such that, for every $u \in H^1(\Omega)$ and $B \in \mathcal{R}$,

$$L^2(\Omega)\text{-}\mathrm{lm}_e\Big[\int_\Omega |\mathrm{grad}u|^2 dx + I_{\{v=0 \text{ on } T_n \cap B\}}(u)\Big]$$

$$= \int_\Omega |\mathrm{grad}u|^2 + \int_B a(x)\tilde{u}(x)^2 d\mu(x),$$

where $\mu \in H^{-1}(\Omega)^+$ and $a:\Omega \to \bar{\mathbb{R}}^+$ is μ-measurable.

Conversely, any such functional can be obtained as an upper limit by taking an appropriate sequence $\{T_n;\ n \in \mathbb{N}\}$.

Comments: The above theorem gives the description of the exact closure of the class of constraints $\{u = 0 \text{ on } T\}$, $T \subset \Omega$: from the quadratic property of the functionals

$$u \to \int_\Omega |\mathrm{grad}u|^2 dx + I_{\{v=0 \text{ on } T_n\}}(u),$$

which is preserved by epi-limit, it follows that the limit integrand j is quadratic too! Since we know we can obtain any measure μ (cf. Theorem 2.36) it follows that the more general form of such a limit constraint is

$$\int_\Omega a(x)\tilde{u}(x)^2 d\mu(x).$$

We notice that by taking $a = +\infty$ on $\mathrm{supp}(\mu) = T$ and zero elsewhere, we obtain the indicator functional of the convex constraint set $K_T = \{u \in H^1 / u = 0 \text{ on } T\}$.

Let us end this section by describing a striking application of Theorem 2.38 in optimal control design.

D. Application to a problem in optimal control design

In some optimal design problems it may happen that there does not exist an optimal domain. This difficulty is often overcome by requiring the domains to satisfy some additional properties (connectedness, boundedness of the curvature,...), the problem in this restricted class having now an optimal solution which is a domain.

In some typical situations, these restrictions are not natural: in optimization of structures one may easily imagine that, for example, the resistance of a material increases as its components are more and more intimately mixed. The solution has to be sought in a larger class called, depending on the situation, a "fuzzy", "relaxed" or "homogenized" domain.

Let us describe a model situation derived from the "condensator problem" where Theorem 2.38 provides the right compactness result to prove existence of an optimal "relaxed" control.

Let Ω be a bounded open set in $\mathbb{R}^N$, $f \in L^2(\Omega)$, and U a positive constant. For any subset D of Ω, D is the control, and the corresponding state u_D is the solution of the following state equation:

$$\begin{cases} -\Delta u = f \text{ on } \Omega \setminus D \\ u = U \text{ on } D \\ u = 0 \text{ on } \partial\Omega . \end{cases} \tag{2.113}$$

The corresponding cost J(D) is equal to

$$J(D) = \int_{\Omega \setminus D} |\text{grad} u_D|^2 dx + N \int_\Omega |u_D(x) - z_d(x)|^2 dx$$

where $N > 0$ and $z_d \in L^2(\Omega)$ are given.

The optimal control design problem is

$$\inf_{D \in \mathcal{D}} J(D), \tag{2.114}$$

where $\mathcal{D}$ is the admissible class of control domain D. For simplicity, let us assume that $\mathcal{D}$ is the class of all compact subsets of Ω.

Noticing that u_D minimizes over $H_o^1(\Omega)$ the functional

$$F(u,D) = \int_\Omega |\text{grad} u|^2 dx + I_{\{u=U \text{ on } D\}} - \int_\Omega fu \, dx,$$

by considering a minimizing sequence of domain $\{D_n; n \in \mathbb{N}\}$ for (2.114) and applying the compactness Theorem 2.38, we easily obtain the following result.

<u>PROPOSITION 2.39</u> The following equality holds

$$\inf_{D \in \mathcal{D}} J(D) = \min_{\mu \in \tilde{\mathcal{D}}} \tilde{J}(\mu), \tag{2.115}$$

where $\tilde{\mathcal{D}} = \{\mu = a\nu/\nu \in H^{-1}(\Omega)^+, \ a:\Omega \to \bar{\mathbb{R}}^+ \text{ is } \nu\text{-measurable}\}$

$$\tilde{J}(\mu) = \int_\Omega |\mathrm{grad}\, u_\mu|^2 dx + \int_\Omega (\tilde{u}_\mu - U)^2 d\mu + N \int_\Omega |\tilde{u}_\mu - z_d|^2 dx$$

and u_μ is the solution of the minimization problem

$$\min_{v \in H_0^1(\Omega)} \{ \int_\Omega |\mathrm{grad}\, v|^2 dx + \int_\Omega (\tilde{v} - U)^2 d\mu - \int_\Omega f v \, dx \}.$$

Comments: The solution has to be sought in the enlarged class $\tilde{\mathcal{D}}$, the optimal control being now a measure μ. When taking $\mu = a\, dx$, where $a = +\infty$ on D and zero elsewhere, we find $u_\mu = u_D$, i.e. the class $\tilde{\mathcal{D}}$ of "relaxed controls" does contain the initial class of domains $\mathcal{D}$.

Let us notice that in order to write the infimum over $\mathcal{D}$ as a minimum over a larger class $\tilde{\mathcal{D}}$, one has to relax *both* the state equation and the cost functional.

The above result is interesting too from a numerical point of view since it allows us to replace the initial control problem where the control is a domain by a simpler one: it is sufficient, indeed, in order to realize the same infimum (2.114), to take the control in any "dense" subset of $\tilde{\mathcal{D}}$.

Typically, by considering density measures $\mu = h\, dx$ with $h \in L^\infty(\Omega)$, we obtain

$$\inf_{D \in \mathcal{D}} J(D) = \min_{\mu \in \tilde{\mathcal{D}}} \tilde{J}(\mu) = \inf_{h \in L^\infty(\Omega)^+} \tilde{J}(h),$$

where given $h \in L^\infty(\Omega)^+$ the new control, the corresponding state u_h is a solution of

$$\begin{cases} -\Delta u + h(u-U) = f \text{ on } \Omega \\ u = 0 \text{ on } \partial\Omega . \end{cases} \quad (2.116)$$

The cost $\tilde{J}(h)$ is given by

$$\tilde{J}(h) = \int_\Omega |\mathrm{grad}\, u_h|^2 dx + \int_\Omega |u_h - U|^2 h \, dx + N \int_\Omega |u_h - z_d|^2 dx. \quad (2.117)$$

The new state equation is bilinear with respect to (h,u). This formulation is close to the penalized approach of Gonzalez de la Paz [1].

When considering a minimizing sequence $(h_n;\, n \in \mathbb{N})$ for the relaxed control problem $\inf_{h \in L^\infty(\Omega)} \tilde{J}(h)$, the set where h_n tends to $+\infty$ yields the optimal domain, while on the complementary set the optimal domain takes a fuzzy form. Let us end these considerations by noticing the following striking result: When taking $f = 0$ and

$$\mathcal{D}_V = \{D \subset \Omega \text{ such that meas } (D) = V\},$$

where $V > 0$ is a positive constant (this is exactly the "condensator problem") there exists indeed an optimal domain D which is a solution of

$$\min_{D \in \mathcal{D}_V} J(D). \quad \square$$

2.5 EPI-CONVERGENCE OF MONOTONE SEQUENCES OF FUNCTIONS

We have already pointed out (cf. Example 1.1.1) that epi-convergence is a concept different from pointwise convergence. Nevertheless, there is an important case where the two notions coincide: it is the case of monotonically (increasing or decreasing) sequences of functions. This is why monotone approximation methods such as penalization, viscosity, barrier methods, etc. are successful. The epi-convergence approach allows us to give, in a unified way, a blitz proof of the convergence of such monotone approximation schemes.

2.5.1 Epi-convergence of increasing sequences of functions. Penalization Method

Let us first state the following result:

THEOREM 2.40 Let X be an abstract space and $(F^n)_{n\in\mathbb{N}}$ a monotone increasing sequence of functions from X into $\bar{\mathbb{R}}$. For every topology τ on X, the sequence $(F^n)_{n\in\mathbb{N}}$ is τ-epi convergent and its limit is equal to

$$\tau\text{-lm}_e F^n = \sup_{n\in\mathbb{N}} (c\ell_\tau F^n). \tag{2.118}$$

Proof of Theorem 2.40 Since the sequence $(F^n)_{n\in\mathbb{N}}$ is increasing, so, for every V belonging to $N_\tau(x)$, is the sequence $(\inf_{u\in V} F^n(u))_{n\in\mathbb{N}}$. Consequently,

$$\liminf_n \inf_{u\in V} F^n(u) = \limsup_n \inf_{u\in V} F^n(u).$$

This being true for every $V \in N_\tau(x)$, taking the supremum with respect to V, we obtain

$$\tau\text{-}li_e F^n = \tau\text{-}ls_e F^n,$$

which is the definition of τ-epi convergence of the sequence $(F^n)_{n\in\mathbb{N}}$. Let us now determine its τ-epi-limit and prove the equality

$$\tau\text{-}lm_e F^n = \sup_{n\in\mathbb{N}} c\ell_\tau F^n.$$

By definition of the τ-epi limits and by the increasing property of the sequence $(F^n)_{n\in\mathbb{N}}$, for every $x \in X$,

$$\tau\text{-}lm_e F^n(x) = \sup_{V\in N_\tau(x)} \lim_{n\to+\infty} \inf_{u\in V} F^n(u)$$

$$= \sup_{V\in N_\tau(x)} \sup_{n\in\mathbb{N}} \inf_{u\in V} F^n(u).$$

Exchanging the two suprema (which is always permitted) we obtain

$$\tau\text{-}lm_e F^n(x) = \sup_{n\in\mathbb{N}} \sup_{V\in N_\tau(x)} \inf_{u\in V} F^n(u). \tag{2.119}$$

We now recall (cf. Proposition 2.2) the equality

$$c\ell_\tau F^n(x) = \sup_{V\in N_\tau(x)} \inf_{u\in V} F^n(u),$$

which combined with (2.119) yields:

$$\forall x \in X \quad \tau\text{-}lm_e F^n(x) = \sup_{n\in\mathbb{N}} c\ell_\tau F^n(x). \quad \square$$

We notice that this formula is compatible with the τ-lower semicontinuity of the τ-epi limit: a supremum of τ-lower semicontinuous functions is still lower semicontinuous. $\square$

Let us now consider the geometric interpretation of the preceding result. For an increasing sequence of functions, the corresponding sequence of subsets of $X \times \mathbb{R}$, formed by the epigraphs is decreasing; since epi-convergence is equivalent to Kuratowski convergence of the epigraphs (for the product topology of $X \times \mathbb{R}$), the following result gives another proof and the geometric interpretation of Theorem 2.40.

PROPOSITION 2.41 Let $(C^n)_{n\in\mathbb{N}}$ be a decreasing sequence of sets in a topological space (X,τ). Then, it converges in the Kuratowski sense, to

$$\tau\text{-Lm}C^n = \bigcap_{n\in\mathbb{N}} c\ell_\tau(C^n). \tag{2.120}$$

Proof of Proposition 2.41 From the decreasing property of the sequence $(C^n)_{n\in\mathbb{N}}$, for every $n \in \mathbb{N}$

$$\bigcup_{k\geq n} C^k = C^n$$

and

$$\bigcap_{k\geq n} C^k = \bigcap_{k\in\mathbb{N}} C^k.$$

In order to prove this last equality, just notice that the right-hand side is included in the left-hand side and that for every $n \in \mathbb{N}$ and every $k_o \in \mathbb{N}$,

$$C^{k_o} \supset C^{\sup(k_o,n)} \supset \bigcap_{k\geq n} C^k.$$

From definitions (1.110) and (1.111) of the limit sets $\tau\text{-Li}C^n$ and $\tau\text{-Ls}C^n$ and the above equalities, it follows

$$\tau\text{-Li}C^n = c\ell_\tau \bigcup_{n\in\mathbb{N}} \bigcap_{k\geq n} c\ell_\tau C^k = c\ell_\tau \bigcap_{n\in\mathbb{N}} c\ell_\tau C^n = \bigcap_{n\in\mathbb{N}} c\ell_\tau C^n$$

$$\tau\text{-Ls}C^n = \bigcap_{n\in\mathbb{N}} c\ell_\tau\Big(\bigcup_{k\geq n} C^k\Big) = \bigcap_{n\in\mathbb{N}} c\ell_\tau C^n.$$

Hence $\tau\text{-Li}C^n = \tau\text{-Ls}C^n = \bigcap_{n\in\mathbb{N}} c\ell_\tau C^n$, which is the definition of set-convergence of the sequence $(C^n)_{n\in\mathbb{N}}$ and gives formula (2.120). □

From the monotone epi-convergence Theorem 2.40 and variational properties of epi-convergence, we derive the following result, well adapted to a large number of applications.

PROPOSITION 2.42 Let $(F^n)_{n\in\mathbb{N}}$ be an increasing sequence of functions from a space X into $\bar{\mathbb{R}}$. Let $(x_n)_{n\in\mathbb{N}}$ be a sequence of ε_n-minimizers of the problems $\inf_{u\in X} F^n(u)$, which is assumed to be τ-relatively compact in X, for a topology τ on X. Then,

$$\inf_{u\in X} F^n(u) \quad \uparrow_{(n \uparrow +\infty)} \quad \min_{u\in X} (\sup_{n\in\mathbb{N}} c\ell_\tau F^n)(u); \tag{2.121}$$

Moreover, every τ-cluster point of such a minimizing sequence is actually a minimizer of the limit problem.

Proof of Proposition 2.42 and comments Since the sequence $(F^n)_{n\in\mathbb{N}}$ is increasing, so is the sequence $(\inf_{u\in X} F^n(u))_{n\in\mathbb{N}}$. Thus, $\lim_{n\to+\infty} (\inf_{u\in X} F^n(u))$ exists. In general,

$$\lim_{n\to+\infty} (\inf_{u\in X} F^n(u)) = \sup_{n\in N} \inf_{u\in X} F^n(u) \leq \inf_{u\in X} \sup_{n\in\mathbb{N}} F^n(u).$$

The problem of determining under what conditions the equality holds is a minimax problem and, is open in its full generality. We are going to give an answer under the hypotheses of Proposition 2.42.

Since the sequence $(F^n)_{n\in\mathbb{N}}$ is increasing, from Theorem 2.40, for every topology τ on X,

$$\tau\text{-}\mathrm{lm}_e F^n = \sup_{n\in\mathbb{N}} (c\ell_\tau F^n).$$

By assumption, the set $K = \{x_n\}$ is τ-relatively compact and satisfies

$$\forall n \in \mathbb{N} \quad K \subset \varepsilon_n\text{-argmin } F^n.$$

From the variational Theorem 2.11,

$$\lim_{n\to+\infty} \inf_{u\in X} F^n(u) = \min_{u\in X} (\sup_{n\in\mathbb{N}} c\ell_\tau F^n)(u).$$

and every τ-cluster point of a minimizing sequence actually minimizes the function $\sup_{n\in\mathbb{N}} (c\ell_\tau F^n)$.

In other words, if $(F^n)_{n\in\mathbb{N}}$ is an increasing sequence of τ-lower semi-continuous functions on (X,τ) such that for some $n_o \in \mathbb{N}$, F^{n_o} is τ-inf compact, then

$$\sup_{n\in\mathbb{N}} \inf_{u\in X} F^n(u) = \inf_{u\in X} \sup_{n\in\mathbb{N}} F^n(u).$$

We stress the fact that the topology τ is not given a priori, but comes from the analysis of the coerciveness and inf-compactness properties of the sequence $(F^n)_{n\in\mathbb{N}}$. The conclusion of Proposition 2.42 holds if, for example, for some $x_o \in X$ $\sup_{n\in\mathbb{N}} F^n(x_o) < +\infty$, and, if the sequence $(F^n)_{n\in\mathbb{N}}$ is minorized (at least for large n) by a τ-inf compact function G. □

An important application of the monotone epi-convergence Theorem 2.40 is the *penalization method*. Let us illustrate it on some examples.

EXAMPLE 2.43 Consider the non-linear optimization problem

$$\underset{x\in S}{\text{minimize}}\ g_o(x)$$

where $S = \{x \in \mathbb{R}^p / g_i(x) \leq 0,\ i = 1,\dots,m / g_i(x) = 0,\ i = m+1,\dots,\bar{m}\}$ is the set of the constraints, also called the set of feasible solutions. For $i = 0,1,\dots,\bar{m}$, the g_i are continuous real-valued functions defined on $\mathbb{R}^p$.

Let us introduce a penalty function $p:]0,+\infty[\times \mathbb{R}^p \to [0,+\infty[$ associated with the (closed) constraint S (cf. Attouch & Wets [1]):

$$\begin{cases} p \text{ is continuous, non-negative and finite on }]0,+\infty[\times \mathbb{R}^p. \\ \forall\theta \in]0,+\infty[,\quad \forall x \in S \quad p(\theta,x) = 0. \\ \forall x \in \mathbb{R}^p\setminus S \quad \theta \to p(\theta,x) \text{ increases to } +\infty. \end{cases}$$

All common (exterior) penalty functions satisfy these conditions; for example

$$p(\theta,x) = \theta \sum_{i=1}^{m} [\max(0,g_i(x))]^\alpha + \theta \sum_{i=m+1}^{\bar{m}} |g_i(x)|^\beta \quad \alpha \geq 1,\ \beta \geq 1.$$

Then, taking $(\theta_\nu)_{\nu\in\mathbb{N}}$ a sequence strictly increasing with ν to $+\infty$, and defining

$$f_\nu(x) = g_o(x) + p(\theta_\nu,x),$$

the sequence of continuous functions $(f_\nu)_{\nu\in\mathbb{N}}$ increases and hence epi-converges (for the usual topology) to f defined by

$$f(x) = \begin{cases} g_o(x) & \text{if } x \in S \\ +\infty & \text{elsewhere.} \end{cases}$$

From the variational Theorem 2.11 it follows that any cluster point of a minimizing sequence $(x_\nu)_{\nu\in\mathbb{N}}, \forall\nu \in \mathbb{N}\ x_\nu \in \varepsilon_\nu\text{-Argmin } f_\nu\ (\varepsilon_\nu \underset{\nu}{\to} 0)$, actually minimizes f, i.e. solves the initial problem.

The advantage of such an exterior penalization method is that the approximation problems "$\inf_{x\in\mathbb{R}^p} f_\nu(x)$" are minimization on the whole space $\mathbb{R}^p$ of continuous functions - there are no more constraints on x; on the other hand, the disadvantage of such a method is that the solutions of the approximating problems $\{x_\nu \in \text{Argmin } f_\nu\}$ are no longer in general in the set S of feasible solutions.

EXAMPLE 2.44 Let us consider the variational problem

$$\underset{\{u \geq g \text{ on } \Omega\}}{\text{Min}} \frac{1}{2}\int_\Omega |\text{grad}u|^2 dx - \int_\Omega fu\, dx$$

where Ω is a bounded open subset in $\mathbb{R}^N$, f belongs to $L^2(\Omega)$ and $g \in L^2(\Omega)$ is an "obstacle"; the associated set S of constraints is given by

$$S = \{u \in H_0^1(\Omega)/u(x) \geq g(x) \text{ a.e. on } \Omega\} .$$

We assume that S is non-void, then S is a closed convex non-empty set in $H_0^1(\Omega)$. From a classical argument, the above minimization problem admits a unique solution; a natural way to approach it by a penalization method is to consider for every $\varepsilon > 0$

$$\underset{u\in H_0^1(\Omega)}{\text{Min}} \{F^\varepsilon(u) - \int_\Omega fu\, dx\},$$

where

$$F^\varepsilon(u) = \frac{1}{2}\int_\Omega |\text{grad}u|^2 dx + \frac{1}{\varepsilon}\int_\Omega (u-g)^{-2}(x)dx.$$

The functionals $(F^\varepsilon)_{\varepsilon>0}$ are convex lower semicontinuous (and hence weakly lower semicontinuous) on $H_0^1(\Omega)$.

$\forall\varepsilon > 0$ $F^\varepsilon \geq G$ where $G(u) = \frac{1}{2}\int_\Omega |\text{grad}u|^2 dx$ is weakly inf-compact on $H_0^1(\Omega)$.

The sequence $(F^\varepsilon)_{\varepsilon>0}$ increases to F, defined by

$$F(u) = \begin{cases} \frac{1}{2}\int_\Omega |\text{grad}u|^2 dx & \text{if } u \geq g \text{ a.e., } u \in H_0^1(\Omega) \\ +\infty & \text{elsewhere.} \end{cases}$$

The hypotheses of Proposition 2.42 are satisfied. Therefore, the solutions $(u_\varepsilon)_{\varepsilon>0}$ of the penalized problems converge (as $\varepsilon \to 0$) in $w\text{-}H_0^1(\Omega)$ (and in fact in s-$H_0^1(\Omega)$ to u, the solution of the initial problem.

In this situation, consideration of the approximating problems is particularly interesting in order to obtain regularity results on the solution of the initial problem (cf. Brezis & Kinderlehrer [1]: One establishes estimations on the approximating problems, independently of $\varepsilon > 0$, then goes to the limit as $\varepsilon \to 0$.

EXAMPLE 2.45 Let us consider the non-standard optimal control problem: (cf. J.L. Lions [1], Haraux & Murat [1] [2]). The state equation is given by

$$(\mathcal{E}) \quad \left| \begin{array}{l} -\Delta z - z^3 = v \text{ in } \Omega \\ z = 0 \text{ on } \partial\Omega \\ z \in L^6(\Omega),\ v \in L^2(\Omega) \end{array} \right.$$

The control is $v \in L^2(\Omega)$ and the state is $z \in L^6(\Omega)$. The major difficulty of this situation is that, for v given in $L^2(\Omega)$, there exists in general an infinity of solutions z of $(\mathcal{E})$; so we introduce $\mathcal{E}$, the graph of $(\mathcal{E})$,

$$\mathcal{E} = \{(v,z) \in L^2(\Omega) \times L^6(\Omega)/(v,z) \in (\mathcal{E})\}.$$

Let U_{ad} be a closed convex subset of $L^2(\Omega)$ such that $U_{ad} \times (L^6(\Omega)) \cap \mathcal{E} \neq \emptyset$. The cost function J is

$$J(v,z) = 1/6\ \|z-z_d\|^6_{L^6(\Omega)} + N/2\ \|v\|^2_{L^2(\Omega)}.$$

The optimal control problem

$$\operatorname*{Inf}_{\substack{v \in U_{ad} \\ (v,z) \in \mathcal{E}}} J(v,z)$$

is written as an optimization problem on the pairs (v,z), v and z being linked by state equation $(\mathcal{E})$; in order to approach it by a penalization method of the constraint $(v,z) \in \mathcal{E}$, it is natural to introduce the penalty function:

$$p(\varepsilon,v,z) = \frac{1}{2\varepsilon}\ \|\Delta z + z^3 + v\|^2_{L^2(\Omega)};$$

the problems

$$\operatorname{Inf}_{\begin{cases} v\in U_{ad} \\ z\in L^2(\Omega)\end{cases}} J(v,z) + 1/2\varepsilon \ \|\Delta z + z^3 + v\|^2_{L^2(\Omega)}$$

clearly satisfy the conditions of Proposition 2.42: the functions

$$F^{\varepsilon}(v,z) = 1/6 \quad \|z-z_d\|^6_{L^6(\Omega)} + N/2 \ \|v\|^2_{L^2(\Omega)} + I_{U_{ad}}(v) + 1/2\varepsilon\|\Delta z + z^3 + v\|^2_{L^2(\Omega)}$$

are lower semicontinuous on $L^2(\Omega) \times L^6(\Omega)$ for the weak topology (this follows from a compactness argument) uniformly inf-compact for this topology, and the sequence $(F^{\varepsilon})_{\varepsilon>0}$ increases to F given by

$$F(v,z) = \begin{cases} J(v,z) \text{ if } v \in U_{ad} \text{ and } (v,z) \in E \\ +\infty \quad \text{otherwise.}\end{cases}$$

The penalized problems clearly admit solutions $(v_{\varepsilon}, z_{\varepsilon})$ and from Proposition 2.42, every weak cluster point in $L^2(\Omega) \times L^6(\Omega)$ of the bounded sequence $(v_{\varepsilon}, z_{\varepsilon})_{\varepsilon>0}$ is actually a solution of the initial optimal control problem.

The interest of this method in this situation is that it is easy to write the optimality conditions for the penalized problems.

Then one can go to the limit, as $\varepsilon \to 0$, and obtain the optimality conditions for the initial problem. (This last limit process is difficult, and so far has been done only under some topological assumptions on U_{ad}). □

2.5.2 Epi-convergence of decreasing sequences of functions. Barrier and viscosity methods

The following theorem, in some sense the dual of Theorem 2.40, states the epi-convergence of every decreasing sequence of functions.

THEOREM 2.46 Let $(F^n)_{n\in\mathbb{N}}$ be a monotone decreasing sequence of functions from X, a general space, into $\bar{\mathbb{R}}$. Then, for every topology τ on X, the sequence $(F^n)_{n\in\mathbb{N}}$ is τ-epi convergent, and its limit is equal to

$$\tau\text{-}\mathrm{lm}_e F^n = c\ell_{\tau}(\inf_{n\in\mathbb{N}} F^n) = c\ell_{\tau}(\inf_{n\in\mathbb{N}} c\ell_{\tau} F^n). \tag{2.122}$$

Proof of Theorem 2.46 The same monotonicity argument as in Theorem 2.40 tells us that

$$\forall x \in X,\quad \tau\text{-}li_e F^n(x) = \tau\text{-}ls_e F^n(x) = \sup_{V \in N_\tau(x)} \lim_{n \to +\infty} \inf_{u \in V} F^n(u).$$

Thus, the sequence $(F^n)_{n \in \mathbb{N}}$ is τ-epi convergent and

$$\forall x \in X \quad \tau\text{-}lm_e F^n(x) = \sup_{V \in N_\tau(x)} \lim_{n \to +\infty} \inf_{u \in V} F^n(u). \tag{2.123}$$

Since the limit with respect to $n \in \mathbb{N}$ (by the decreasing property of the sequence $(F^n)_{n \in \mathbb{N}}$) is equal to the infimum with respect to $n \in \mathbb{N}$, (2.123) becomes

$$\forall x \in X \quad \tau\text{-}lm_e F^n(x) = \sup_{V \in N_\tau(x)} \inf_{n \in \mathbb{N}} \inf_{u \in V} F^n(u). \tag{2.124}$$

Exchanging the two inf operations in (2.124), which is always permitted,

$$\forall x \in X \quad \tau\text{-}lm_e F^n(x) = \sup_{V \in N_\tau(x)} \inf_{u \in V} (\inf_{n \in \mathbb{N}} F^n)(u).$$

From the formulation of the $c\ell_\tau$ operation given in Proposition 2.2,

$$\tau\text{-}lm_e F^n = c\ell_\tau (\inf_{n \in \mathbb{N}} F^n).$$

The last equality of (2.122) follows from the fact that the two sequences $(F^n)_{n \in \mathbb{N}}$ and $(c\ell_\tau F^n)_{n \in \mathbb{N}}$ have the same τ-epi limit. □

Let us now give the geometric interpretation (by consideration of the epigraphs $(\text{epi } F^n)_{n \in \mathbb{N}}$) of the above theorem.

<u>THEOREM 2.47</u> Let $(C^n)_{n \in \mathbb{N}}$ be an increasing sequence of subsets of a topological space (X, τ). Then, it converges in Kuratowski sense and its τ-limit set, $\tau\text{-}LmC^n$ is equal to

$$\tau\text{-}LmC^n = c\ell_\tau (\bigcup_{n \in \mathbb{N}} C^n) = c\ell_\tau \bigcup_{n \in \mathbb{N}} (c\ell_\tau C^n). \tag{2.125}$$

<u>Proof of Theorem 2.47</u> From the increasing property of the sequence $(C^n)_{n \in \mathbb{N}}$, for every $n \in \mathbb{N}$

$$\bigcap_{k \geq n} c\ell_\tau (C^k) = c\ell_\tau C^n$$

and

$$\bigcup_{k \geq n} C^k = \bigcup_{k \in \mathbb{N}} C^k.$$

From the definitions (1.110) and (1.111) of the limit-sets $\tau\text{-Li}C^n$ and $\tau\text{-Ls}C^n$ and the above equalities it follows

$$\tau\text{-Li}C^n = c\ell_\tau \bigcup_{n\in\mathbb{N}} \bigcap_{k\geq n} c\ell_\tau C^k = c\ell_\tau \bigcup_{n\in\mathbb{N}} c\ell_\tau C^n$$

and

$$\tau\text{-Ls}C^n = \bigcap_{n\in\mathbb{N}} c\ell_\tau \big(\bigcup_{k\geq n} C^k\big) = c\ell_\tau \bigcup_{n\in\mathbb{N}} C^n.$$

So $\tau\text{-Li}C^n \supset \tau\text{-Ls}C^n$, which is the definition of (Kuratowski) set-convergence of the sequence $(C^n)_{n\in\mathbb{N}}$ and

$$\tau\text{-Lm}C^n = c\ell_\tau \bigcup_{n\in\mathbb{N}} C^n = c\ell_\tau \bigcup_{n\in\mathbb{N}} c\ell_\tau C^n. \quad \square$$

Let us now describe in the following corollary how to apply Theorem 2.46 to optimization problems.

PROPOSITION 2.48 Let $(F^n)_{n\in\mathbb{N}}$ be a decreasing sequence of functions from X, an abstract space into $\bar{\mathbb{R}}$. Then,

$$\inf_{u\in X} F^n(u) \underset{(n\to+\infty)}{\downarrow} \inf_{u\in X} (\inf_{n\in\mathbb{N}} F^n)(u). \tag{2.126}$$

Moreover, if a minimizing sequence $(x_n)_{n\in\mathbb{N}}$ $(\forall n \in \mathbb{N}\ x_n \in \varepsilon_n\text{-Argmin } F^n,\ \varepsilon_n \underset{n}{\to} 0)$ is τ-relatively compact in X, for a topology τ on X, then every τ-cluster point of such a sequence minimizes $c\ell_\tau(\inf F^n)$ on X:

$$\tau\text{-Ls}(\varepsilon_n\text{-Argmin } F^n) \subset \text{Argmin } c\ell_\tau(\inf_n F^n). \tag{2.127}$$

Proof of Proposition 2.48 and comments The sequence $(F^n)_{n\in\mathbb{N}}$ being decreasing, so is the sequence $(\inf_{u\in X} F^n(u))_{n\in\mathbb{N}}$. Thus,

$$\lim_{n\to+\infty} (\inf_{u\in X} F^n(u)) = \inf_{n\in\mathbb{N}} \inf_{u\in X} F^n(u).$$

Exchanging the two infima, we obtain

$$\inf_{u\in X} F^n(u) \downarrow \inf_{u\in X} (\inf_n F^n)(u),$$

which is (2.126). Let us consider a minimizing sequence $(x_n)_{n\in\mathbb{N}}$, $x_n \in \varepsilon_n$-Argmin F^n, which is assumed τ-relatively compact in X. From Theorem 2.46, which states

$$\tau\text{-lm}_e F^n = c\ell_\tau(\inf_{n\in\mathbb{N}} F^n)$$

and the variational Theorem 2.11, every τ-cluster point x of the sequence $(x_n)_{n\in\mathbb{N}}$ actually minimizes $c\ell_\tau(\inf_{n\in\mathbb{N}} F^n)$:

$$\inf_{u\in X} F^n(u) \downarrow \min_{u\in X} \; c\ell_\tau(\inf_{n\in\mathbb{N}} F^n)(u) = c\ell_\tau(\inf_{n\in\mathbb{N}} F^n)(x).$$

As in Proposition 2.42, we stress the fact that the topology τ is not given and that, in applications, the difficulty is to determine a topology τ for which the minimizing sequences are τ-relatively compact, and then to determine the τ-closure of $\inf_{n\in\mathbb{N}} F^n$. □

Let us now give some applications of the above convergence results to the barrier method, then, to the viscosity approximation method, and finally to a "perturbation" problem in optimal control theory.

EXAMPLE 2.49 The *barrier method* (cf. Attouch & Wets [1]). Let us consider the non-linear optimization problem in $\mathbb{R}^p$

$$\begin{cases} \text{minimize } g_o(x) \text{ with constraint } x \in S, \text{ where} \\ S = \{x \in \mathbb{R}^p / g_i(x) < 0 \quad i = 1,2,\ldots,m\}. \end{cases} \tag{2.128}$$

We assume that the functions $\{g_i : \mathbb{R}^p \to \mathbb{R}; i = 0,1,2,\ldots,m\}$ are continuous.
Define

$$f(x) = \begin{cases} g_o(x) \text{ if } x \in S \\ +\infty \text{ otherwise} \end{cases}$$

and

$$f_\nu(x) = g_o(x) + q(\theta_\nu, x)$$

where the $\theta_\nu > 0$ are strictly increasing to $+\infty$ with ν and

$$q:]0,+\infty[\times \mathbb{R}^p \to]0,+\infty]$$

is continuous, finite if $x \in \text{int } S$, $+\infty$ otherwise and, if $x \in \text{int } S$, $\theta \to q(\theta,x)$ is strictly decreasing to zero.

Such a function q is called a barrier function. The most commonly used barrier functions are

$$q(\theta,x) = -\theta^{-1} \sum_{i=1}^{m} [\min(0,g_i(x))]^{-1}$$

$$q(\theta,x) = \theta^{-2} \sum_{i=1}^{m} [\min(0,g_i(x))]^{-2}$$

$$q(\theta,x) = -\theta^{-1} \sum_{i=1}^{m} \log [\min(0.5, -g_i(x))] \quad (\text{with } \log a = -\infty \text{ if } a \leq 0)$$

Let us now study the sequence $(f_\nu)_{\nu\in\mathbb{N}}$; when ν increases to $+\infty$, from properties of the barrier function q, the sequence f_ν decreases to the function f_o defined by

$$f_o(x) = \begin{cases} g_o(x) & \text{if } x \in \text{int } S \\ +\infty & \text{otherwise.} \end{cases}$$

From Theorem 2.46, the epi-limit (for the usual topology τ of $\mathbb{R}^n$) of the sequence f_ν is equal to $c\ell f_o$. Let us prove that

$$c\ell f_o = \begin{cases} g_o(x) & \text{if } x \in c\ell(\text{int } S) \\ +\infty & \text{otherwise.} \end{cases}$$

Clearly,

$$f_o \geq g_o + I_{\{c\ell(\text{int } S)\}},$$

where I_C is the indicator function of the set C. The second member of this inequality being lower semicontinuous, it follows that

$$c\ell f_o \geq g_o + I_{\{c\ell(\text{int } S)\}}.$$

In order to obtain the converse inequality, let us verify, equivalently, that

$$\forall x \in c\ell(\text{int } S) \quad g_o(x) \geq c\ell f_o(x).$$

Let $x_n \in \text{int } S$ such that $x_n \to x$; by definition of f_o, for every $n \in \mathbb{N}$,

$$f_o(x_n) = g_o(x_n).$$

Thus, g_o being continuous,

$$g_o(x) = \liminf_n f_o(x_n) \geq c\ell f_o(x).$$

So we can conclude that the sequence $(f_\nu)_{\nu \in \mathbb{N}}$ epi-converges to f if and only if $f = c\ell f_o$, i.e.

$$S = c\ell(\text{int } S). \tag{2.129}$$

When (2.129) is satisfied, since f_ν epi-converges to f, it follows from Theorem 2.11 that if for each $\nu \in \mathbb{N}$, $x_\nu \in \varepsilon_\nu$-Argmin f_ν ($\varepsilon_\nu \to 0$) and x is a cluster point of the sequence $(x_\nu)_{\nu \in \mathbb{N}}$, then x minimizes f, i.e. it solves (2.128). □

<u>EXAMPLE 2.50</u> <u>The viscosity method</u> Let us illustrate, on the classical minimal surface problem in calculus of variations, the viscosity method. Let $\Omega \subset \mathbb{R}^N$ be a bounded open set and for $g \in L^1(\partial\Omega)$ given on the boundary $\partial\Omega$ of Ω (which is assumed to be regular) let us consider the minimal surface problem

$$\inf_{\begin{cases} u \in W^{1,1}(\Omega) \\ u|_{\partial\Omega} = g \end{cases}} \int_\Omega \sqrt{1 + |Du|^2(x)}\,dx \tag{2.130}$$

(cf. Ekeland & Teman [1]).

A direct attack on this problem is quite difficult, essentially because of the lack of reflexivity of the space $W^{1,1}$ (in which the minimizing sequence are bounded). So, the idea is to approach (2.130) by

$$\inf_{\substack{u \in H^1(\Omega) \\ u|_{\partial\Omega} = g}} \int_\Omega \sqrt{1 + |Du|^2(x)}\,dx + \frac{\varepsilon}{2} \cdot \int_\Omega (u^2 + |Du|^2)\,dx. \tag{2.131}$$

For every $\varepsilon > 0$, problem (2.131) has a unique solution $u_\varepsilon \in H^1(\Omega) = W^{1,2}(\Omega)$ (by reflexivity of the space $H^1(\Omega)$). By standard arguments, we can obtain uniform estimates on the approximating solutions $(u_\varepsilon)_{\varepsilon>0}$. The simplest one is that the sequence $(u_\varepsilon)_{\varepsilon>0}$ is bounded in $W^{1,1}(\Omega)$ and hence relatively compact in $L^1(\Omega)$. Then we notice that the sequence of functionals $(F^\varepsilon)_{\varepsilon>0}$ defined on $L^1(\Omega)$ by

$$F^{\varepsilon}(u) = \begin{cases} \int_{\Omega} \sqrt{1 + |Du|^2}\, dx + \varepsilon/2 \int_{\Omega} (u^2 + |Du|^2) dx \text{ if } u \in H^1(\Omega),\ u|_{\partial\Omega} = g \\ + \infty \text{ otherwise} \end{cases}$$

decreases to

$$F(u) = \begin{cases} \int_{\Omega} \sqrt{1 + |Du|^2}\, dx \text{ if } u \in H^1(\Omega),\ u|_{\partial\Omega} = g \\ + \infty \text{ otherwise} \end{cases}$$

and that problem (2.131) can be written:

$$\inf_{u \in L^1(\Omega)} F^{\varepsilon}(u).$$

From Proposition 2.48 it follows that every $L^1(\Omega)$-cluster point of the sequence $(u_{\varepsilon})_{\varepsilon>0}$ minimizes on $L^1(\Omega)$ the functional, $L^1(\Omega)$-$c\ell F$, the $L^1(\Omega)$-lower semicontinuous regularization of F. One can prove that the $L^1(\Omega)$ closure of F is in fact equal to the $L^1(\Omega)$ closure of the initial minimal surface functional given on $W^{1,1}(\Omega)$ (and not on $H^1(\Omega)$). Finally every $L^1(\Omega)$-cluster point of the sequence $(u_{\varepsilon})_{\varepsilon>0}$ actually minimizes the "relaxed" problem

$$\min_{u \in L^1(\Omega)} [c\ell_{L^1(\Omega)} (\int_{\Omega} \sqrt{1 + |Du|^2}(x) dx + I_{\mathfrak{U}}(u))]$$

with $\mathfrak{U} = \{u \in W^{1,1}(\Omega);\ u|_{\partial\Omega} = g\}$.

One can prove (cf. Ekeland & Teman [1], De Giorgi [3], Miranda [2]) that

$$c\ell_{L^1(\Omega)} [\int_{\Omega} \sqrt{1 + |Du|^2} dx + I_{\mathfrak{U}}(u)] = \begin{cases} \int_{\Omega} \sqrt{1 + |D_a u|^2} + \int_{\Omega} |D_s u| + \int_{\partial\Omega} |u-g| d\Gamma & u \in BV(\Omega) \\ + \infty \text{ if } u \in L^1(\Omega) \setminus BV(\Omega) \end{cases}$$

where $BV(\Omega)$ denotes the space of bounded variation functions on Ω. Noticing that the solutions of the above problem are in fact in $W^{1,1}_{loc}$ (Miranda [2]) we obtain that the limit problem is

$$\min_{u \in BV(\Omega) \cap W^{1,1}_{loc}(\Omega)} [\int_{\Omega} \sqrt{1 + |Du|^2}\, dx + \int_{\partial\Omega} |u-g| d\Gamma] \tag{2.132}$$

We notice that in the "relaxed problem" (2.132), contrary to the initial problem where the constraint functional $I_{\mathcal{U}}$ takes only the values 0 and $+\infty$, the relaxed constraint functional $u \to \int_{\partial\Omega} |u-g| d\Gamma$ may take all the values between 0 and $+\infty$! (Compare with Example 1.1.2, the "cloud of ice"). □

EXAMPLE 2.51 A minimal cost control problem The following situation, studied by J.-L Lions [1], Shih Shu Chang [1], illustrates the difficulty, when applying Proposition 2.48, of obtaining, in certain cases, compactness information on the minimizing sequences. Let A be a second order elliptic operator, take $A = -\Delta$ for simplicity, on a bounded regular open set Ω in $\mathbb{R}^N$. We denote by $\delta(x-b)$ the Dirac measure at $b \in \Omega$. The state equation is

$$\begin{cases} \frac{\partial y}{\partial t} - \Delta y = v(t)\ \delta(x-b) \text{ on } \Omega \times (0,T) \\ y = 0 \text{ on } \Sigma = \partial\Omega \times (0,T) \\ y(x,0) = 0 \text{ on } \Omega. \end{cases}$$

The control is v and the corresponding state is $y(v)$. The cost function $J_\varepsilon(\cdot)$ is defined for every $\varepsilon > 0$ by

$$J_\varepsilon(v) = \int_\Omega |y(x,T,v) - z_d|^2 dx + \varepsilon \int_0^T v^2(t)dt,$$

where $z_d \in L^2(\Omega)$.

The optimal control problem is

$$\operatorname{Inf}_{v \in U_{ad}} J_\varepsilon(v) \tag{2.133}$$

where U_{ad} is the set of admissible constraints.

The minimal cost control problem is the study of the limit control problem when ε goes to zero, that is, when the cost of the control goes to zero. Introducing L, the linear operator $v \to L(v) = y(T,v)$, the difficulty is that in the present situation, L is not continuous and everywhere defined from $H = L^2(0,T)$ into $\mathcal{H} = L^2(\Omega)$; its domain is

$$U = \{v \in L^2(0,T);\ \ y(T,v) \in L^2(\Omega)\}.$$

With this notation, the problem (2.133) can be written

$$\text{Inf}_{v \in U_{ad}} \{ \|L(v) - z_d\|^2_{\mathcal{H}} + \varepsilon \|v\|^2_H \}.$$

Since L has a closed graph in $H \times \mathcal{H}$, it follows that if U_{ad} is a closed convex subset of U (equipped with the graph norm of L) then, for every $\varepsilon > 0$, the minimization problem (2.133) has a unique solution u_ε. Let us establish estimations on $\{u_\varepsilon; \varepsilon \to 0\}$; first

$$\sup_{\varepsilon>0} \|Lu_\varepsilon\|_{\mathcal{H}} < +\infty. \tag{2.134}$$

In order to obtain a bound on $\{u_\varepsilon; \varepsilon \to 0\}$ in H we have to assume that

$$\inf_{v \in U_{ad}} \int_\Omega |y(x,T,v) - z_d|^2 dx \tag{2.135}$$

has at least a solution, that is to say the set S of solutions of (2.135) is non-void. Then, denoting $v_0 \in S$, by definition of u_ε,

$$\|Lu_\varepsilon - z_d\|^2_{\mathcal{H}} + \varepsilon \|u_\varepsilon\|^2_H \leq \|Lv_0 - z_d\|^2_{\mathcal{H}} + \varepsilon \|v_0\|^2_H$$

and, by definition of v_0,

$$\leq \|Lu_\varepsilon - z_d\|^2_{\mathcal{H}} + \varepsilon \|v_0\|^2_H.$$

Therefore,

$$\forall v_0 \in S \quad \|u_\varepsilon\|_H \leq \|v_0\|_H. \tag{2.136}$$

From (2.134) and (2.136) the sequence $\{u_\varepsilon; \varepsilon \to 0\}$ remains bounded in U (equipped with the norm of the graph) which is a Hilbert space. Hence, it is weakly relatively compact in U. We can now apply Proposition 2.48 :

Every cluster point (in weak U) of the sequence $\{u_\varepsilon; \varepsilon \to 0\}$ actually minimizes on U_{ad} the functional

$$c\ell_{w-U} \|L(\cdot) - z_d\|^2 = \|L(\cdot) - z_d\|^2$$

(the function $v \to \|L(v) - z_d\|^2$ is convex continuous on U and hence weakly lower semicontinuous).

Moreover, from (2.136) it follows that if $\bar{u}$ is a limit point of the sequence $(u_\varepsilon)_{\varepsilon>0}$ it satisfies

$$\|\bar{u}\|_H < \|v_o\|_H \quad \text{for every } v_o \text{ belonging to S.}$$

Since S is closed convex in H, this implies the weak (and in fact the strong) convergence of the whole sequence $(u_\varepsilon)_{\varepsilon>0}$ to $\bar{u}$, with

$$\|\bar{u}\|_H = \min_{v\in S} \|v\|_H . \tag{2.137}$$

On the other hand,

$$\text{Inf}_{v\in U_{ad}} J_\varepsilon(v) \downarrow \text{Inf}_{v\in U_{ad}} \int_\Omega |y(x,T,v)-z_d|^2 dx.$$

In other words, the approximating solutions converge to the solution of the minimal $L^2(0,T)$-norm of this last minimization problem. □

2.6 COMPARISON OF EPI-LIMITS FOR DIFFERENT TOPOLOGIES

2.6.1 Epi-convergence and pointwise convergence

We have already noticed that, in general τ-epi-convergence and pointwise convergence are different notions. In the preceding section we pointed out an important case where these two notions coincide - the case of monotone convergence. It is quite natural to ask the following question: What are the relations between epi and pointwise convergence? Let us first notice that pointwise convergence is a τ-epi-convergence, by taking precisely for τ the discrete topology (that is the topology for which all subsets are open). Let us denote by τ_d the discrete topology on space X. Then, for any sequence $\{F^n; n = 1,2,...\}$ of functions from X into $\bar{\mathbb{R}}$,

$$\tau_d\text{-li}_e F^n(x) = \sup_{V\in N_d(x)} \liminf_n \inf_{u\in V} F^n(u)$$

$$= \liminf_n F^n(x)$$

(since the supremum is realized by taking $V = \{x\}$). Similarly,

$$\tau_d\text{-ls}_e F^n(x) = \limsup_n F^n(x).$$

Thus,

$$\tau_d\text{-li}_e F^n(x) = \tau_d - \text{ls}_e F^n(x)$$

holds if and only if

$$\liminf_n F^n(x) = \limsup_n F^n(x).$$

In other words,

PROPOSITION 2.52 Epi-convergence for the discrete topology is equivalent to pointwise convergence.

Thus, the initial problem turns to be a special case of the more general one which consists of comparing the τ-limits for distinct topologies. We have already stressed the importance of this problem when studying perturbations of epi-convergent sequences (Section 2.4.2). It also occurs in the study of the Mosco convergence (see the next chapter) where the two topologies are respectively equal to the strong and the weak topology of a reflexive Banach space.

A first and quite elementary answer to the above question is given by the following proposition. We recall that a topology τ_1 is said to be finer than a topology τ_2 if any τ_2-open set is also a τ_1 open set, i.e. $\mathcal{O}_{\tau_1} \supset \mathcal{O}_{\tau_2}$. (We then write $\tau_1 > \tau_2$).

PROPOSITION 2.53 Let τ_1 and τ_2 be two topologies on a space X, and let us assume that τ_2 is finer than τ_1, $\tau_1 < \tau_2$. For any sequence of functions $(F^n)_{n\in\mathbb{N}}$ from X into $\bar{\mathbb{R}}$, the following inequalities hold:

$$\begin{cases} \tau_1\text{-li}_e F^n \leqslant \tau_2\text{-li}_e F^n \\ \tau_1\text{-ls}_e F^n \leqslant \tau_2\text{-ls}_e F^n. \end{cases} \tag{2.138}$$

As a consequence, if $\tau_1\text{-lm}_e F^n$ and $\tau_2\text{-lm}_e F^n$ exist,

$$\tau_1\text{-lm}_e F^n \leqslant \tau_2\text{-lm}_e F^n. \tag{2.139}$$

Proof of Proposition 2.53 Let us denote, for every $x \in X$, by $\mathcal{O}_i(x) = \mathcal{O}_{\tau_i}(x)$ the τ_i-open set containing x (i = 1,2). By assumption, $\mathcal{O}_1(x) \subset \mathcal{O}_2(x)$. Thus,

$$\sup_{V\in\mathcal{O}_1(x)} \liminf_n \inf_{u\in V} F^n(u) \leqslant \sup_{V\in\mathcal{O}_2(x)} \liminf_n \inf_{u\in V} F^n(u);$$

that is,

$$\tau_1\text{-}li_e F^n(x) \leq \tau_2\text{-}li_e F^n(x).$$

Similarly,

$$\tau_1\text{-}ls_e F^n(x) \leq \tau_2\text{-}ls_e F^n(x).$$

Another way of deriving these inequalities (in the metrizable case) is to notice that if $\tau_1 < \tau_2$ there are more sequences converging to x for the topology τ_1 than for the topology τ_2. Therefore

$$\inf_{\{x_n \xrightarrow{\tau_1} x\}} \liminf_n F^n(x_n) \leq \inf_{\{x_n \xrightarrow{\tau_2} x\}} \liminf_n F^n(x_n). \quad \square$$

Since the discrete topology is the finest topology that one can define on a space X, as a consequence of Propositions 2.52 and 2.53 we obtain

COROLLARY 2.54 Let $(F^n)_{n\in\mathbb{N}}$ be a sequence of functions from a topological space (X,τ) into $\bar{\mathbb{R}}$. Then, for every $x \in X$

$$\begin{cases} \tau\text{-}li_e F^n(x) \leq \liminf_n F^n(x) \\ \tau\text{-}ls_e F^n(x) \leq \limsup_n F^n(x) \end{cases} \tag{2.140}$$

Thus, if $\tau\text{-}lm_e F^n$ and $\lim_n F^n$ both exist

$$\tau\text{-}lm_e F^n \leq \lim_n F^n. \tag{2.141}$$

The above results ask for a few comments:

REMARKS 2.55

(a) In general, inequalities (2.140) are strict: we observed that fact, for example, on the homogenization formulae (example 1.1.1). Let us describe a finite dimensional quite elementary example. Take

$$\{F^n : \mathbb{R} \to \mathbb{R}: n = 1,2,...\} \tag{2.142}$$

as follows (See Fig. 2.8):

$$F^n(x) = \begin{cases} 0 \text{ for } x < 0 \text{ or } x > 2/n \\ -nx \text{ for } 0 < x < 1/n \\ nx-2 \text{ for } 1/n < x < 2/n \end{cases}$$

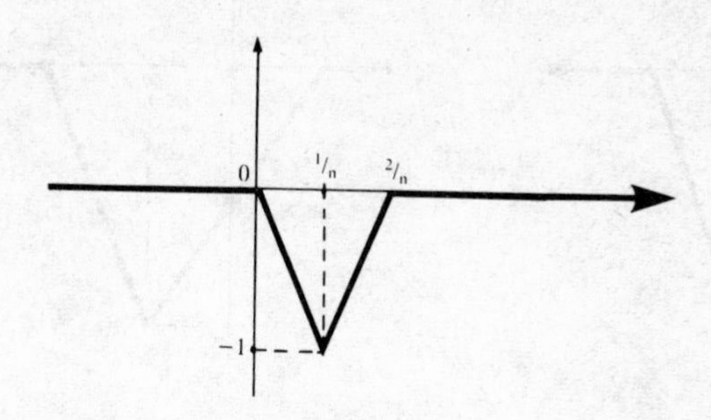

Figure 2.8

Clearly the pointwise limit of the sequence $\{F^n; n \in \mathbb{N}\}$ is the zero function. But, the epi-limit of the sequence $\{F^n; n \in \mathbb{N}\}$ ($\mathbb{R}$ being equipped with the usual topology) is the function F equal to zero on $\mathbb{R} \setminus \{0\}$ and equal to -1 at x = 0! So

$$-1 = \text{lm}_e F^n(0) < \lim F^n(0) = 0.$$

(b) Since $\tau\text{-li}_e F^n$ and $\tau\text{-ls}_e F^n$ are τ-lower semicontinuous (Theorem 2.1) inequalities (2.140) and (2.141) can be sharpened:

$$\begin{aligned} &\tau\text{-li}_e F^n \leq c\ell_\tau(\liminf F^n) \\ &\tau\text{-ls}_e F^n \leq c\ell_\tau(\limsup F^n) \qquad (2.143) \\ &\tau\text{-lm}_e F^n \leq c\ell_\tau(\lim F^n) \end{aligned}$$

But inequalities (2.143), as shown by the above example, are still strict in general.

(c) In all preceding examples, pointwise limit and epi limits both exist. Is this a general phenomenon? In other words, does the existence of one limit imply the existence of the other? The answer is no. Let us justify it on the following examples: Take $F^n: \mathbb{R} \to \mathbb{R}$ defined by Fig. 2.9.

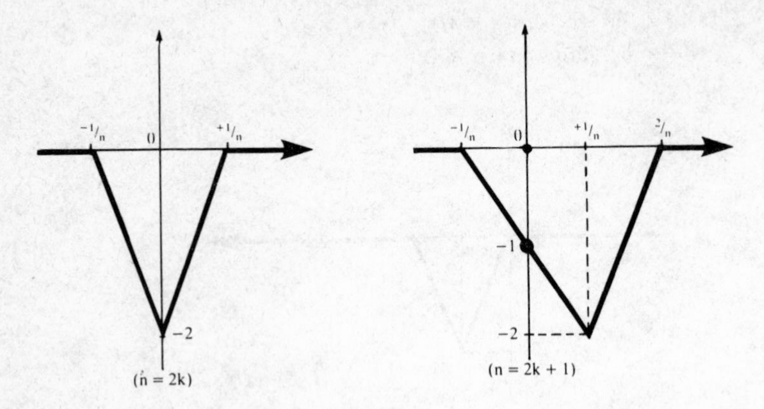

Figure 2.9

Then the sequence $\{F^n;\ n \in \mathbb{N}\}$ does not converge pointwise at zero since

$$\liminf_n F^n(0) = -2$$

$$\limsup_n F^n(0) = -1.$$

But it does epi-converge at zero: for every open set $\mathcal{O}$ containing zero and for n large enough

$$\inf_{u \in V} F^n(u) = -2.$$

Therefore,

$$\tau\text{-}li_e F^n(0) = \tau\text{-}ls_e F^n(0) = -2$$

and

$$\tau\text{-}lm_e F^n(0)$$

exists and is equal to -2. □

Let us finally give an example where the pointwise limit exists, but where the τ-epi limit does not:

The idea is to build two sequences $(F_1^n)_{n \in \mathbb{N}}$ and $(F_2^n)_{n \in \mathbb{N}}$ which converge pointwise to the same limit, but τ-epi converge to different limits. Then, take

$$F^n = \begin{cases} F_1^n \text{ if } n \text{ odd} \\ F_2^n \text{ if } n \text{ even.} \end{cases}$$

For example, take functions as in Example (2.142) with

$$F_1^n(1/n) = -1$$

and

$$F_2^n(1/n) = -2.$$

Clearly, the sequences $\{F_1^n; n \in \mathbb{N}\}$ and $\{F_2^n; n \in \mathbb{N}\}$ pointwise converge to zero, but

$$lm_e F_1^n(0) = -1 \neq lm_e F_2^n(0) = -2.$$

Thus

$$li_e F^n(0) = -2 \neq ls_e F^n(0) = -1$$

and the sequence $\{F^n; n \in \mathbb{N}\}$ is not epi-convergent at $x = 0$!

We can summarize this section by saying that, in a general setting, the only relations which tie the epi and the pointwise convergence are inequalities (2.140). In order to get a more precise result one has to do some restrictive assumptions on the family of functions $\{F^n; n = 1,2,...\}$. In the next section we study a general condition called τ/σ equi-lower-semicontinuity assuring the coincidence of the notions of τ-epi-convergence and σ-epi-convergence, for two distinct topologies τ and σ on a space X.

2.6.2 τ/σ equi-lower-semicontinuity

This notion, introduced by Salinetti & Wets [1] (1977) and developed by these authors in collaboration with Dolecki [1] (1981) stems from the following theorem:

THEOREM 2.56 Let X be a space equipped with two topologies τ and σ which are comparable.

Let $(F^n)_{n\in\mathbb{N}}$ be a sequence of functions from X into $\bar{\mathbb{R}}$ and assume that the sequence $(F^n)_{n\in\mathbb{N}}$ epi-converges for one of the two topologies. Then the following statements are equivalent:

(i) the sequence $(F^n)_{n\in\mathbb{N}}$ epi-converges for the other topology to the same limit:

$$\tau\text{-lm}_e F^n = \sigma\text{-lm}_e F^n$$

(ii) $\forall x \in D = \text{dom}(\tau\text{-li}_e F^n)$, $\forall \varepsilon > 0$, $\forall W \in N_\sigma(x)$ $\exists V \in N_\tau(x)$ $\exists H \in \mathbb{N}$ s.t:

$$\forall n \in H \quad \inf_{y\in V} F^n(y) > \inf_{y\in W} F^n(y) - \varepsilon$$

(where σ denotes the stronger and τ the weaker of the two topologies).

(iii) $\begin{cases} \tau\text{-li}_e F^n = \sigma\text{-li}_e F^n & \text{and} \\ \tau\text{-ls}_e F^n = \sigma\text{-ls}_e F^n. \end{cases}$

Proof of Theorem 2.56

(i) ⇒ (ii) Let us assume that (ii) is not true: there exists x belonging to $D = \text{dom}\,(\tau\text{-li}_e F^n)$, $\varepsilon_o > 0$ and W_o belonging to $N_\sigma(x)$ satisfying:

$$\forall V \in N_\tau(x) \quad \forall H \in \mathbb{N} \quad \exists n \in H \text{ such that } \inf_{y\in V} F^n(y) < \inf_{y\in W_o} F^n(y) - \varepsilon_o.$$

For each $V \in N_\tau(x)$ this last property allows us to construct a subsequence $(n_k)_{k\in\mathbb{N}}$ (depending on V) such that

$$\forall k \in \mathbb{N} \inf_{y\in V} F^{n_k}(y) < \inf_{y\in W_o} F^{n_k}(y) - \varepsilon_o.$$

Taking the upper limit with respect to k, this last inequality becomes

$$\limsup_k \inf_{y\in V} F^{n_k}(y) < \limsup_k \inf_{y\in W_o} F^{n_k}(y) - \varepsilon_o.$$

Since we have no additional assumptions on the subsequence $(n_k)_{k\in\mathbb{N}}$, all we can say is

$$\inf_{(n_k)_{k\in\mathbb{N}}} \limsup_k \inf_{y\in V} F^{n_k}(y) < \sup_{(n_k)_{k\in\mathbb{N}}} \limsup_k \inf_{y\in W_o} F^{n_k}(y) - \varepsilon_o$$

i.e.

$$\liminf_n \inf_{y\in V} F^n(y) < \limsup_n \inf_{y\in W_o} F^n(y) - \varepsilon_o.$$

Since this is true for any $V \in N_\tau(x)$, we finally obtain

$$\sup_{V \in N_\tau(x)} \liminf_n \inf_{y \in V} F^n(y) \leq \sup_{W \in N_\sigma(x)} \limsup_n \inf_{y \in W} F^n(y) - \varepsilon_o,$$

$$\tau\text{-}li_e F^n(x) \leq \sigma\text{-}ls_e F^n(x) - \varepsilon_o. \tag{2.144}$$

By assumption (i), $\tau\text{-}li_e F^n(x) = \sigma\text{-}ls_e F^n(x)$ and $x \in D$ implies $\tau\text{-}li_e F^n(x) < +\infty$. Comparing with inequality (2.144), we clearly get the contradiction. Let us notice that, in preceding proof, topologies τ and σ play a symmetric role. In (ii) we stated only the significant result since the other one, obtained by exchanging τ and σ is automatically satisfied.

(ii) ⇒ (iii) Let us prove that, without any assumptions on the sequence $(F^n)_{n \in \mathbb{N}}$, we have the following implication:

$$\text{(ii)} \Rightarrow \begin{cases} \sigma\text{-}li_e F^n \leq \tau\text{-}li_e F^n \\ \sigma\text{-}ls_e F^n \leq \tau\text{-}ls_e F^n. \end{cases}$$

Let us fix $H \in \mathcal{N}$, $W \in N_\sigma(x)$ and $\varepsilon > 0$. From hypothesis (ii), with W and $\varepsilon > 0$ are associated $V_W^\varepsilon \in N_\tau(x)$, $H_W^\varepsilon \in \mathcal{N}$ such that

$$\forall n \in H_W^\varepsilon \quad \inf_{y \in W} F^n(y) \leq \inf_{y \in V_W^\varepsilon} F^n(y) + \varepsilon \quad \text{(we take } x \in D).$$

This implies

$$\inf_{\substack{n \in H \\ y \in W}} F^n(y) \leq \inf_{\substack{n \in H \cap H_W^\varepsilon \\ y \in W}} F^n(y) \leq \inf_{\substack{n \in H \cap H_W^\varepsilon \\ y \in V_W^\varepsilon}} F^n(y) + \varepsilon$$

and since $H \cap H_W^\varepsilon \in \mathcal{N}$ ($\mathcal{N}$ is a filter)

$$\leq \sup_{\substack{H \in \mathcal{N} \\ V \in N_\tau(x)}} \inf_{\substack{n \in H \\ y \in V}} F^n(y) + \varepsilon .$$

This being true for every $H \in \mathcal{N}$, $W \in N_\sigma(x)$ and $\varepsilon > 0$,

$$\sup_{\substack{H \in \mathcal{N} \\ W \in N_\sigma(x)}} \inf_{\substack{n \in H \\ y \in W}} F^n(y) \leq \sup_{\substack{H \in \mathcal{N} \\ V \in N_\tau(x)}} \inf_{\substack{n \in H \\ y \in V}} F^n(y)$$

i.e.

$$\sigma\text{-}li_e F^n(x) \leqslant \tau\text{-}li_e F^n(x).$$

If $x \notin D = \text{dom}\,(\tau\text{-}li_e F^n)$, then $\tau\text{-}li_e F^n(x) = +\infty$ and the preceding inequality is still satisfied.

When proving the other inequality $\sigma\text{-}ls_e F^n \leqslant \tau\text{-}ls_e F^n$, one has just to adapt the preceding argument noticing that, from Remark 1.37,

$$\tau\text{-}ls_e F^n(x) = \sup_{\substack{V \in N_\tau(x) \\ H \in \ddot{\mathbb{N}}}} \inf_{\substack{y \in V \\ n \in H}} F^n(y).$$

So, one has just to replace $\mathbb{N}$ by its grill $\ddot{\mathbb{N}}$ and notice that: If $H \in \ddot{\mathbb{N}}$ and $H^\varepsilon \in \mathbb{N}$, then $H \cap H^\varepsilon$ still belongs to $\ddot{\mathbb{N}}$. In order to achieve the proof of (iii), we notice that since $\tau < \sigma$,

$$\tau\text{-}li_e F^n \leqslant \sigma\text{-}li_e F^n$$

$$\tau\text{-}ls_e F^n \leqslant \sigma\text{-}ls_e F^n \qquad \text{(cf. Proposition 2.53)}$$

and finally the equalities (iii) hold.

(iii) $\Rightarrow$ (i) This implication is a clear consequence of the definition of epi-convergence, that is, $li_e = ls_e$. □

The preceding theorem tells us that property (ii) is a minimal assumption in order that epi-convergences for two topologies τ and σ coincide: it is the "τ/σ equi-lower semicontinuity" property:

DEFINITION 2.57 Let X be a space equipped with two topologies τ and σ. A sequence of functions $(F^n)_{n \in \mathbb{N}}$ from X into $\bar{\mathbb{R}}$ is said to be τ/σ equi-lower-semicontinuous if:

$$\forall x \in D = \text{dom}\,(\tau\text{-}li_e F^n)\ \forall \varepsilon > 0\ \ \forall W \in N_\sigma(x),\ \exists V \in N_\tau(x)\ \exists H \in \mathbb{N} \tag{2.145}$$

such that

$$\forall n \in H \qquad \inf_{y \in V} F^n(y) \geqslant \inf_{y \in W} F^n(y) - \varepsilon.$$

This definition calls for a few comments. The justification of this terminology appears clearly when studying the case (important with respect to applications) where σ is the discrete topology (every set is open for this topology): in this case Definition 2.57 becomes (we omit the reference to $\sigma = \tau_d$ = discrete topology).

DEFINITION 2.58 Let (X,τ) be a topological space and $(F^n)_{n\in\mathbb{N}}$ a sequence of functions from X into $\bar{\mathbb{R}}$. The sequence $(F^n)_{n\in\mathbb{N}}$ is said to be τ-equi lower semicontinuous if

$$\forall x \in D = \text{dom } (\tau\text{-li}_e F^n) \ \forall \varepsilon > 0 \ \exists V \in N_\tau(x) \ \exists H \in \mathbb{N}$$

such that

$$\forall n \in H \inf_{y\in V} F^n(y) \geq F^n(x) - \varepsilon.$$

Noticing (cf. Proposition 2.52) that the τ_d-epi-convergence (τ_d = discrete topology) is nothing but the pointwise convergence, Theorem 2.56 becomes:

COROLLARY 2.59 Let (X,τ) be a topological space and $(F^n)_{n\in\mathbb{N}}$ a sequence of functions from X into $\bar{\mathbb{R}}$. Any pair formed by two of the following three statements implies the third one.

(i) $F = \tau\text{-lm}_e F^n$

(ii) $F = \lim F^n$ (i.e. $\forall x \in X \ F(x) = \lim_{n\to+\infty} F^n(x)$: pointwise convergence)

(iii) The sequence $(F^n)_{n\in\mathbb{N}}$ is τ-equi lower semicontinuous.

The following formulations of lower semicontinuity justify the terminology:

PROPOSITION 2.60 Let (X,τ) be a topological space and $F:X \to \bar{\mathbb{R}}$ a real-extended valued function. F is τ-lower semicontinuous iff one of the following equivalent properties is satisfied:

(i) $\forall x \in D = \text{dom } (c\ell_\tau F), \ \forall \varepsilon > 0, \ \exists V \in N_\tau(x)$ such that

$$\inf_{y\in V} F(y) \geq F(x) - \varepsilon.$$

(ii) $$\begin{cases} \forall x \in \text{dom } F \ \forall \varepsilon > 0 \ \exists V \in N_\tau(x) \text{ such that} \\ \inf_{y\in V} F(y) \geq F(x) - \varepsilon. \\ \forall x \notin \text{dom } F \ \forall a > 0 \ \exists V \in N_\tau(x) \text{ such that } \inf_{y\in V} F(y) \geq a. \end{cases}$$

We notice that property (i) of Proposition 2.60 is nothing but the notion of τ-equi lower semicontinuity, when considering the sequence $F^n = F \ \forall n \in \mathbb{N}$.

2.6.3 Example

EXAMPLE 2.61 Let $\mathcal{F}_M$ be the family of functionals

$$\mathcal{F}_M = \{F: H^1(\Omega) \to \mathbb{R}^+ / F(u) = \int_\Omega j(x, Du(x))dx \text{ with } j: \Omega \times \mathbb{R}^N \to \mathbb{R}^+ \atop (x,z) \to j(x,z)$$

measurable in x, convex continuous in z and satisfying $0 \leq j(x,z) \leq M(1 + |z|^2)\}$ where Ω is a bounded open subset of $\mathbb{R}^N$ and M belongs to $\mathbb{R}^+$. Let us prove that the family $\mathcal{F}_M$ satisfies a $L^1(\Omega)/L^\infty(\Omega)$ equi-lsc property:

$\forall u \in H^1(\Omega)\ \forall \varepsilon > 0\ \forall \eta > 0\ \ \exists \rho(\eta, \varepsilon, u) > 0$ such that for every $F \in \mathcal{F}_M$

$$\inf_{\{\|v-u\|_{L^1(\Omega)} \leq \rho(\eta)\}} F(u) \geq \inf_{\{\|v-u\|_{L^\infty(\Omega)} \leq \eta\}} F(u) - \varepsilon. \tag{2.146}$$

Let us argue by contradiction and assume that (2.146) is false. Then, there exists $u \in H^1(\Omega)$, $\varepsilon_0 > 0$, $\eta_0 > 0$ such that, for every $\rho > 0$, there exists $F_\rho \in \mathcal{F}_M$ satisfying

$$\inf_{\{\|v-u\|_{L^1(\Omega)} \leq \rho\}} F_\rho(v) < \inf_{\{\|v-u\|_{L^\infty(\Omega)} \leq \eta_0\}} F_\rho(v) - \varepsilon_0.$$

This implies the existence for every $\rho > 0$ of $v_\rho \in H^1(\Omega)$ satisfying

$$\begin{cases} \|v_\rho - u\|_{L^1(\Omega)} \leq \rho \\ F_\rho(v_\rho) < \inf_{\{\|v-u\|_{L^\infty(\Omega)} \leq \eta_0\}} F_\rho(v) - \varepsilon_0. \end{cases} \tag{2.147}$$

For every $\alpha \in \mathbb{R}^+$, let us introduce $T_\alpha : \mathbb{R} \to \mathbb{R}$ the contraction defined by (see Fig. 2.10):

$$T_\alpha(r) = \begin{cases} r & \text{if } |r| \leq \alpha \\ \alpha & \text{if } r \geq \alpha \\ -\alpha & \text{if } r \leq -\alpha . \end{cases}$$

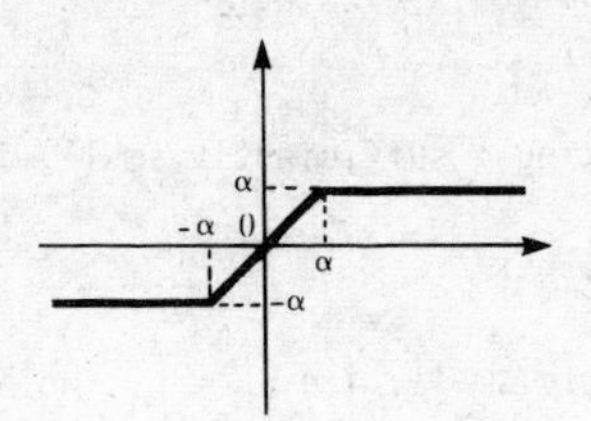

Figure 2.10

Let us consider $w_\rho = u + T_{\alpha_\rho}(v_\rho - u)$ (with $\alpha_\rho \to 0$ when $\rho \to 0$, to be precise):

$$\|w_\rho - u\|_{L^\infty(\Omega)} = \|T_{\alpha_\rho}(v_\rho - u)\|_{L^\infty(\Omega)} \leq \alpha_\rho . \tag{2.148}$$

Moreover,

$$Dw_\rho = Du + T'_{\alpha_\rho}(v_\rho - u)(Dv_\rho - Du)$$

$$= T'_{\alpha_\rho}(v_\rho - u)Dv_\rho + [1 - T'_{\alpha_\rho}(v_\rho - u)]Du.$$

Since T_ρ is a monotone contraction, $0 \leq T'_{\alpha_\rho} \leq 1$; by convexity of j_ρ,

$$j_\rho(x, Dw_\rho(x)) \leq T'_{\alpha_\rho}(v_\rho - u)j_\rho(x, Dv_\rho(x) + (1 - T'_{\alpha_\rho}(v_\rho - u)(x))j_\rho(x, Du(x))$$

$$\leq j_\rho(x, Dv_\rho(x)) + (1 - T'_{\alpha_\rho}(v_\rho - u))j_\rho(x, Du(x)).$$

Integrating over Ω, we get

$$F_\rho(w_\rho) \leq F_\rho(v_\rho) + \int_\Omega [1 - T'_{\alpha_\rho}(v_\rho - u)]j_\rho(x, Du(x))dx.$$

Using that $0 \leq j_\rho(x,z) \leq M(1 + |z|^2)$ and definition of T_{α_ρ}, we get

$$F_\rho(w_\rho) \leq F_\rho(v_\rho) + M \int_{\{|v_\rho - u| \geq \alpha_\rho\}} (1 + |Du|^2)dx. \tag{2.149}$$

Since $\|v_\rho - u\|_{L^1(\Omega)} \leq \rho$, meas $\{|v_\rho - u| \geq \alpha_\rho\} \leq \rho/\alpha_\rho$; taking $\alpha_\rho = \sqrt{\rho}$, for example, we obtain meas $\{|v_\rho - u| \geq \alpha_\rho\} \xrightarrow[\rho \to 0]{} 0$; since $u \in H^1(\Omega)$, $|Du|^2$ belongs to $L^1(\Omega)$ and

$$\lim_{\rho\to 0} \int_{\{|v_\rho - u| > \alpha_\rho\}} (1 + |Du|^2)dx = 0. \tag{2.150}$$

From (2.149) and (2.150), taking ρ sufficiently small, i.e. $\rho < \rho_0$, we obtain

$$F_\rho(w_\rho) < F_\rho(v_\rho) + \varepsilon_0/2.$$

From (2.148) for ρ sufficiently small, i.e. $\rho < \rho_1$, we have

$$||w_\rho - u||_{L^\infty(\Omega)} < \eta_0 \quad \text{(since } \alpha_\rho \to 0 \text{ when } \rho \to 0).$$

From the two last inequalities

$$\forall \rho,\ 0 < \rho < \inf(\rho_0, \rho_1) \quad \inf_{\{||v-u||_{L^\infty(\Omega)} < \eta_0\}} F_\rho(v) < F_\rho(w_\rho) < F_\rho(v_\rho) + \varepsilon_0/2$$

which combined with (2.147) implies

$$\forall 0 < \rho < \inf(\rho_0, \rho_1) \inf_{\{||v-u||_{L^\infty(\Omega)} < \eta_0\}} F_\rho(v) < \inf_{\{||v-u||_{L^\infty(\Omega)} < \eta_0\}} F_\rho(v) - \varepsilon_0/2$$

a clear contradiction. □

<u>COROLLARY 2.62</u> Let $\{F^\varepsilon; \varepsilon > 0\}$ be the sequence of functionals of the homogenization problem (cf. Section 1.1.1),

$$F^\varepsilon(u) = \int_\Omega j(\frac{x}{\varepsilon}, \text{grad}u(x))dx.$$

Then,

$$\forall u \in H^1(\Omega) \quad s\text{-}L^1(\Omega)\text{-}\text{lm}_e F^\varepsilon(u) = s\text{-}L^\infty(\Omega)\text{-}\text{lm}_e F^\varepsilon(u) = F^{\text{hom}}(u).$$

This is a direct consequence of the above L^1/L^∞ equi-lsc property and of Theorem 1.20. (cf. application to Example 2.18).

2.7 MOREAU-YOSIDA PROXIMAL APPROXIMATION

In this section, we introduce and study a basic tool in optimization theory: the *Moreau-Yosida proximal approximation*.

This approximation enjoys nice properties with respect to minimization problems. It is a "variational notion of approximation".

In [1] (1965) Moreau introduced, for a general convex lower semicontinuous function $F:V \to]-\infty, +\infty]$ defined on a real Hilbert space V, the notion of proximal approximation: by definition, for every $u \in V$, $\text{prox}_F(u)$ is the point where the function

$$\phi(v) = F(v) + \frac{1}{2} \|u-v\|_V^2$$

reaches its minimum.

The terminology is justified by the fact that, when F is equal to I_K, the indicator function of a closed convex set K, the proximal approximation is nothing but the projection on K.

Moreau noticed, too, that the value of this infimum enjoys nice properties. More generally, for every $\lambda > 0$, one can define

$$F_\lambda(u) = \inf_{v \in V} \{F(v) + \frac{1}{2\lambda} \|u-v\|_V^2\}.$$

The function F_λ is called the Moreau-Yosida approximation of index λ of F. The reference to Yosida is justified by the fact that the Fréchet derivative of F_λ is nothing but the Yosida approximation of index λ of the maximal monotone operator ∂F (cf. Brezis [1])

$$\partial(F_\lambda) = (\partial F)_\lambda$$

where ∂, the subdifferential operation, extends the notion of Fréchet derivative to the convex, lower semicontinuous functions.

Many authors have contributed to the development of this concept in the framework of convex analysis (Attouch [4], Aubin [1],[2], Benilan [1], Brezis [1], Damlamian [4], Moreau [1], [2], Rockafellar [1], Sonntag [1], Wets [3], Fougeres [1]).

Recently it emerged that this concept is of more general significance and can be developed in a very general setting, without any convexity assumption on the function F (cf. Attouch [8], Attouch & Wets [3], De Giorgi [2]).

2.7.1 Definition-Properties

DEFINITION 2.63 Let (X,d) be a metric space and $F:X \to \bar{\mathbb{R}}$ a real extended valued function.

For every $\lambda > 0$, the Moreau-Yosida proximal approximation of index λ of F is the function $F_\lambda:X \to \mathbb{R}$ equal to

$$\forall x \in X \quad F_\lambda(x) = \inf_{u\in X} \{F(u) + \frac{1}{2\lambda} d^2(u,x)\}. \tag{2.151}$$

We are going to prove that, without any regularity assumption on F, the functions $(F_\lambda)_{\lambda>0}$ are locally Lipschitz and that the sequence $\{F_\lambda ; \lambda \downarrow 0\}$ converges increasingly to $c\ell_\tau F$ (where τ is the topology induced by d) (Theorem 2.64).

Then, we shall prove that all the epi-convergence properties of a sequence $(F^n)_{n\in\mathbb{N}}$ of functions can be re-expressed (in a simpler way) in terms of the pointwise convergence of their Moreau-Yosida approximates (Theorem 2.65).

At this point, we would like to stress that most of the results we establish are of a topological nature and do not really depend on the form of the perturbation term used in formula (2.151). Rather than the square of the distance, we could have used locally Lipschitz functions of the distance or even functions not directly related to the distance (cf. R. Wets [3]). Our choice is justified by the fact that in the Hilbertian case, for this choice of the perturbation term, we have additional properties for the resolvents.

We shall sometimes write d_τ for the metric d on X, in order to refer to the topology τ induced by d, and, emphasize the topological aspects of the question, being studied.

<u>THEOREM 2.64</u> Let $F:X \to]-\infty, +\infty]$, where X is equipped with a metric d_τ, a real-extended valued function satisfying:

$$\begin{cases} F \not\equiv +\infty \\ \exists r > 0 \quad \exists x_0 \in X \text{ such that: } \forall x \in X \quad F(x) + r(d^2(x,x_0) + 1) > 0. \end{cases} \tag{2.152}$$

Then, the family $(F_\lambda)_{\lambda>0}$ of Moreau-Yosida approximates of F (defined by (2.151)) satisfies the following properties (i) and (ii):

(i) For every $x \in X$, the sequence $(F_\lambda(x))_{\lambda>0}$ increases to $c\ell_\tau F(x)$, the τ-closure of F at x (τ is the topology induced by d) as λ decreases to zero:

$$c\ell_\tau F(x) = \lim_{\lambda\downarrow 0} F_\lambda(x) = \sup_{\lambda>0} F_\lambda(x). \tag{2.153}$$

Consequently, if F is τ-lower semicontinuous

$$F = \sup_{\lambda>0} F_\lambda(x).$$

(ii) For every $0 < \lambda \le 1/8r$, F_λ is real (finite) valued, locally Lipschitz with respect to x: there exists $C:\mathbb{R}^+ \times \mathbb{R}^+ \to \mathbb{R}^+$ bounded on bounded subsets such that

$$\forall \lambda > 0 \ \forall x_1, x_2 \in X \quad |F_\lambda(x_1) - F_\lambda(x_2)| \le \frac{1}{\lambda} d(x_1, x_2).C(d(x_1, x_o); d(x_2, x_o)). \tag{2.154}$$

Moreover, C depends only and in a continuous way of r and $F(x_o)$, given in (2.152).

Proof of Theorem 2.64

(i) By definition (2.151) of F_λ,

$$\forall u \in X \quad F_\lambda(x) \le F(u) + \frac{1}{2\lambda} d^2(u,x).$$

Taking $u = x$, we obtain $F_\lambda(x) \le F(x)$; this being true for any $\lambda > 0$

$$\sup_{\lambda>0} F_\lambda \le F. \tag{2.155}$$

Let us prove that, under assumptions (2.152), for every $\lambda > 0$, F_λ is finite valued:

From (2.152)

$$F_\lambda(x) = \inf_{u\in X} \{F(u) + \frac{1}{2\lambda} d^2(u,x)\} \ge \inf_{u \in X} \{-rd^2(u,x_o) - r + \frac{1}{2\lambda} d^2(u,x)\}.$$

Using the inequality $d^2(u,x_o) \le 2d^2(u,x) + 2d^2(x,x_o)$, we obtain

$$\forall 0 < \lambda \le 1/4r, \quad \forall x \in X \quad F_\lambda(x) \ge -2rd^2(x,x_o) - r. \tag{2.156}$$

On the other hand, using again (2.152) and introducing $x_o \in X$ such that $F(x_o) < +\infty$,

$$F_\lambda(x) \le F(x_o) + \frac{1}{2\lambda} d^2(x_o,x).$$

The two above inequalities imply that F_λ is a real (finite) valued function. Returning to the definition of F_λ, since the infimum is finite, for every $\lambda > 0$ there exists x_λ belonging to X which satisfies

$$F_\lambda(x) \le F(x_\lambda) + \frac{1}{2\lambda} d^2(x_\lambda,x) \le F_\lambda(x) + \lambda. \tag{2.157}$$

We are now able to prove the reverse inequality to (2.155):

$$c\ell_\tau F \leq \sup_{\lambda>0} F_\lambda.$$

If $\sup_{\lambda>0} F_\lambda(x) = +\infty$, this is clear; so let us take $x \in X$ such that $\sup_{\lambda>0} F_\lambda(x) < +\infty$. From (2.157)

$$F(x_\lambda) + \frac{1}{2\lambda} d^2(x_\lambda, x) \leq \sup_{\lambda>0} F_\lambda(x) + \lambda.$$

From hypothesis (2.152), it follows that

$$-rd^2(x_\lambda, x_o) - r + \frac{1}{2\lambda} d^2(x_\lambda, x) \leq \sup_{\lambda>0} F_\lambda(x) + \lambda,$$

and

$$d^2(x_\lambda, x) \leq 2\lambda[\sup_{\lambda>0} F_\lambda(x) + \lambda + r + 2rd^2(x, x_o)] + 4\lambda rd^2(x_\lambda, x).$$

$$\forall 0 < \lambda < \frac{1}{8r} \quad d^2(x_\lambda, x) \leq 4\lambda[\sup_{\lambda>0} F_\lambda(x) + \lambda + r + 2rd^2(x, x_o)].$$

Thus,

$$d(x_\lambda, x) \xrightarrow[\lambda \to 0]{} 0, \text{ i.e. } x_\lambda \xrightarrow[\lambda \to 0]{\tau} x.$$

From the second inequality of (2.157) and the positivity of d^2,

$$F(x_\lambda) \leq \sup_{\lambda>0} F_\lambda(x) + \lambda.$$

Let λ go to zero; since $x_\lambda \xrightarrow{\tau} x$, by definition of $c\ell_\tau F$,

$$c\ell_\tau F(x) \leq \sup_{\lambda>0} F_\lambda(x).$$

Let us assume for now that we have proved that the Moreau-Yosida approximates F_λ are locally Lipschitz; they are consequently τ-continuous (τ is the topology induced by d) and $\sup_{\lambda>0} F_\lambda$ as a supremum of τ-continuous functions is τ-lower semicontinuous.

So (2.155) can be made precise: $\sup_{\lambda>0} F_\lambda \leq c\ell_\tau F$; combined with $c\ell_\tau F \leq \sup_{\lambda>0} F_\lambda$ this yields the equality

$$c\ell_\tau F = \sup_{\lambda>0} F_\lambda.$$

(ii) Let us now prove the locally Lipschitz properties of Moreau-Yosida approximates: Since F is not identically equal to $+\infty$, let us denote $x_o \in X$ such that $F(x_o) < +\infty$. Clearly, the minorization assumption (2.152) do not depend on the point x_o chosen in X (one has just to adapt the constant r) and, without restriction, we can take a point x_o such that $F(x_o) < +\infty$. For any $0 < \lambda \leq 1/8r$, and $\varepsilon > 0$, let us consider a point x_λ^ε (which depends on $x \in X$, $\lambda > 0$, $\varepsilon > 0$) satisfying

$$F_\lambda(x) \leq F(x_\lambda^\varepsilon) + \frac{1}{2\lambda} d^2(x_\lambda^\varepsilon, x) \leq F_\lambda(x) + \varepsilon. \tag{2.158}$$

From (2.158) and the quadratic minorization assumption (2.152),

$$-r(d^2(x_\lambda^\varepsilon, x_o) + 1) + \frac{1}{2\lambda} d^2(x_\lambda^\varepsilon, x) \leq F_\lambda(x) + \varepsilon.$$

On the other hand, by definition of F_λ,

$$F_\lambda(x) \leq F(x_o) + \frac{1}{2\lambda} d^2(x, x_o).$$

Combining these two last inequalities, we obtain

$$d^2(x_\lambda^\varepsilon, x) \leq 2\lambda[F(x_o) + \varepsilon + r(d^2(x_\lambda^\varepsilon, x_o) + 1)] + d^2(x, x_o).$$

It follows that for any $0 < \lambda < \frac{1}{8r}$,

$$d^2(x_\lambda^\varepsilon, x) \leq 4\lambda[F(x_o) + \varepsilon + r + 2rd^2(x, x_o)] + 2d^2(x, x_o). \tag{2.159}$$

Let us consider two points x and $\hat{x}$ in X and, for any $0 < \lambda \leq \frac{1}{8r}$, $\varepsilon > 0$, x_λ^ε and $\hat{x}_\lambda^\varepsilon$ corresponding points (defined by (2.152)).

By definition of F_λ,

$$F_\lambda(\hat{x}) \leq F(x_\lambda^\varepsilon) + \frac{1}{2\lambda} d^2(x_\lambda^\varepsilon, \hat{x})$$

and by definition of x_λ^ε

$$\leq F_\lambda(x) + \varepsilon + \frac{1}{2\lambda} [d^2(x_\lambda^\varepsilon, \hat{x}) - d^2(x_\lambda^\varepsilon, x)]$$

$$\leq F_\lambda(x) + \varepsilon + \frac{1}{2\lambda} d(x, \hat{x})[2d(x_\lambda^\varepsilon, x) + d(x, \hat{x})].$$

Using (2.159), this last inequality becomes: for any $0 < \lambda \leq \frac{1}{8r}$, $\varepsilon > 0$,

$$F_\lambda(\hat{x}) - F_\lambda(x) \leq \varepsilon + \frac{1}{2\lambda} d(x,\hat{x}) [d(x,x_0) + d(\hat{x},x_0) + 4d^2(x,x_0)$$

$$+ 8\lambda(F(x_0) + \varepsilon + r) + 16\lambda r d^2(x,x_0)].$$

Letting ε go to zero, and interchanging the role of x and $\hat{x}$, we obtain:

$$\forall 0 < \lambda \leq \frac{1}{8r}, \ \forall x,\hat{x} \in X \quad |F_\lambda(x)-F_\lambda(\hat{x})| \leq \frac{1}{\lambda} d(x,\hat{x}) C(d(x,x_0);d(\hat{x},x_0))$$

with

$$C(d(x,x_0),d(\hat{x},x_0)) = [d(x,x_0) + d(\hat{x},x_0) + 2(1+4\lambda r)(d^2(x,x_0) + d^2(\hat{x},x_0))$$

$$+ 4\lambda(F(x_0) + r)]$$

$$\leq [d(x,x_0)+d(\hat{x},x_0)+3(d^2(x,x_0)+d^2(\hat{x},x_0))+ \frac{1}{2r}(F(x_0)+r)],$$

which is bounded when $d(x,x_0)$ and $d(\hat{x},x_0)$ remain bounded. Let us observe that C depends continuously on the parameters r and $F(x_0)$.

2.7.2 Moreau-Yosida approximation and epi-convergence

In the next theorem, we give the formulation in terms of Moreau-Yosida approximates of the τ-epi limits of a sequence of functions:

THEOREM 2.65 Let (X,τ) be a topological metrizable space, d_τ a distance inducing the topology τ. For every function $F:X \to \bar{\mathbb{R}}$, and $\lambda > 0$ we denote by F_λ its Moreau-Yosida approximate of index λ:

$$\forall x \in X \quad F_\lambda(x) = \inf_{u \in X} \{F(u) + \frac{1}{2\lambda} d_\tau^2(u,x)\}.$$

Let $(F^n)_{n\in\mathbb{N}}$ be a sequence of functions from X into $]-\infty, +\infty]$ satisfying: there exist $r \geq 0$, $x_0 \in X$, such that for every $n \in \mathbb{N}$ and $x \in X$

$$F^n(x) + r(d^2(x,x_0) + 1) \geq 0 . \tag{2.161}$$

Then, the following equalities hold: for every $x \in X$,

$$\tau\text{-}li_e F^n(x) = \sup_{\lambda>0} \ \liminf_n F_\lambda^n(x) \tag{2.162}$$

$$\tau\text{-}ls_e F^n(x) = \sup_{\lambda>0} \ \limsup_n F_\lambda^n(x). \tag{2.163}$$

Proof of Theorem 2.65 Let us prove (2.162) and begin with the following inequality:

$$a = \sup_{\lambda>0} \liminf_n F_\lambda^n(x) \geq \tau\text{-}li_e F^n(x). \tag{2.164}$$

If $a = +\infty$, there is nothing to prove; so let us assume that a is finite. For every $\lambda > 0$, $a \geq \liminf_n F_\lambda^n(x)$. Take a sequence $(\lambda_m)_{m\in\mathbb{N}}$ decreasing to zero, then

$$a \geq \limsup_m \liminf_n F_{\lambda_m}^n(x).$$

From diagonalization Lemma 1.17, there exists an increasing map $n \to m(n)$ such that

$$a \geq \liminf_n F_{\lambda_{m(n)}}^n(x).$$

Let us denote $\lambda_n = \lambda_{m(n)}$; the sequence $(\lambda_n)_{n\in\mathbb{N}}$ decreases to zero as $n \to +\infty$.

By definition (2.157) of x_λ,

$$\lambda_n + F_{\lambda_n}^n(x) \geq F^n(x_{\lambda_n}) + \frac{1}{2\lambda_n} d^2(x_{\lambda_n}, x).$$

Combining these two last inequalities, and using x_n to denote x_{λ_n}, we obtain

$$a \geq \liminf_n [F^n(x_n) + \frac{1}{2\lambda_n} d^2(x_n, x)].$$

By definition of $\liminf_n$, there exists a subsequence n_k such that

$$a \geq \lim_k [F^{n_k}(x_{n_k}) + \frac{1}{2\lambda_{n_k}} d^2(x_{n_k}, x)] \geq \liminf_k F^{n_k}(x_{n_k}).$$

On the other hand, from the quadratic minorization assumption, the sequence $(x_k)_{k\in\mathbb{N}}$, with $x_k = x_{n_k}$, satisfies, for k sufficiently large,

$$+\infty > a + 1 \geq -r[d^2(x_k, x_o) + 1] + \frac{1}{2\lambda_{n_k}} d^2(x_k, x).$$

This clearly implies $d(x_k, x) \xrightarrow[k\to+\infty]{} 0$, i.e. $x_k \xrightarrow[k\to+\infty]{\tau} x$. Finally,

$$a \geq \liminf_k F^{n_k}(x_k) \geq \inf_{\begin{cases}(n_k)_{k\in\mathbb{N}} \\ x_k \xrightarrow{\tau} x\end{cases}} \liminf_k F^{n_k}(x_k).$$

From Theorem 1.13 and Remark 1.19, this last expression is equal to $\tau\text{-li}_e F^n(x)$ (it is the sequential formulation of this quantity in a metrizable space). This proves the first inequality (2.164).

Let us now prove the opposite inequality

$$\sup_{\lambda>0} \liminf_n F_\lambda^n(x) \leq \tau\text{-li}_e F^n(x) = \inf_{\begin{cases}(n_k)_{k\in\mathbb{N}} \\ x_k \xrightarrow[k]{\tau} x\end{cases}} \liminf_k F^{n_k}(x_k).$$

Equivalently, let us prove that for every $\lambda > 0$, for every subsequence $\{n_k;\ k \in \mathbb{N}\}$ and every convergent sequence $x_k \xrightarrow{\tau} x$,

$$\liminf_n F_\lambda^n(x) \leq \liminf_k F^{n_k}(x_k).$$

$$\begin{aligned}\liminf_n F_\lambda^n(x) &\leq \liminf_k F_\lambda^{n_k}(x) \\ &\leq \liminf_k \inf_{u\in X} \{F^{n_k}(u) + \frac{1}{2\lambda} d_\tau^2(u,x)\} \\ &\leq \liminf_k \{F^{n_k}(x_k) + \frac{1}{2\lambda} d_\tau^2(x_k,x)\} \\ &\leq \liminf_k \ F^{n_k}(x_k) \quad (\text{since } d_\tau(x_k,x) \xrightarrow[k\to+\infty]{} 0),\end{aligned}$$

which completes the proof of (2.162).

The proof of (2.163) is obtained in a similar way; for completeness let us describe it briefly and prove first

$$b = \sup_{\lambda>0} \limsup_n F_\lambda^n(x) \geq \tau\text{-ls}_e F^n(x).$$

We can assume without restriction that $b < +\infty$; taking a sequence $(\lambda_m)_{m\in\mathbb{N}}$ decreasing to zero,

$$b = \limsup_m \limsup_n F_{\lambda_m}^n(x);$$

from the diagonalization Corollary 1.16, there exists a mapping $n \to m(n)$ increasing to $+\infty$ such that

$$b \geq \limsup_n F_{\lambda_n}^n(x) \quad (\text{where we denote } \lambda_n = \lambda_{m(n)}).$$

For n sufficiently large, $F^n_{\lambda_n}(x)$ is finite ($b < +\infty$); considering $x_n = x_{\lambda_n}$ defined by

$$\lambda_n + F^n_{\lambda_n}(x) \geq F^n(x_n) + \frac{1}{2\lambda_n} d^2_\tau(x_n,x),$$

we get

$$b \geq \limsup_n \{F^n(x_n) + \frac{1}{2\lambda_n} d^2_\tau(x_n,x)\}.$$

By the same argument as before, we get, using the quadratic minorization assumption on the $(F^n)_{n\in\mathbb{N}}$, (2.161), that x_n τ-converges to x and thus

$$b \geq \limsup_n F^n(x_n) \geq \inf_{\{x_n \xrightarrow{\tau} x\}} \limsup_n F^n(x_n) = \tau\text{-}ls_e F^n(x).$$

The opposite inequality

$$\sup_{\lambda>0} \limsup_n F^n_\lambda(x) \leq \tau\text{-}ls_e F^n(x) = \inf_{\{x_n \xrightarrow{\tau} x\}} \limsup_n F^n(x_n)$$

is equivalent to:

$$\forall\lambda > 0, \forall x_n \xrightarrow{\tau} x \quad \limsup_n F^n_\lambda(x) \leq \limsup_n F^n(x_n),$$

which clearly follows from the inequality:

$$F^n_\lambda(x) \leq F^n(x_n) + \frac{1}{2\lambda} d^2_\tau(x_n,x). \quad \square$$

REMARK 2.66 The first part of Theorem 2.64 can also be obtained as a consequence of Theorem 2.65. Taking $F^n \equiv F$, from (2.162) or (2.163), noticing that $\tau\text{-}li_e F^n = \tau\text{-}ls_e F^n = c\ell_\tau F$ (cf. Corollary 2.3), it follows that

$$c\ell_\tau F = \sup_{\lambda>0} F_\lambda . \quad \square$$

Using the preceding formulations of epi limits (Theorem 2.65) we obtain a new characterization of epi-convergence.

COROLLARY 2.67 Let $(F^n)_{n\in\mathbb{N}}$ be a sequence of functions from (X,τ), a topological metrizable space, into $]-\infty, +\infty]$ satisfying (2.161). Let d_τ be a distance inducing the topology τ and $(F^n_\lambda)_{\substack{n\in\mathbb{N}\\ \lambda>0}}$ the corresponding Moreau-Yosida approximates.

Then, the sequence $(F^n)_{n\in\mathbb{N}}$ τ-epi-converges to F at x if and only if

$$\sup_{\lambda>0} \limsup_n F^n_\lambda(x) = \sup_{\lambda>0} \liminf_n F^n_\lambda(x) = F(x). \tag{2.165}$$

Consequently if, for every $\lambda > 0$, the sequence $(F^n_\lambda(x))_{n\in\mathbb{N}}$ converges in $\mathbb{R}$, then the sequence $(F^n)_{n\in\mathbb{N}}$ τ-epi converges at x and

$$\tau\text{-}\mathrm{lm}_e F^n(x) = \sup_{\lambda>0} \lim_n F^n_\lambda(x). \quad \square \tag{2.166}$$

Let us now prove the "resolvent equation" which ties the Moreau-Yosida approximates one to the other:

PROPOSITION 2.68 Let H be a Hilbert space and $F:H \to \bar{\mathbb{R}}$ a real extended valued function. Then for every $\lambda > 0$, $\mu > 0$ and every $x \in H$,

$$(F_\lambda)_\mu(x) = F_{\lambda+\mu}(x). \tag{2.167}$$

Proof of Proposition 2.68 The proof follows directly from the definition of Moreau-Yosida approximation, after interchanging some infimum operations:

$$\begin{aligned}(F_\lambda)_\mu(x) &= \inf_{u\in H} \{F_\lambda(u) + \frac{1}{2\mu}\|x-u\|^2_H\} \\ &= \inf_{u\in H}\inf_{v\in H} \{F(v) + \frac{1}{2\lambda}\|u-v\|^2_H + \frac{1}{2\mu}\|x-u\|^2_H\} \\ &= \inf_{v\in H} [F(v) + \inf_{u\in H} \{\frac{1}{2\lambda}\|u-v\|^2_H + \frac{1}{2\mu}\|x-u\|^2\}].\end{aligned}$$

Let us compute

$$A = \inf_{u\in H} \{\frac{1}{2\lambda}\|v-u\|^2_H + \frac{1}{2\mu}\|x-u\|^2_H\}.$$

The infimum is realized at $\underline{u}$ satisfying

$$\frac{1}{\lambda}(\bar{u}-v) + \frac{1}{\mu}(\bar{u}-x) = 0;$$

that is,

$$\bar{u} = \frac{\lambda x + \mu v}{\lambda + \mu} .$$

Using this in A, we obtain

$$A = \frac{1}{2(\lambda+\mu)} ||x-v||_H^2 .$$

Finally,

$$(F_\lambda)_\mu(x) = \inf_{v \in H} \{F(v) + \frac{1}{2(\lambda+\mu)} ||x-v||_H^2\} = F_{\lambda+\mu}(x). \quad \square \tag{2.168}$$

As a by product, it follows from the continuity of $F_\lambda(\cdot)$ and from (2.153) that $\lim_{\mu \to 0} F_{\lambda+\mu}(x) = F_\lambda(x)$ i.e. $\lambda \to F_\lambda(x)$ is continuous. In fact, it is locally Lipschitz on $]0,+\infty[$.

Let us now give another, rather elegant proof of the abstract compactness Theorem 2.22. It relies on the characterization (Theorem 2.65) of the epi-limits via Moreau-Yosida approximates. In fact, we are going to prove a *stronger compactness result:*

PROPOSITION 2.69 Let (X,τ) be a metrizable separable space and d_τ a distance inducing the topology τ on X.

Let $F^n : X \to \bar{\mathbb{R}}$ be a sequence of real extended valued functions which satisfies the following uniform minorization property:

$$\text{there exists } r \geq 0 \text{ and } x_o \in X : F^n(x) + r(d^2(x,x_o) + 1) \geq 0$$
$$\text{for all } n \in \mathbb{N} \text{ and } x \in X. \tag{2.169}$$

Then one can extract a subsequence $(F^{n_k})_{k \in \mathbb{N}}$ such that

$$\forall x \in X, \forall \lambda > 0 \quad \lim_{k \to +\infty} F_\lambda^{n_k}(x) \text{ exists.} \tag{2.170}$$

Before proving Proposition 2.69, *let us show how it implies the compactness theorem:*

We first notice that (cf. Corollary 2.67) the convergence (2.170) of the Moreau-Yosida approximates, implies the τ-epi convergence of the extracted subsequence F^{n_k}.

We then have to explain how, given a general sequence $(F^n)_{n \in \mathbb{N}}$, $F^n : X \to \bar{\mathbb{R}}$ one can reduce to the preceding situation (2.169). Let us describe two methods:

(a) The first one is to use truncations and introduce, for every $\ell \in \mathbb{N}$,

$$F^{n,\ell} = \sup (F^n, -\ell).$$

The sequence $(F^{n,\ell})_{n\in\mathbb{N}}$, for ℓ fixed, satisfies condition (2.169). By a diagonalization argument and using (2.170) one can extract a subsequence $(n_k)_{k\in\mathbb{N}}$ such that

$$\forall \ell \in \mathbb{N} \quad \tau\text{-}\mathrm{lm}_e F^{n_k,\ell} = F_\ell \text{ exists.}$$

Introducing

$$F = \inf_{\ell\in\mathbb{N}} F_\ell,$$

we claim that

$$F = \tau\text{-}\mathrm{lm}_e F^{n_k}.$$

Let us write (for simplicity) F^k instead of F^{n_k} and first verify that

$$F \geq \tau\text{-}\mathrm{ls}_e F^k. \tag{2.171}$$

For every integer ,

$$F^{k,\ell} \geq F^k.$$

Thus,

$$\tau\text{-}\mathrm{lm}_e F^{k,\ell} = F_\ell \geq \tau\text{-}\mathrm{ls}_e F^k.$$

Taking the infimum with respect to ℓ, we obtain (2.171).

Let us now verify that

$$\tau\text{-}\mathrm{li}_e F^k \geq F$$

i.e. that, for every convergent sequence $x_k \xrightarrow{\tau} x$

$$\liminf_k F^k(x_k) \geq F(x).$$

For each ℓ, since $\tau\text{-}\mathrm{lm}_e F^{k,\ell} = F_\ell$,

$$\liminf_{k} F^{k,\ell}(x_k) \geq F_{\ell}(x).$$

Therefore,

$$\liminf_{\ell} \; \liminf_{k} F^{k,\ell}(x_k) \geq F(x).$$

From the diagonalization Lemma 1.15, there exists a strictly increasing mapping $k \to \ell(k)$ such that

$$\liminf_{k} F^{k,\ell(k)}(x_k) \geq \liminf_{\ell} \; \liminf_{k} F^{k,\ell}(x_k)$$

$$\geq F(x).$$

So

$$\liminf_{k} \sup \{F^k(x_k), -\ell(k)\} \geq F(x),$$

which clearly implies

$$\liminf_{k} F^k(x_k) \geq F(x).$$

(b) The second method relies on the following lemma of independent interest.

LEMMA 2.70 Let $\phi:\bar{\mathbb{R}} \to [a,b]$ be an increasing homeomorphism. Then for every sequence $(F^n)_{n\in\mathbb{N}}$ of functions from (X,τ) a general topological space into $\bar{\mathbb{R}}$ the following equalities hold:

$$\phi(\tau\text{-}li_e F^n) = \tau\text{-}li_e \, \phi(F^n)$$

$$\phi(\tau\text{-}ls_e F^n) = \tau\text{-}ls_e \, \phi(F^n).$$

Thus,

$$F = \tau\text{-}lm_e F^n$$

is equivalent to

$$\phi(F) = \tau\text{-}lm_e \, \phi(F^n).$$

Given a sequence $F^n:X \to \bar{\mathbb{R}}$, we apply the compactness theorem to the sequence

$\phi(F^n):X\to [a,b]$ (we are in the situation of (2.169) since the sequence $\phi(F^n)$ is uniformly bounded!):

$$\exists n_k \text{ such that } \tau\text{-}\mathrm{lm}_e\phi(F^{n_k})\text{exists.}$$

From the above lemma,

$$\tau\text{-}\mathrm{lm}_e F^{n_k} \text{ exists and is equal to } \phi^{-1}(\tau\text{-}\mathrm{lm}_e\ \phi(F^{n_k})).\ \square$$

Let us now give the proof of Proposition 2.69 Let $(x_p)_{p\in\mathbb{N}}$ be a dense denumerable subset of (X,τ) and $(\lambda_m)_{m\in\mathbb{N}}$ a sequence of strictly positive numbers decreasing to zero.

For each $(m,p)\in\mathbb{N}\times\mathbb{N}$, the sequence $(F^n_{\lambda_m}(x_p))_{n\in\mathbb{N}}$ is bounded:

$$-2rd^2(x_p,x_o)-r < F^n_{\lambda_m}(x_p) < F^n(x_{on}) + \frac{1}{2\lambda_m}d^2(x_{on},x_p). \tag{2.172}$$

The left inequality follows from (2.156). The right one relies on the following remark: given a sequence $(F^n)_{n\in\mathbb{N}}$ satisfying (2.169), either

$$\forall x\in X,\quad \forall\lambda>0\ \lim_{n\to+\infty} F^n_\lambda(x) = +\infty,$$

(or there exists a subsequence $(n_k)_{k\in\mathbb{N}}$ and a bounded sequence $(x_k)_{k\in\mathbb{N}}$ in X such that $\sup_{k\in\mathbb{N}} F^{n_k}(x_k) < +\infty$. (2.173)

Thus (up to the extraction of a subsequence) property (2.172) is satisfied. By a standard diagonalization argument we can extract an(other) subsequence, which we still denote by $(n_k)_{k\in\mathbb{N}}$, such that

$$\forall(m,p)\in\mathbb{N}\times\mathbb{N}\quad \lim_{k\to+\infty} F^{n_k}_{\lambda_m}(x_p) \text{ exists.} \tag{2.174}$$

From Theorem 2.64, for every $m\in\mathbb{N}$, the Moreau-Yosida approximates $(F^{n_k}_{\lambda_m}(\cdot))_{k\in\mathbb{N}}$ satisfy an equi-Lipschitz property. (The constant C in formula (2.154) can be taken as independent of k, because of (2.173) and of the uniform minorization assumption (2.169).) This, combined with (2.174), implies that for every $m\in\mathbb{N}$, for every $x\in X$,

$$\lim_{k\to+\infty} F^{n_k}_{\lambda_m}(x) \text{ exists.}$$

Using the equi-Lipschitz dependence of the Moreau-Yosida approximates F_λ^n with respect to λ, by the same type of argument (take $(\lambda_m)_{m\in\mathbb{N}}$ dense in $\mathbb{R}^+$), we finally obtain

$$\forall\lambda > 0,\ \forall x \in X,\ \lim_{k\to+\infty} F_\lambda^{n_k}(x) \text{ exists.} \quad \square$$

REMARK 2.71

(a) Notice that the family of the Moreau-Yosida approximates $(F_\lambda^n)_{n\in\mathbb{N}}$ may be pointwise convergent for every $\lambda > 0$, i.e.

$$\forall\lambda > 0\ \ \forall x \in X\ \ \lim_{n\to+\infty} F_\lambda^n(x) = F^\lambda(x) \text{ exists,}$$

(which implies the τ-epi convergence of the sequence $(F^n)_{n\in\mathbb{N}}$ (Corollary 2.67) to $F = \sup_{\lambda>0} F^\lambda$) *but*, in general the limit family $(F^\lambda)_{\lambda>0}$ is no longer the Moreau-Yosida approximation of a given function. (In particular $F_\lambda \neq F^\lambda$!).

This can be verified on the following example. Take $X = L^2(0,1)$, τ the strong topology of X and

$$F^n(u) = \int_0^1 a_n(t)u^2(t)dt,$$

where the a_n are measurable and satisfy $0 < \lambda_o \leqslant a_n(t) \leqslant \Lambda_o < +\infty$ for almost every $t \in (0,1)$ and all integers $n \in \mathbb{N}$. Let us assume that

$$a_n \xrightarrow[n\to+\infty]{} a\ \sigma(L^\infty,L^1),$$

for example a_n oscillates more and more rapidly as n goes to $+\infty$, between two values α and β.

An elementary computation yields

$$F_\lambda^n(u) = \inf_{v\in L^2(0,1)} \int_0^1 \{a_n(t)v^2(t) + \frac{1}{2\lambda}(v(t)-u(t))^2\}dt$$

$$= \int_0^1 \frac{a_n(t)}{1 + 2\lambda a_n(t)} u^2(t)dt.$$

Thus

$$\lim_{n\to+\infty} F_\lambda^n(u) = F^\lambda(u) \text{ exists,}$$

where

$$F^{\lambda}(u) = \int_0^1 (\text{w-}\lim_n \frac{a_n}{1+2\lambda a_n}) u^2(t)\, dt \quad (\text{by w- we denote } \sigma(L^\infty, L^1)). \quad (2.175)$$

It follows from Corollary 2.67 that the sequence (F^n) τ-epi converges to

$$F = \sup_{\lambda>0} F^{\lambda},$$

that is

$$F(u) = \int_0^1 (\text{w-}\lim_n a_n) u^2(t)dt$$

(this could have been obtained by a direct argument!). By the same computation as above,

$$F_{\lambda}(u) = \int_0^1 \frac{(\text{w-}\lim a_n)}{1 + 2\lambda(\text{w-}\lim a_n)} u^2(t)dt.$$

One can easily verify that

$$\text{w-}\lim_n \left(\frac{a_n}{1 + 2\lambda a_n}\right) \neq \frac{\text{w-}\lim_n a_n}{1 + 2\lambda(\text{w-}\lim_n a_n)} \quad (2.176)$$

(the only functions $\phi: \mathbb{R} \to \mathbb{R}$ such that $\phi(\text{w-}\lim a_n) = \text{w-}\lim \phi(a_n)$ for any such weakly converging sequences $(a_n)_{n\in\mathbb{N}}$ are the affine functions. Here ϕ is homographic!) In the above example, one can verify that (2.176) holds whenever $\alpha \neq \beta$.

Equivalently,

$$F^{\lambda} \neq F_{\lambda}. \quad (2.177)$$

There is in fact an inequality which links these quantities in a general way:

$$\forall \lambda > 0 \quad F_{\lambda} \geqslant F^{\lambda}. \quad (2.178)$$

Since $F = \tau\text{-lm}_e F^n$, for every $u \in X$, there exists $u_n \xrightarrow{\tau} u$ such that $F^n(u_n) \to F(u)$. By definition of F^n_{λ}

$$\forall x \in X \quad F^n_{\lambda}(x) \leqslant F^n(u_n) + \frac{1}{2\lambda} d^2_{\tau}(u_n, x)$$

Letting n go to $+\infty$,

$$F^{\lambda}(x) = \lim_{n} F^n_{\lambda}(x) \leq F(u) + \frac{1}{2\lambda} d^2_{\tau}(u,x).$$

This being true for any $u \in X$,

$$F^{\lambda}(x) \leq \inf_{u \in X} \{F(u) + \frac{1}{2\lambda} d^2_{\tau}(u,x)\} = F_{\lambda}(x).$$

(b) In the next chapter, we shall prove that, in the case of convex functions on a reflexive Banach space X, pointwise convergence of the Moreau-Yosida approximates F^n_{λ} to the corresponding approximates of F, that is

$$\forall \lambda > 0, \quad \forall x \in X \quad F^n_{\lambda}(x) \xrightarrow[(n \to +\infty)]{} F_{\lambda}(x),$$

is equivalent to Mosco convergence (i.e. epi-convergence for *both* strong and weak topologies of X) of the sequence F^n. In the preceding example

$$s\text{-}X\text{-}\mathrm{lm}_e F^n(u) = \int_0^1 \text{w-lim } a_n \, u^2(x)du$$

$$w\text{-}X\text{-}\mathrm{lm}_e F^n(u) = \int_0^1 \frac{1}{\text{w-lim } \frac{1}{a_n}} u^2(x)dx.$$

The same reasons which explain why $s\text{-}\mathrm{lm}_e F^n \neq w\text{-}\mathrm{lm}_e F^n$ (i.e. the sequence $(a_n)_{n \in \mathbb{N}}$ converges weakly and not strongly, i.e.

$$\text{w-lim } a_n \neq \frac{1}{\text{w-lim } \frac{1}{a_n}})$$

explain also why the family $(F^{\lambda})_{\lambda > 0}$ is no longer the Moreau-Yosida approximation of some F.

(c) We have already proved (cf. Corollary 2.67) that

(i) $\forall \lambda > 0 \quad \forall x \in X \quad \lim_{n \to +\infty} F^n_{\lambda}(x)$ exists

implies

(ii) $\tau\text{-}\mathrm{lm}_e F^n$ exists,

(where τ is the topology associated with the distance used in the definition of F^n_{λ}). It is quite natural to study the reverse statement: does (ii) imply (i)? the answer is no: take

$$F^n(u) = \int_0^1 a_n(x)\dot{u}^2(x)dx,$$

where

$$a_n = \begin{cases} b_n & \text{if } n \text{ odd}, \\ c_n & \text{if } n \text{ even}, \end{cases}$$

such that w-lim b_n = w-lim c_n but

$$\text{w-lim} \frac{b_n}{1 + 2\lambda b_n} \neq \text{w-lim} \frac{c_n}{1 + 2\lambda c_n}$$

(cf. (2.175)).

2.8 TOPOLOGY OF EPI-CONVERGENCE

In this section, we study the existence of a topology on the family of real extended valued functions defined on a topological space (X,τ), which induces the convergence in the τ-epigraph sense.

In Section 1.4, we established the equivalence between τ-epi convergence of a sequence of functions and set-convergence of the corresponding epigraphs in the product space $X \times \mathbb{R}$ (equipped with the product topology $\tau \times d_o$). Moreover, as noticed in Corollary 2.7, when studying the τ-epi convergence of a sequence $(F^n)_{n\in\mathbb{N}}$ one can, without restriction, assume the F^n to be τ-lower semicontinuous, since $\tau\text{-li}_e F^n = \tau\text{-li}_e(c\ell_\tau F^n)$ and $\tau\text{-ls}_e F^n = \tau\text{-ls}_e(c\ell_\tau F^n)$. Noticing that a function is τ-lower semicontinuous iff its epigraph is τ-closed, the problem can be formulated in the following equivalent way: Let us study the existence of a topology $\mathfrak{T}$ on the closed subsets of a topological space (Y,σ) which induces convergence in the Kuratowski sense (the set convergence).

Taking $Y = X \times \mathbb{R}$, $\sigma = \tau \times d_o$, by restriction of the topology $\mathfrak{T}$ to the family of the closed epigraphs of Y, we shall obtain the topology of the τ-epi convergence.

We first consider the general problem of the existence of such a topology $\mathfrak{T}$ then give a positive answer in the case of a locally compact Hausdorff space X. Most theoretical results found in this section are extracted from Choquet (1947-48) [1], Michael (1951) [1], from the book of Kuratowski (1958)

[1], and from the recent presentation of Dolecki, Salinetti and Wets (1983) [1]. The only structure used in this section, as in the whole chapter, is the topological one. In the convex case, we shall make precise and enrich a number of these results (refer to next chapter).

Finally, we show on two examples, highly oscillatory potentials and stochastic homogenization, how the topological point of view (by contrast with the "convergence" point of view) allows a sharp approach.

2.8.1 Existence of a topology including epi-convergence. The general situation

Given (X,τ) a general topological space we have introduced on sequences of functions from X into $\bar{\mathbb{R}}$ (the theory can be readily extended to *filtered sequences*) a convergence notion called τ-epi convergence. Following C. Berge [1], we recall the four properties which characterize the convergences which are associated to a topology: (by $\vdash$ we denote a subsequence)

(i) $F^\nu \to F, \quad F^\mu \vdash F^\nu \Rightarrow F^\mu \to F.$

(ii) $F^\nu \equiv F \Rightarrow F^\nu \to F.$

(iii) $F^\nu \to F$ when

$$\forall F^\mu \vdash F^\nu \quad \exists F^\gamma \vdash F^\mu : \quad F^\gamma \to F$$

(iv) If $F^\nu \to F$ and, for every $\nu \in \mathbb{N}$, $F^\nu_\mu \to F^\nu$, one can extract from the $\{F^\nu_\mu\}$ a filtered sequence converging to F.

PROPOSITION 2.72 Epi convergence satisfies properties (i), (ii), (iii).

The difficulty comes from the diagonalization property (iv). We shall be able to verify it (in fact a posteriori), and hence derive the topological character of the epi convergence only in some particular cases: the locally compact case will be examined in this section, the case of convex functions in the next chapter.

Property (i) is a consequence of the following inequality: if $F^\mu \vdash F^\nu$,

$$\tau\text{-}li_e F^\nu \leq \tau\text{-}li_e F^\mu \leq \tau\text{-}ls_e F^\mu \leq \tau\text{-}ls_e F^\nu.$$

Property (ii) follows from Theorem 2.1:

$$F^\nu \equiv F \Rightarrow \tau\text{-}\mathrm{lm}_e F^\nu = c\ell_\tau F.$$

At this point, we stress the fact that we have to work with closed functions.

- Let us verify (iii) and first prove that

$$\forall x_\nu \xrightarrow{\tau} x \ \liminf_\nu F^\nu(x_\nu) \geqslant F(x).$$

By definition of the lower limit, there exists a subsequence $\mu \vdash \nu$ such that

$$\liminf_\nu F^\nu(x_\nu) = \lim_\mu F^\mu(x_\mu).$$

By assumption, there exists a subsequence $F^\gamma \vdash F^\mu$ such that $F = \tau\text{-}\mathrm{lm}_e F^\gamma$. Thus,

$$\lim_\mu F^\mu(x_\mu) = \lim_\gamma F^\gamma(x_\gamma) \geqslant F(x).$$

Let us now verify that, for every $x \in X$ there exists a sequence (x_ν) such that:

$$x_\nu \xrightarrow{\tau} x \text{ and } F^\nu(x_\nu) \to F(x).$$

Via the one to one bicontinuous correspondence $F \to \mathrm{epi}\, F$ between closed functions and closed sets (refer to Theorems 1.36 and 1.39), it is equivalent to verify the following implication: for any filtered sequence $(K^\nu)_{\nu \in \mathbb{H}}$ of closed sets,

$$(\forall K^\mu \vdash K^\nu \ \exists K^\gamma \vdash K^\mu : K = \mathrm{Lm} K^\gamma) \Rightarrow (K \subset \mathrm{Li} K^\nu)$$

(the convergence being taken in Kuratowski sense: it is set convergence).

Let us argue by contradiction and assume that there exists $x_o \in K$ such that $x_o \notin \mathrm{Li} K^\nu = \bigcap_{\mu \vdash \nu} \mathrm{Ls} K^\mu$ (this last equality follows from (2.44)).

Then there exists $\mu \vdash \nu$ such that $x_o \notin \mathrm{Ls} K^\mu$. By assumption, there exists $K^\gamma \vdash K^\mu$ such that $K = \mathrm{Lm} K^\gamma$. From $K^\gamma \vdash K^\mu$, we derive $\mathrm{Ls} K^\gamma \subset \mathrm{Ls} K^\mu$, and from $x_o \notin \mathrm{Ls} K^\mu$ we obtain

$$x_o \notin \mathrm{Ls} K^\gamma.$$

On the other hand, since $K = \mathrm{Lm} K^\gamma = \mathrm{Ls} K^\gamma$ and $x_o \in K$, we have

$$x_o \in \mathrm{Ls} K^\gamma,$$

a clear contradiction. □

Because of its practical importance (it is the base of the proof of epi-convergence via compactness argument) let us formulate property (iii) in terms of epi-convergence: the following statement follows from the above argument and from the compactness Theorem 2.22.

PROPOSITION 2.73 Let (X,τ) be a second countable topological space and $\{F^n; n \in \mathbb{N}\}$ a sequence of functions from X into $\bar{\mathbb{R}}$. If the sequence $\{F^n; n \in \mathbb{N}\}$ possesses a unique limit point F for the τ-epi convergence then it τ-epi converges to F.

2.8.2 The locally compact case

When (X,τ) is a locally compact space, we are going to prove the existence of a topology inducing τ-epi convergence. As indicated in the introduction, we can formulate the problem in an equivalent way in terms of the existence of a topology inducing set convergence on the closed subsets of (Y,σ) (where $Y = X \times \mathbb{R}$). We denote respectively by

$\mathcal{F}_\sigma$ the hyperspace of closed subsets C of (Y,σ).

$\mathbb{K}_\sigma$ the hyperspace of compact subsets K of (Y,σ).

$\mathbb{O}_\sigma$ the hyperspace of open subsets $\mathbb{O}$ of (Y,σ).

The following definition introduces a topology on $\mathcal{F}_\sigma$: this notion works in a general topological space. But, as we shall see, it induces set convergence only in some particular cases: the locally compact case is considered in this section.

DEFINITION 2.74 The topology $\mathbb{E}_\sigma$, called the topology of set convergence on $\mathcal{F}_\sigma$ (also called the Kuratowski topology) is generated by the subbase of open sets

$$\{C^K; K \in \mathbb{K}_\sigma\} \quad \text{and} \quad \{C_{\mathbb{O}}; \mathbb{O} \in \mathbb{O}_\sigma\}$$

where for any $Q \subset Y$

$$C^Q = \{C \in \mathcal{F}_\sigma \ / \ C \cap Q = \emptyset\}$$

and

$$C_Q = \{C \in \mathcal{F}_\sigma / C \cap Q \neq \emptyset\}.$$

In other words, C^K is the family of the closed subsets of (Y,σ) which avoid the compact K, and $C_{\mathcal{O}}$ is the family of the closed subsets of (Y,σ) which meet the open set $\mathcal{O}$.

This definition calls for a few comments: we first notice that an open set of $\mathcal{E}_\sigma$ is a family of closed subsets of (Y,σ)! By definition,

$$\{C^K; K \in \mathcal{K}_\sigma\} \cup \{C_{\mathcal{O}}; \mathcal{O} \in \mathcal{O}_\sigma\}$$

is a subbase for the topology $\mathcal{E}_\sigma$, so the class of subsets

$$\{C^K \cap C_{\mathcal{O}_1} \cap C_{\mathcal{O}_2} \cap \ldots \cap C_{\mathcal{O}_n} / K \in \mathcal{K}_\sigma ; \mathcal{O}_i \in \mathcal{O}_\sigma \; i = 1,2,\ldots,n; n \in \mathbb{N}\}$$

is a base for the topology $\mathcal{E}_\sigma$ (every open set of $\mathcal{E}_\sigma$ can be obtained as the union of members of this class).

THEOREM 2.75 Let (Y,σ) be a locally compact topological space and $(C^n)_{n\in\mathbb{N}}$ a sequence of closed subsets of Y. Then, $C = \sigma\text{-Lm}C^n$ if and only if C^n converges to C for the topology $\mathcal{E}_\sigma$. In other words, convergence for the topology $\mathcal{E}_\sigma$ is the Kuratowski convergence (also called set convergence).

Proof of Theorem 2.75

(a) Let us prove that:

$C \subset \sigma\text{-Li}C^n$ if and only if, to every $\mathcal{O} \in \mathcal{O}_\sigma$ such that $C \cap \mathcal{O} \neq \emptyset$, there corresponds $H_{\mathcal{O}} \in \mathcal{N}$ such that: $\forall n \in H_{\mathcal{O}} \; C^n \cap \mathcal{O} \neq \emptyset$. (2.179)

Let us first prove the part "if":

Suppose that x belongs to C. Then, for every $\mathcal{O} \in \mathcal{O}_\sigma(x)$, $\mathcal{O} \cap C \neq \emptyset$. From the hypothesis, this implies that

$$\forall \mathcal{O} \in \mathcal{O}_\sigma(x) \quad \exists H_{\mathcal{O}} \in \mathcal{N}: \forall n \in H_{\mathcal{O}} \; C^n \cap \mathcal{O} \neq \emptyset.$$

Take $H' \in \ddot{\mathcal{N}}$ the grill of $\mathcal{N}$; then, H' meets every $H \in \mathcal{N}$ and hence

$$\forall \mathcal{O} \in \mathcal{O}_\sigma(x) \quad \forall H' \in \ddot{\mathbb{N}} \quad (\bigcup_{n \in H'} C^n) \cap \mathcal{O} \neq \emptyset;$$

equivalently

$$\forall H' \in \ddot{\mathbb{N}} \quad x \in c\ell_\sigma(\bigcup_{n \in H'} C^n), \text{ or, } x \in \bigcap_{H' \in \ddot{\mathbb{N}}} c\ell_\sigma(\bigcup_{n \in H'} C^n)$$

which, from Proposition 1.32, means $x \in \sigma\text{-Li}C^n$; so $C \subset \sigma\text{-Li}C^n$.

Suppose now that $C \subset \tau\text{-Li}C^n$; then, $C \cap \mathcal{O} \neq \emptyset$ implies that

$$\mathcal{O} \cap (\bigcap_{H' \in \ddot{\mathbb{N}}} c\ell_\sigma(\bigcup_{n \in H'} C^n)) \neq \emptyset$$

which, in turn, implies:

$$\forall H' \in \ddot{\mathbb{N}} \quad \mathcal{O} \cap (\bigcup_{n \in H'} C^n) \neq \emptyset.$$

Equivalently, for every $H' \in \ddot{\mathbb{N}}$ there exists $n_{H'} \in H'$ such that

$$\mathcal{O} \cap C^{n_{H'}} \neq \emptyset.$$

Take $H = \bigcup_{H' \in \ddot{\mathbb{N}}} \{n_{H'}\}$; then H belongs to the grill of $\ddot{\mathbb{N}}$, i.e. to $\mathbb{N}$, since $\mathbb{N}$ is a filter. So,

$$\forall n \in H \quad C^n \cap \mathcal{O} \neq \emptyset,$$

which completes the proof of (2.179).

(b) Let us now prove that

$C \supset \sigma\text{-Ls}C^n$, if and only if to every $K \in \mathbb{K}_\sigma$ such that $K \cap C = \emptyset$,

there corresponds $H_K \in \mathbb{N}$ such that:

$$\forall n \in H_K \quad C^n \cap K = \emptyset. \tag{2.180}$$

Let us prove the part "if"; take $x \in \sigma\text{-Ls}C^n$, from Proposition 1.32

$$\forall H \in \mathbb{N} \quad x \in c\ell_\sigma(\bigcup_{n \in H} C^n).$$

Suppose that $x \notin C$. The set C being closed and (Y,σ) being locally compact, there exists a compact neighbourhood K of x, $K \ni x$, such that

$$K \cap C = \emptyset.$$

By assumption, with $K \in \mathbb{K}_\sigma$ we can associate $H_K \in \mathbb{N}$ such that

$$\forall n \in H_K \quad C^n \cap K = \emptyset \text{ i.e. } K \cap (\bigcup_{n \in H_K} C^n) = \emptyset.$$

Thus $x \notin c\ell_\sigma(\bigcup_{n \in H_K} C^n)$; this contradicts the assumption that $x \in \sigma\text{-Ls}C^n$. So $x \in C$, and $C \supset \sigma\text{-Ls}C^n$, which proves the part "if" of (2.180). Let us now prove the part "only if" and assume that $C \supset \sigma\text{-Ls}C^n$. Take $K \in \mathbb{K}_\sigma$ such that $K \cap C = \emptyset$ and argue by contradiction: let us assume that for every $H \in \mathbb{N}$ there exists $n_H \in H$ such that

$$C^{n_H} \cap K \neq \emptyset.$$

Taking $H' = \bigcup_{H \in \mathbb{N}} \{n_H\}$, we have $H' \in \ddot{\mathbb{N}}$ and $\forall n \in H' \quad C^n \cap K \neq \emptyset$. The set K being σ-compact, $\{C^n \cap K; n \in H'\}$ admit at least a cluster point $x \in K$. Then, for every $H \in \mathbb{N}$,

$$x \in c\ell_\sigma(\bigcup_{n \in H} C^n) \cap K$$

and consequently, $x \in \sigma\text{-Ls}C^n \cap K$, i.e. $\sigma\text{-Ls}C^n \cap K \neq \emptyset$. Since $C \supset \sigma\text{-Ls}C^n$ $C \cap K \neq \emptyset$ which contradicts the hypothesis.

(c) Let us now interpret properties (2.179) and (2.180) in terms of topology $\mathbb{E}_\sigma$:

$$\begin{cases} C \subset \sigma\text{-Li}C^n \iff \forall C_{\mathbb{O}} \ni C \quad \exists H_{\mathbb{O}} \in \mathbb{N} \quad \forall n \in H_{\mathbb{O}} \quad C_{\mathbb{O}} \ni C^n. \\ C \supset \sigma\text{-Ls}C^n \iff \forall C^K \ni C \quad \exists H_K \in \mathbb{N} \quad \forall n \in H_K \quad C^K \ni C^n. \end{cases}$$

$$C = \sigma\text{-Lm}C^n \text{ iff } \sigma\text{-Ls}C^n \subset C \subset \sigma\text{-Li}C^n;$$

using (2.179), (2.180) and the fact that the $\{C_{\mathbb{O}}; C^K; \mathbb{O} \in \mathbb{O}_\sigma\ K \in \mathbb{K}_\sigma\}$ form an open subbase of $\mathbb{E}_\sigma$, we obtain that $(C^n)_{n \in \mathbb{N}}$ converges to C for $\mathbb{E}_\sigma$ iff $C = \sigma\text{-Lm}C^n$. □

Let us now study properties of the topology $\mathbb{E}_\sigma$ and give in the present context the topological interpretation of compactness Theorems 2.22 and 2.23.

THEOREM 2.76 Let (Y,σ) be a Hausdorff, locally compact topological space. Then $(\mathcal{F}_\sigma, \mathcal{E}_\sigma)$, the hyperspace of the closed subsets of (Y,σ) equipped with the topology of set convergence, is Hausdorff and compact.

Proof of Theorem 2.76

1. Let us prove first that, if (Y,σ) is Hausdorff, so is $(\mathcal{F}_\sigma, \mathcal{E}_\sigma)$. Let C^1 and C^2 be two distinct closed subsets of Y. Then, there exists some y that belongs to C^1 and not to C^2 (or vice versa).

At this point, let us use the following result: a Hausdorff, locally compact topological space is normal (i.e. one can separate two disjoint closed subsets) and hence, regular (i.e. one can separate a point from a disjoint closed subset). This can be viewed rapidly by introduction of the Alexandrov compactification $\hat{Y}$ of Y. Since $\hat{Y}$ is compact Hausdorff, it is normal; the normal property being hereditary and topological, Y is also normal.

The space Y being locally compact and regular, there exist two open sets $\mathcal{O}^1$ and $\mathcal{O}^2$ such that:

$$y \in \mathcal{O}^1 \text{ and } \mathcal{O}^1 \text{ is relatively compact}$$

$$C^2 \subset \mathcal{O}^2 \text{ and } \mathcal{O}^1 \cap \mathcal{O}^2 = \emptyset.$$

This implies that $K^1 = c\ell_\sigma \mathcal{O}^1$ is compact and disjoint from C^2. Hence, $C^1 \in C_{\mathcal{O}^1}$ and $C^2 \in C^{K^1}$; noticing that $C_{\mathcal{O}^1} \cap C^{K^1} = \emptyset$, we can separate two distinct points of the topological space $(\mathcal{F}_\sigma, \mathcal{E}_\sigma)$, which is therefore Hausdorff.

2. Let us now prove that $(\mathcal{F}_\sigma, \mathcal{E}_\sigma)$ is compact. To this end, we use the Alexander's characterization of compactness in terms of the finite intersection property (for a subbase). Noticing that the family $\{C^K; C_{\mathcal{O}}; K \in \mathcal{K}_\sigma; \mathcal{O} \in \mathcal{O}_\sigma\}$ is a subbase of $\mathcal{E}_\sigma$ and that the complements of C^K and $C_{\mathcal{O}}$ are respectively C_K and $C^{\mathcal{O}}$, the problem can be reduced to: Suppose that

$$\bigcap_{i\in I} C_{K_i} \bigcap_{j\in J} C^{\mathcal{O}^j} = \emptyset, \text{ where } K_i \in \mathcal{K}_\sigma,\ \mathcal{O}^j \in \mathcal{O}_\sigma \quad (2.181)$$

and I and J are arbitrary index sets. We have to show that the family of sets $\{C_{K_i}, C^{\mathcal{O}^j}; i \in I, j \in J\}$ contains a finitely indexed subfamily which has an empty intersection.

Let us observe that $\bigcap_{j\in J} C^{\mathcal{O}^j} = C^{\mathcal{O}}$ with $\mathcal{O} = \bigcup_{j\in J} \mathcal{O}^j$, which as a union of open sets is still open. So

$$\Longleftrightarrow C^{\mathcal{O}} \cap \left(\bigcap_{i\in I} C_{K_i} \right) = \emptyset;$$

let us interpret this property in a simpler way. The following sentences are equivalent to (2.181):

$$\forall C \in \mathcal{F}_\sigma \text{ satisfying } C \in C^{\mathcal{O}},\ C \in \complement\left(\bigcap_{i\in I} C_{K_i} \right) = \bigcup_{i\in I} C^{K_i}.$$

$$\forall C \in \mathcal{F}_\sigma \text{ satisfying } C \cap \mathcal{O} = \emptyset,\ \exists i \in I \text{ such that } C \cap K_i = \emptyset. \tag{2.182}$$

$$\text{There exists an index } i_o \in I \text{ such that } K_{i_o} \subset \mathcal{O}. \tag{2.183}$$

The last equivalence is a consequence of the following remark: If (2.182) holds, noticing that $\mathcal{O} \in \mathcal{O}_\sigma$, then $Y \setminus \mathcal{O} \in \mathcal{F}_\sigma$ and $(Y\setminus\mathcal{O}) \cap \mathcal{O} = \emptyset$. So there exists $i_o \in I$ such that $(Y\setminus\mathcal{O}) \cap K_{i_o} = \emptyset$ i.e. $K_{i_o} \subset \mathcal{O}$ and (2.183) holds. Conversely, if $K_{i_o} \subset \mathcal{O}$ and $C \cap \mathcal{O} = \emptyset$ then $C \cap K_{i_o} = \emptyset$.

From the Heine-Borel property, K_{i_o} being compact, from the open cover $\mathcal{O} = \bigcup_{j\in J} \mathcal{O}^j$ one can extract a finite subcover

$$K_{i_o} \subset \mathcal{O}^{j_1} \cup \mathcal{O}^{j_2} \cup \ldots \cup \mathcal{O}^{j_\ell}.$$

From the above equivalence, reinterpreting this inclusion in terms of the C_K and $C^{\mathcal{O}}$, we obtain

$$C_{K_{i_o}} \cap C^{\mathcal{O}^{j_1}} \cap C^{\mathcal{O}^{j_2}} \cap \ldots \cap C^{\mathcal{O}^{j_\ell}} = \emptyset$$

which ends the proof of the finite intersection property. □

PROPOSITION 2.77 Let (Y,σ) be a Hausdorff, locally compact, second countable topological space. Then $(\mathcal{F}_\sigma, \mathcal{E}_\sigma)$, the hyperspace of the closed subsets of (Y,σ) equipped with the topology of set convergence, is second countable.

Proof of Proposition 2.77 The topological space (Y,σ) being second countable, let us denote by $(\mathcal{O}^n)_{n\in\mathbb{N}}$ a countable base of open subsets of Y. From local compactness of (Y,σ), we can take the $(\mathbb{O}^n)_{n\in\mathbb{N}}$ to be relatively compact.

For every $n \in \mathbb{N}$, take $K^n = c\ell_\sigma \mathbb{O}^n$ and so define a countable family $(K^n)_{n\in\mathbb{N}}$ of compact subsets of (Y,σ).

Let us prove that the family $\{C_{\mathbb{O}^n}; n \in \mathbb{N}\} \cup \{C^{K^n}; n \in \mathbb{N}\}$ is still a subbase (and hence a countable subbase) of $(\mathcal{F}_\sigma, \mathcal{E}_\sigma)$. The family whose elements are the finite intersections of such subsets will be still countable and form a base of $(\mathcal{F}_\sigma, \mathcal{E}_\sigma)$.

Let us denote by $\mathfrak{T}$ the topology generated by the class $\{C_{\mathbb{O}^n}; C^{K^n}; n \in \mathbb{N}\}$. Clearly, $\mathfrak{T} \subset \mathcal{E}_\sigma$; let us prove the opposite inclusion:

(a) Let $\mathbb{O} \in \mathcal{O}_\sigma$ and $C \in C_{\mathbb{O}}$ i.e. $C \cap \mathbb{O} \neq \emptyset$. By definition of a base, $\mathbb{O}$ is equal to a union of elements of the family $(\mathbb{O}^n)_{n\in\mathbb{N}}$:

$$\mathbb{O} = \bigcup_{k\in\mathbb{N}} \mathbb{O}^{n_k}.$$

Since $C \cap \mathbb{O} \neq \emptyset \Rightarrow \exists k_o \in \mathbb{N}$ such that $C \cap \mathbb{O}^{n_{k_o}} \neq \emptyset$; noticing that $(\mathbb{O}^{n_{k_o}} \subset \mathbb{O}) \Rightarrow (C_{\mathbb{O}^{n_{k_o}}} \subset C_{\mathbb{O}})$, we derive that, for every $C \in C_{\mathbb{O}}$, there exists $k_o \in \mathbb{N}$ such that $C \in C_{\mathbb{O}^{n_{k_o}}} \subset C_{\mathbb{O}}$. Hence, $C_{\mathbb{O}}$ can be written as a union of open sets of $\mathfrak{T}$, and for every $\mathbb{O} \in \mathcal{O}_\sigma$, $C_{\mathbb{O}} \in \mathfrak{T}$.

(b) Let $K \in \mathcal{K}_\sigma$ and $C \in C^K$, i.e. $C \cap K = \emptyset$.

The topological space (Y,σ) being locally compact, Hausdorff is normal and thus the two closed subsets C and K can be separated: there exists $\mathbb{O}^1$ and $\mathbb{O}^2$, two disjoint open sets such that

$$C \subset \mathbb{O}^1 \text{ and } K \subset \mathbb{O}^2.$$

Since $K \subset \mathbb{O}^2$ and K is compact, by definition of a base of open sets, and the Heine-Borel property, there exists a finite number of elements $\mathbb{O}^{n_1},\ldots,\mathbb{O}^{n_\ell}$ of the base $(\mathbb{O}^n)_{n\in\mathbb{N}}$ such that

$$K \subset \bigcup_{i=1}^{\ell} \mathbb{O}^{n_i} \subset \mathbb{O}^2.$$

This implies that $C \cap c\ell_\sigma(\bigcup_{i=1}^{\ell} \mathbb{O}^{n_i}) = \emptyset$. Denoting $K^{n_i} = c\ell_\sigma \mathbb{O}^{n_i}$, $C \cap (\bigcup_{i=1}^{\ell} K^{n_i}) = \emptyset$.

Equivalently, $C \in \bigcap_{i=1}^{\ell} C^{K^{n_i}}$; moreover, since $K \subset \bigcup_{i=1}^{\ell} K^{n_i}$, we have $\bigcap_{i=1}^{\ell} C^{K^{n_i}} \subset C^K$;

finally $C \in \bigcap_{i=1}^{\ell} C^{K^{n_i}} \subset C^K$; so C^K can be written as a union of finite intersections of elements of the family $\{C^{K^n}; n \in \mathbb{N}\}$. This implies that $C^K \in \mathfrak{C}$ which, combined with $C_\emptyset \in \mathfrak{C}$, achieves the proof of Proposition 2.77. □

Let us now interpret the preceding results in terms of epi convergence. Let us assume that (X,τ) is a Hausdorff, locally compact topological space. Let Sc_τ denote the space of the τ-lower semicontinuous functions from (X,τ) into $\bar{\mathbb{R}}$. Let $Y = X \times \mathbb{R}$, equipped with the product topology $\sigma = \tau \times d_o$ (d_o is the usual topology of $\mathbb{R}$) and $\mathcal{F}_\sigma$ be the hyperspace of closed subsets of Y. The application $F \longmapsto \text{epi}(F)$ is a one to one correspondence between Sc_τ and the subclass of $\mathcal{F}_\sigma$ formed by the closed epigraphs of Y.

We denote by $\mathfrak{E}_\sigma$ the topology induced by the topology $\mathfrak{E}_\sigma = \mathfrak{E}_{\tau \times d}$ on the closed epigraphs of $(Y, \tau \times d_o)$ and call it the topology of epi-convergence on Sc_τ (this justifies the notation $\mathfrak{E}_\tau$).

THEOREM 2.78 Let (X,τ) be a Hausdorff locally compact topological space. The space Sc_τ of the τ-lower semicontinuous functions from (X,τ) into $\bar{\mathbb{R}}$, equipped with the epi-convergence topology $\mathfrak{E}_\tau$, is a compact, Hausdorff topological space.

For every sequence of functions $(F^n)_{n\in\mathbb{N}}$ of Sc_τ, the following equivalence holds:

$$F^n \xrightarrow{\mathfrak{E}_\tau} F \iff F = \tau\text{-}\mathrm{lm}_e F^n.$$

Moreover, if (X,τ) is second countable, so is the topological space $(Sc_\tau, \mathfrak{E}_\tau)$.

Proof of Theorem 2.78 The properties of being Hausdorff and second countable are hereditary. The only thing we have to prove is that the family of the closed epigraphs in $Y = X \times \mathbb{R}$ ($\sigma = \tau \times d_o$) is closed in $\mathcal{F}_\sigma$ for the topology $\mathfrak{E}_\sigma$ of set convergence. As a closed subset of a compact space, it will still be compact. Let us denote by Epi_σ the family of the closed epigraphs of $Y = X \times \mathbb{R}$, and take $C \in \mathcal{F}_\sigma \setminus \text{Epi}_\sigma$; from the characteristic property of epigraphs, there exists $x \in X$ and $a < b$ such that $(x,a) \in C$ but $(x,b) \notin C$. The space X being locally compact Hausdorff, so is $X \times \mathbb{R}$; hence, the space $X \times \mathbb{R}$ is regular and we can separate (x,b) from the closed set C:

$$\exists \mathcal{O}^1 \in \mathcal{O}_\sigma \quad \exists \mathcal{O}^2 \in \mathcal{O}_\sigma \quad \mathcal{O}^1 \cap \mathcal{O}^2 = \emptyset \text{ and } \mathcal{O}^1 \ni (x,b), \quad \mathcal{O}^2 \supset C.$$

Since $\mathbb{O}^2 \supset C \ni (x,a)$ there exists $\mathbb{O}_2 \in \mathcal{O}_\tau(x)$ and $\varepsilon_2 > 0$ such that

$$\mathbb{O}_2 \times]a-\varepsilon_2, a + \varepsilon_2[\subset \mathbb{O}^2.$$

Since $\mathbb{O}^1 \ni (x,b)$ there exists $\mathbb{O}_1 \in \mathcal{O}_\tau(x)$ and $\varepsilon_1 > 0$ such that

$$\mathbb{O}_1 \times]b-\varepsilon_1, b + \varepsilon_1[\subset \mathbb{O}^1.$$

Moreover, we can assume (from local compactness of X), $\mathbb{O}_1$ and $\mathbb{O}_2$ relatively compact; take $\mathbb{O} = \mathbb{O}_1 \cap \mathbb{O}_2 \in \mathcal{O}_\tau(x)$, $\varepsilon = \inf(\varepsilon_1,\varepsilon_2,b-a) > 0$ and $K = c\ell_\tau \mathbb{O} \in \mathbb{K}_\sigma$; we claim that the set

$$G = C^{K\times\{b\}} \cap C_{\mathbb{O} \times]a-\varepsilon,a+\varepsilon[}$$

is an open neighbourhood of C in $\mathcal{F}_\sigma$ that does not contain any epigraph:

(α) $C \in G$: since $(x,a) \in C$ and $\mathbb{O} \in \mathcal{O}_\tau(x)$ $\mathbb{O}\times]a-\varepsilon,a+\varepsilon[\cap C \neq \emptyset$.

Since $\mathbb{O} \times \{b\}$ does not intersect a neighbourhood of C, we still have $c\ell_\tau \mathbb{O} \times \{b\} = K \times \{b\}$ has an empty intersection with C.

(β) G does not contain any epigraph: let E be an epigraph which belongs to $C_{\mathbb{O}\times]a-\varepsilon,a+\varepsilon[}$. Then, $\exists x \in \mathbb{O}$ and $\alpha \in]a-\varepsilon,a+\varepsilon[$ such that $(x,\alpha) \in E$; since $b - a > \varepsilon$, $b > \alpha$ and $(x,b) \in E \cap K \times \{b\}$, i.e. $E \notin C^{K\times\{b\}}$. Thus $\mathcal{F}_\sigma \setminus \mathrm{Epi}_\sigma$ is open, or, equivalently, Epi_σ is closed. □

<u>COROLLARY 2.79</u> Let (X,τ) be a locally compact, Hausdorff, second countable space; then, Sc_τ, the space of the τ-lower semicontinuous functions from X into $\bar{\mathbb{R}}$, equipped with the topology $\mathbb{E}_\tau$ of τ-epi convergence, is compact Hausdorff and second countable and hence *metrizable*.

<u>Proof of Corollary 2.79</u> This is a direct consequence of Theorem 2.78 and of Urysohn's metrization theorem. Let us recall the statement of this theorem: every second countable normal, Hausdorff space is metrizable.

We have just to notice that a compact Hausdorff space is normal. □

Urysohn's metrization theorem does not give us a constructive method of obtaining a metric inducing the topology $\mathbb{E}_\tau$ of the set convergence on Sc_τ; let us briefly examine this question.

First, let us notice that a locally compact, Hausdorff topological space is normal and hence the space (X,τ), which is assumed normal and second countable, is metrizable. So, without restriction, we can assume that (X,τ) is metrizable, second countable, locally compact or equivalently metrizable, separable and locally compact.

In the important particular case, $X = \mathbb{R}^m$ equipped with the usual topology, we can describe fairly simply a metric inducing the epi convergence topology. Let us denote by d_o a distance on $X = \mathbb{R}^m$ inducing τ_o, the usual topology (for example take d_o the Euclidian metric).

Given a function $F:X \to]-\infty,+\infty]$, we denote by F_λ its Moreau-Yosida approximation (refer to Section 2.7): for any $\lambda > 0$ and $x \in X$

$$F_\lambda(x) = \inf_{y \in X} \{F(y) + \frac{1}{2\lambda} d_o^2(y,x)\}.$$

In the following proposition epi convergence of a sequence of functions from $\mathbb{R}^m$ into $]-\infty,+\infty]$ is characterized in terms of pointwise convergence of their Moreau-Yosida approximations; then, we derive a metric on Sc_τ inducing the epi convergence topology $\mathcal{E}_\tau$.

PROPOSITION 2.80 Let $(F^n)_{n\in\mathbb{N}}$, F, be a sequence of closed functions from $X = \mathbb{R}^m$ into $\bar{\mathbb{R}}$ such that there exists $r \in \mathbb{R}$ and a bounded sequence $(x_{on})_{n\in\mathbb{N}}$ in X such that

$$\begin{cases} \forall n \in \mathbb{N} \quad \forall x \in X \quad F^n(x) + r(d_o(x,x_{on}) + 1) \geqslant 0 \\ \sup_{n\in\mathbb{N}} \quad F^n(x_{on}) < +\infty. \end{cases} \tag{2.184}$$

Then the following equivalences hold:

(i) $F = \tau_o\text{-}\mathrm{lm}_e F^n$

(ii) $\forall\lambda > 0 \quad \forall x \in \mathbb{R}^m \quad F_\lambda(x) = \lim_{n\to+\infty} F_\lambda^n(x)$

(iii) $\forall k \in \mathbb{N} \quad \forall i \in \mathbb{N} \quad F_{\lambda_i}(x_k) = \lim_{n\to+\infty} F^n_{\lambda_i}(x_k)$,

where $(x_k)_{k\in\mathbb{N}}$ is a dense denumerable subset of $X = \mathbb{R}^m$ and $(\lambda_i)_{i\in\mathbb{N}}$ is a sequence of strictly positive numbers decreasing to zero.

The proof of Proposition 2.80 follows from the variational Theorem 2.11 and properties of the Moreau-Yosida approximation (Theorems 2.64 and 2.65).

COROLLARY 2.81 Let $X = \mathbb{R}^m$ and for every $r \in \mathbb{R}$, $c \in \mathbb{R}$, $x_0 \in X$, let us denote

$$Sc_{r,c,x_0} = \{F:X \to]-\infty,+\infty]/F \in Sc,\ F(x_0) \leq c,\ F(\cdot) + r(d(\cdot,x_0)+1) \geq 0\}$$

Take $(x_k)_{k\in\mathbb{N}}$ a denumerable dense subset of X and $(\lambda_i)_{i\in\mathbb{N}}$, $\lambda_i > 0$, $\lambda_i \downarrow 0$. For every F and $G \in Sc$ let us define

$$e(F,G) = \sum_{i,k=1}^{+\infty} \frac{1}{2^{i+k}} \frac{|F_{\lambda_i}(x_k)-G_{\lambda_i}(x_k)|}{1 + |F_{\lambda_i}(x_k)-G_{\lambda_i}(x_k)|} .$$

Then, for every $r \in \mathbb{R}$, $c \in \mathbb{R}$, $x_0 \in X$, e is a metric on Sc_{r,c,x_0} inducing the epi convergence topology.

REMARK 2.82

(a) It has been pointed out by Vervaat [1] that the epi convergence topology $\mathcal{E}_\tau$ on Sc_τ can be defined as the weakest topology on Sc_τ making the applications

(i) $F \to \inf_K F$ lower semicontinuous for every τ-compact set K.

(ii) $F \to \inf_{\mathcal{O}} F$ upper semicontinuous for every open set $\mathcal{O} \in \mathcal{O}_\tau$.

As a corollary, one obtains the variational property of epi convergence, that is, when

$$\inf_{\mathcal{O}} F^n = \inf_K F^n \text{ for every } n \in \mathbb{N},\ K\ \tau\text{-compact},$$

then τ-epi convergence of F^n to F implies convergence of the infimum! Epi-convergence topology is also called *inf-vague topology* by Vervaat.

(b) When studying *random closed set* theory, Cressie [1] stresses the fact the topology $\mathcal{E}$ of the set-convergence (Definition 2.74) reflects exactly the way image data in $\mathbb{R}^m$ are analysed. For this reason, he called it the *"hit or miss" topology*.

(c) Measurability properties for multifunctions $\Gamma:S \to (X,\tau)$ can be attacked very simply by considering corresponding functions from S into the hyperspace $\mathcal{F}_\tau$ of closed subsets of X_τ, equipped with the set convergence topology.

For example (cf. Rockafellar [4], Attouch [4]) a *normal integrand*

$$f:S \times X \to \bar{\mathbb{R}}$$

can be viewed as a measurable function $s \in S \to f(s,\cdot)$ from S into Sc_τ the space of closed functions from X into $\bar{\mathbb{R}}$ (equipped with the τ-epi convergence topology). By application of the *Lusin* property, the study of measurability properties can be reduced to that of continuity properties with respect to epi convergence topology. In the convex case, using the continuity of the Fenchel transformation $F \to F^*$ for epi convergence topology (cf. Chapter 3) one immediately obtains that, whenever f is a normal convex integrand, so is f^*! □

REMARK 2.83

(a) When taking $F^n = I_{C^n}$, the indicator function of a closed set C^n in $\mathbb{R}^N$, Proposition 2.80 becomes

For every sequence $\{C^n; n \in \mathbb{N}\}$ of closed subsets of $\mathbb{R}^N$ such that there exists a bounded sequence $\{x_{on}; x_{on} \in C^n$ for every $n \in \mathbb{N}\}$, the following statements are equivalent (i) $\Leftrightarrow$ (i) $\Leftrightarrow$ (iii): (2.185)

(i) $C = \tau\text{-}\mathop{\mathrm{Lm}}_n C^n$;

(ii) for every $x \in \mathbb{R}^N$ $d(x,C) = \lim_n d(x,C^n)$;

(iii) for every bounded subset K of $\mathbb{R}^N$, $\sup_{x \in K} |d(x,C)-d(x,C^n)| = 0$,

where d is a metric introducing the usual topology on $\mathbb{R}^N$.

(b) This last equivalence (i) $\Leftrightarrow$ (iii) suggests that we study the connection between epi convergence and the *Hausdorff metric;* we recall that given two closed non-void subsets C and D of $\mathbb{R}^N$ their Hausdorff distance h(C,D) is defined by

$$h(C,D) = \sup \{\sup_{x \in C} d(x,D); \ \sup_{x \in D} d(x,C)\}. \tag{2.186}$$

When C and D are bounded

$$h(C,D) = \sup_{x \in \mathbb{R}^N} |d(x,C)-d(x,D)|.$$

Thus the Hausdorff metric induces a topology strictly stronger than set convergence. When restricting these two topologies to closed sets contained in a fixed compact subset of $\mathbb{R}^N$, they turn out to be equivalent (cf. Salinetti and Wets [5] for further developments). □

2.8.3 Applications in infinite dimensional cases

Up to now we have been mainly concerned with the finite dimensional case. Indeed, the preceding considerations can be readily extended to the *infinite dimensional case*, when considering classes of functionals satisfying a *uniform inf-compactness property*. Let us illustrate this on model examples 1.1.1 and 1.1.2 and then give two striking applications of this "topological approach". Let us consider, following the notation of Theorem 2.24, the following class of functionals

$$\mathcal{F}_{\lambda,\Lambda} = \{F:H^1(\Omega) \to \mathbb{R}^+/F(u)=\int_\Omega j(x,\text{grad}u(x))dx \tag{2.187}$$

where $j:\mathbb{R}^N \times \mathbb{R}^N \to \mathbb{R}^+$, $(x,z) \to j(x,z)$, is measurable in x, convex continuous in z and satisfies

$$\lambda|z|^2 \leq j(x,z) \leq \Lambda(1 + |z|^2) \text{ for every } z \in \mathbb{R}^N\}.$$

Assuming that $0 < \lambda \leq \Lambda < +\infty$, we claim that the τ-epi convergence on $\mathcal{F}_{\lambda,\Lambda}$ with τ equal to the strong topology of $L^2(\Omega)$ is associated with a topology and that $\mathcal{F}_{\lambda,\Lambda}$ is compact with respect to this topology.

Thanks to the uniform coercivness of functions of the class $\mathcal{F}_{\lambda,\Lambda}$ and to the compact injection (Ω is assumed to be bounded) of $H^1(\Omega)$ into $L^2(\Omega)$ we can easily verify that, for every sequence $\{F^n; n \in \mathbb{N}\}$ in $\mathcal{F}$

$$F = \tau\text{-lm}_e F^n$$
$$\Updownarrow$$
$$F_\lambda(u) = \lim_n F_\lambda^n(u) \text{ for every } \lambda > 0, \text{ for every } u \in L^2(\Omega)$$

where $F_\lambda^n(u) = \inf_{v\in H^1(\Omega)} \{\int_\Omega j^n(x,\text{grad}v(x))dx + \frac{1}{2\lambda}\int_\Omega |u(x)-v(x)|^2dx\}$.
Description of a metric d inducing τ-epi convergence on $\mathcal{F}_{\lambda,\Lambda}$ is then obtained as in Corollary 2.81. Compactness of $\mathcal{F}_{\lambda,\Lambda}$ is a consequence of Theorem 2.24.

A. Highly oscillating potentials

We stressed the fact that the problem of the existence of a topology inducing epi convergence is related to the diagonalization property. Conversely, as soon as we know that such a topology exists (as in the above example) we can use such a diagonalization argument. Take, for example (refer to model Example 1.1.2), the "cloud of ice":

$$F^{\varepsilon}(u) = \int_{\Omega} |\text{grad}u|^2 dx + I_{\{v=0 \text{ on } T_{\varepsilon}\}}(u)$$

where $T_{\varepsilon} = \bigcup_{i \in I(\varepsilon)} T_{\varepsilon}^{i}$, the T_{ε}^{i} are r_{ε}-homothetic of some $T \subset Y =]0,1[^N$ and ε-periodically distributed.

We proved (Theorems 1.3 and 1.27) that when $N = 3$ and $r_{\varepsilon} \sim \varepsilon^3$ = strong topology $L^2(\Omega)$

$$F = \tau\text{-lm}_{e} F^{\varepsilon} \text{ with } F(u) = \int_{\Omega} |\text{grad}u|^2 dx + C \int_{\Omega} u^2 dx$$

and $C = \text{Cap}_{\mathbb{R}^3} T$.

Let us introduce for every $\lambda > 0$ and $\varepsilon > 0$ the following penalized approximation of F^{ε}:

$$F_{\lambda}^{\varepsilon}(u) = \int_{\Omega} |\text{grad}u|^2 dx + \frac{1}{\lambda} \int_{\Omega} \chi_{\varepsilon}(x) u^2(x) dx, \tag{2.188}$$

where $\chi_{\varepsilon}(x)$ is the characteristic function of T_{ε} (i.e. equal to one on T_{ε} and zero elsewhere).

When λ decreases to zero, the sequence $\{F_{\lambda}^{\varepsilon}; \lambda \downarrow 0\}$ increases to F^{ε} and thus, by Theorem 2.40, τ-epi converges to F^{ε} (functionals F^{ε} are τ-lower semicontinuous with respect to τ equal to the strong topology of $L^2(\Omega)$). We have the following diagram, the convergences being taken with respect to the

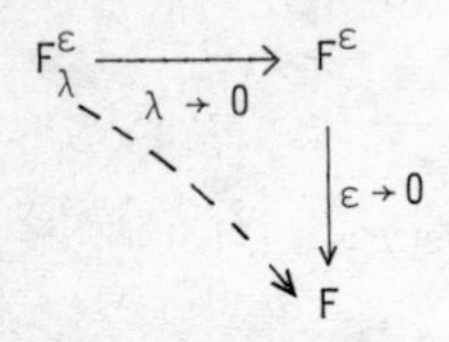

s-$L^2(\Omega)$-epi convergence. Functionals $\{F_{\lambda}^{\varepsilon}; \varepsilon > 0, \lambda > 0\}$ being uniformly coercive and hence uniformly τ-inf-compact, by the above argument, these convergences are associated with a metric. Thus, we can apply a classical diagonalization argument: there exists a map

$$\varepsilon \to \lambda(\varepsilon) \text{ decreasing to zero with } \varepsilon$$

such that

$$F^{\varepsilon}_{\lambda(\varepsilon)} \to F \quad \text{as } \varepsilon \text{ goes to zero.} \tag{2.189}$$

A natural question is: what is $\lambda(\varepsilon)$ equal to?

Indeed following Attouch and Murat [1], one can prove that if <u>$\lambda(\varepsilon) >> \varepsilon^6$</u> *then (2.189) holds.* When $\lambda(\varepsilon) << \varepsilon^6$ then, for every $u \in H^1(\Omega)$

$$\tau\text{-}\mathrm{lm}_e F^{\varepsilon}_{\lambda(\varepsilon)}(u) = \int_\Omega |\mathrm{grad}\,u|^2 dx.$$

When $\lambda(\varepsilon) = k\varepsilon^6$

$$\tau\text{-}\mathrm{lm}_e F^{\varepsilon}_{\lambda(\varepsilon)}(u) = \int_\Omega |\mathrm{grad}\,u|^2 dx + C_k \int_\Omega u^2 dx \tag{2.190}$$

with

$$C_k = \min_{\{w\to 1 \text{ at } \infty\}} \{ \int_\Omega |\mathrm{grad}\,w|^2 dx + k \int_T w^2 dx \}.$$

This is in accordance with (2.189) since when k tends to $+\infty$, C_k tends to $C = \mathrm{Cap}\,T$. The terminology is justified by the fact that functionals $F^{\varepsilon}_{\lambda(\varepsilon)}$ can be written

$$F^{\varepsilon}_{\lambda(\varepsilon)}(u) = \int_\Omega \{ |\mathrm{grad}\,u(x)|^2 + a_\varepsilon(x) u^2(x) \} dx$$

with "potentials" a_ε which are highly oscillating as shown in Fig. 2.11.

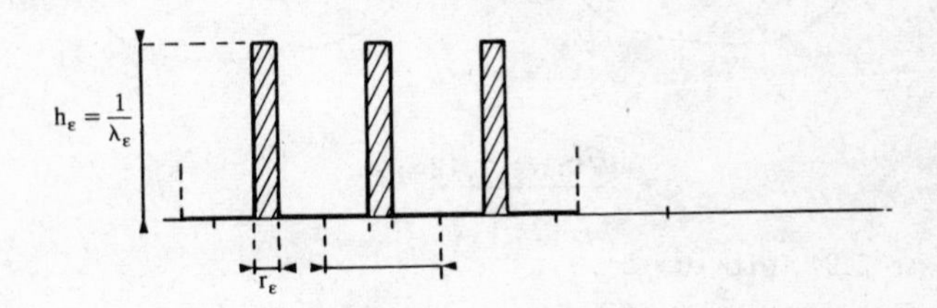

<u>Figure 2.11</u>

More generally in Attouch & Murat [1] one considers the case of highly oscillating potentials and discusses with respect to the two parameters r_ε and h_ε the limit analysis problem. Refer also to A. Brillard [1].

B. <u>Stochastic homogenization</u>

In fact, the periodicity assumption is only a first approximation when describing the structure of composite materials like concrete, alloy, porous or fibred materials. Their real structure is more complicated. A natural extension is to assume the conductivity coefficients to be almost periodic (cf. Kozlov [1]). A more precise approach (including as particular cases the preceding ones) is to consider *conductivity coefficients as random variables*.

Let us describe the stochastic approach of the model in Section 1.1.1. Let $(\Sigma,\mathcal{A},\mu)$ be a probability space; for every $\omega \in \Sigma$ and $u \in H^1(\Omega)$ we define

$$F^{\varepsilon}(\omega,u) = \int_{\Omega} j_{\varepsilon}(x,\omega,\text{grad}\,u(x))dx$$

where $\varepsilon > 0$ describes the tightness of the structure.

Energy functionals $\{F^{\varepsilon};\ \varepsilon \to 0\}$ can be viewed as a stochastic process

$$\omega \xrightarrow{F^{\varepsilon}} F^{\varepsilon}(\omega,\cdot)$$

(see Fig. 2.12) from $(\Sigma,\mathcal{A},\mu)$ into $\mathcal{F} = \mathcal{F}_{\lambda,\Lambda}$, the class of functionals defined

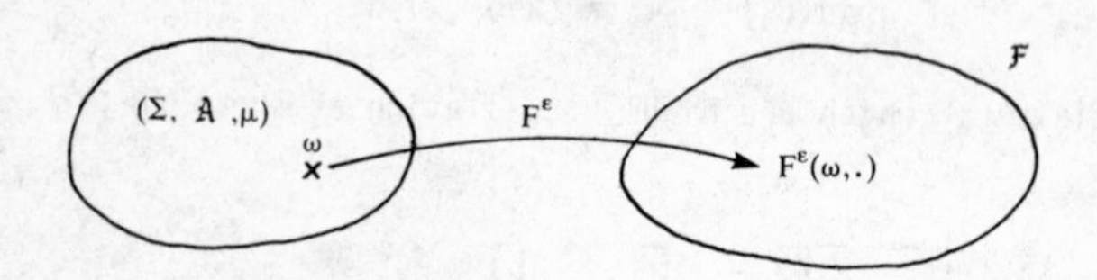

Figure 2.12

in Theorem 2.24 (with $\lambda > 0$).

$\mathcal{F}$ equipped with $L^2(\Omega)$-epi convergence is a compact metric space. The sequence of measures $\{P^{\varepsilon};\ \varepsilon > 0\}$, where P^{ε} is the image measure of μ by F^{ε}, i.e. for every Borel subset A of $\mathcal{F}$, $P^{\varepsilon}(A) = \mu((F^{\varepsilon})^{-1}(A))$, forms a sequence of probability measures on the compact metric space $\mathcal{F}$. Hence, we can extract a converging subsequence $\{P^{\varepsilon_k},\ k \in \mathbb{N}\}$

$$P^{\varepsilon_k} \longrightarrow P \text{ as } k \to +\infty$$

in the following sense:

For every continuous function $f:(\mathcal{F}, \text{epi}) \to \mathbb{R}$

$$\int_{\mathcal{F}} f(F)dP^{\varepsilon_k} \xrightarrow[k]{} \int_{\mathcal{F}} f(F)dP. \tag{2.191}$$

An interesting question is: when is P a Dirac measure, $P = \delta_{F^{hom}}$? It is at this stage that one has to make some stochastic assumptions such as stationariness, which replaces the periodicity assumptions (refer to Papanicolaou & Varadhan [1], Kozlov [2], [3], Modica [3]).

Since, for uniformly τ-inf-compact functions, τ-epi convergence implies convergence of the infimum, the function $f:F \to \inf_{u \in H^1(\Omega)} F(u)$ is continuous on $\mathcal{F}$. From (2.191) it follows that

$$\int_{\Sigma} d\mu(\omega) \min_{u \in H^1(\Omega)} \int_{\Omega} j_{\varepsilon_k}(x,\omega,\text{grad}u(x))dx \xrightarrow[(k \to +\infty)]{}$$

$$\min_{u \in H^1(\Omega)} \int_{\Omega} j_{hom}(x,\text{grad}u(x))dx. \quad \square$$

3 Epi-convergence of convex functions

In the two preceding chapters, the only structure was topological. We now enrich it and assume the space (X,τ) to be a topological vector space and the functions $\{F^n:X \to \bar{\mathbb{R}}\}$ to be convex.

An important feature of epi convergence in this context is to make the Fenchel duality transformation $F \to F^*$ continuous.

The Mosco convergence, which is epi convergence for both the strong and weak topologies of X (X is assumed to be a reflexive Banach space), makes this duality transformation bicontinuous. This explains the central role played by this notion in the study of strong stability properties of appro mation and perturbation schemes in convex optimization.

The bicontinuity of the subdifferential operation $\partial:F \to \partial F$ follows in a natural way, functions F being equipped with the Mosco convergence topolog and subdifferential operators ∂F being equipped with the graph convergence (also called resolvent convergence) topology. In passing, we recall some general results about maximal monotone operators (subdifferentials of clos convex proper functions are particular cases) and explore the convergence some related mathematical objects: generated semi-groups, spectrum, etc.

3.1 CONSERVATION OF CONVEXITY BY EPI-CONVERGENCE

PROPOSITION 3.1 Let (X,τ) be a topological vector space and $\{F^n;\ n \in \mathbb{N}\}$ a sequence of convex functions from X into $\bar{\mathbb{R}}$. Then,

$\tau\text{-ls}_e F^n$ is also a convex function (3.

$\tau\text{-li}_e F^n$ is no longer in general convex. (3.

Therefore, $\tau\text{-lm}_e F^n$, when it exists, is a closed convex function. Taking $F^n = F$ it follows that the closure of a convex function is still convex.

Proof of Proposition 3.1 From a geometric interpretation of epi-convergen (Theorem 1.36)

$$\text{epi } (\tau\text{-}\mathrm{ls}_e F^n) = \tau\text{-Li}(\text{epi } F^n)$$

$$\text{epi } (\tau\text{-}\mathrm{li}_e F^n) = \tau\text{-Ls}(\text{epi } F^n).$$

Noticing that a function F is convex if and only if its epigraph is convex, statements (3.1) and (3.2) are equivalent to the following ones: Let $\{C^n;\ n \in \mathbb{N}\}$ be a sequence of convex sets in a topological space (Y,σ). Then,

$$\sigma\text{-Li}C^n \text{ is also convex} \tag{3.3}$$

$$\sigma\text{-Ls}C^n \text{ is no longer in general convex.} \tag{3.4}$$

Conclusion (3.3) is a straightforward consequence of Definition 1.31 of the lower limit of a sequence of sets $\{C^n;\ n \in \mathbb{N}\}$

$$\sigma\text{-Li}C^n = c\ell_\sigma \bigcup_{n\in\mathbb{N}} \bigcap_{k\geq n} c\ell_\sigma C^k,$$

and of preservation of convexity by closure, intersection and increasing limit. By Definition 1.31 of the upper limit,

$$\sigma\text{-Ls}C^n = \bigcap_{n\in\mathbb{N}} c\ell_\sigma\Big(\bigcup_{k\geq n} C^k\Big).$$

From this formula one cannot derive any convexity property because convexity is not preserved in general by forming unions.

Take for example C_1 and C_2, two closed convex non-empty sets such that $C_1 \cap C_2 = \emptyset$ and $C^n = C_1$ for n odd, $C^n = C_2$ for n even. The above formula yields

$$\sigma\text{-Ls}C^n = C_1 \cup C_2, \text{ which is not convex!}$$

When the sequence $\{F^n;\ n \in \mathbb{N}\}$ epi-converges, then $\tau\text{-}\mathrm{lm}_e F^n = \tau\text{-}\mathrm{ls}_e F^n$ is still convex. Taking $F^n = F$ for every $n \in \mathbb{N}$ and noticing that $\tau\text{-}\mathrm{lm}_e F^n = c\ell_\tau F$, it follows that convexity is preserved by the closure operation.

3.2 EPI-CONVERGENCE AND DUALITY

3.2.1 Definition and classical properties of the Fenchel transformation

DEFINITION-PROPOSITION 3.2 Let (X,τ) be a locally convex topological vector space and $F:X \to]-\infty,+\infty]$ a convex proper function.

The Fenchel conjugate of F is the closed convex proper function $F^*:X^* \to]-\infty, +\infty]$, defined by

$$\forall f \in X^* \quad F^*(f) = \sup_{x \in X} \{\langle f,x \rangle - F(x)\}. \tag{3.5}$$

(by X^* we denote the topological dual space of X; we may more generally work with a duality pairing (X,Y)).

The biconjugate F^{**} of F is equal to $c\ell_\tau F$, the τ-closure of F:

$$F^{**} = c\ell_\tau F. \tag{3.6}$$

Therefore, if F is a closed convex proper function $(X,\tau) \to]-\infty, +\infty]$

$$F^{**} = F. \tag{3.7}$$

(Theorem of Fenchel-Moreau; cf. Brezis [3]).

We denote $\Gamma(X)$ the family of closed convex proper functions from X into $]-\infty, +\infty]$. Equivalently, $\Gamma(X)$ is the family of the functions which can be written as a supremum of continuous affine functions with "slopes" in X^*. The above equality (3.7) implies that the Fenchel transformation $F \to F^*$ *is a one to one mapping from* $\Gamma(X)$ *onto* $\Gamma(X^*)$.

A basic tool in the study of the continuity properties of this transformation is the *Moreau-Yosida approximation* (cf. Section 2.7). Let us make the properties of this approximation precise in the convex setting.

PROPOSITION 3.3 Let X be a normed linear space and $F:X \to]-\infty, +\infty]$ a closed convex proper function (i.e. $F \in \Gamma(X)$). Then, for every $\lambda > 0$, F_λ, the Moreau-Yosida proximal approximation of F

$$\forall x \in X \quad F_\lambda(x) = \inf_{u \in X} \{F(u) + \frac{1}{2\lambda} \|x-u\|_X^2\} \tag{3.8}$$

is a convex continuous function.

The sequence $(F_\lambda)_{\lambda>0}$ converges increasingly to F (as λ decreases to zero). Moreover, the conjugate of F_λ is equal to

$$\forall f \in X^* \quad (F_\lambda)^*(f) = F^*(f) + \frac{\lambda}{2} \|f\|_{X^*}^2. \tag{3.9}$$

Proof of Proposition 3.3 It can be easily derived from the general properties of inf-convolution: Given two functions, $\phi, \psi: X \to]-\infty, +\infty]$, their inf-convolution (denoted by $\phi \nabla \psi$) is the function from X into $]-\infty,+\infty]$ equal to

$$\forall x \in X \qquad \phi \nabla \psi(x) = \inf_{u \in X} \{\phi(u) + \psi(x-u)\}. \tag{3.10}$$

One can verify (cf. Moreau [3]) that when considering strict epigraphs, that is,

$$\text{epi}_s F = \{(\lambda,x) \in \mathbb{R} \times X / \lambda > F(x)\}$$

$$\text{epi}_s(\phi \nabla \psi) = \text{epi}_s(\phi) + \text{epi}_s(\psi). \tag{3.11}$$

The (vectorial) sum of two convex sets being still convex, if ϕ and ψ are convex, so is $\phi \nabla \psi$.

The conjugate of $\phi \nabla \psi$ is always equal to

$$(\phi \nabla \psi)^* = \phi^* + \psi^*. \tag{3.12}$$

This last result does not require any assumption on ϕ or ψ. It just relies on an exchange of supremum as shown by the following computation:

$$\begin{aligned}(\phi \nabla \psi)^*(f) &= \sup_{x \in X} [\langle f,x\rangle - \inf_{u \in X} \{\phi(u) + \psi(x-u)\}] \\ &= \sup_{x \in X} \sup_{u \in X} [\langle f,x\rangle - \phi(u) - \psi(x-u)] \\ &= \sup_{u \in X} \sup_{x \in X} [\langle f,x\rangle - \phi(u) - \psi(x-u)] \\ &= \sup_{u \in X} [-\phi(u) + \sup_{x \in X} \{\langle f,x\rangle - \psi(x-u)\}] \\ &= \phi^*(f) + \psi^*(f).\end{aligned}$$

Taking $\phi = F$, $\psi = \frac{1}{2\lambda} \|\cdot\|_X^2$, and applying (3.12), we obtain (3.9). □

Applying the Fenchel duality transformation to (3.12), one obtains

$$(\phi + \psi)^* = (\phi^* \nabla \psi^*)^{**}.$$

It is important to know when the equality

$$(\phi + \psi)^* = \phi^* \nabla \psi^*$$

holds, or equivalently, when the inf-convolution of two closed convex functions is still closed. The following proposition gives a fairly general answer (cf. J.P. Aubin [1]).

PROPOSITION 3.4 Let $\phi, \psi : X \to]-\infty, +\infty]$ two closed convex proper functions from a reflexive Banach space X into the extended reals. Under the assumption

$$\text{dom } \phi - \text{dom } \psi \text{ is a neighbourhood of the origin,} \tag{3.13}$$

the following equality holds:

$$(\phi + \psi)^* = \phi^* \nabla \psi^*. \tag{3.14}$$

Proof of Proposition 3.4 let us verify that the convex function

$$x^* \to (\phi^* \nabla \psi^*)(x^*) \text{ is closed.}$$

Let $x_n^* \to x^*$; introducing some sequence $\varepsilon_n \to 0$, by definition of $\phi^* \nabla \psi^*$

$$\begin{aligned}(\phi^* \nabla \psi^*)(x_n^*) &= \inf_{u^* \in X^*} \{\phi^*(x_n^* - u^*) + \psi^*(u^*)\} \\ &\geq \phi^*(x_n^* - u_n^*) + \psi^*(u_n^*) - \varepsilon_n.\end{aligned} \tag{3.15}$$

The point is to prove that the sequence $\{u_n^*;\ n \in \mathbb{N}\}$ is bounded in X^*: by assumption, there exists $\rho > 0$ such that

$$\text{dom } \phi - \text{dom } \psi \supset B(0,\rho).$$

Therefore, for every $x \in B(0,\rho)$, there exists $\alpha \in \text{dom } \phi$, $\beta \in \text{dom } \psi$ such that $x = \beta - \alpha$. Let us majorize

$$\begin{aligned}\langle u_n^*, x\rangle &= \langle u_n^*, \beta\rangle - \langle u_n^*, \alpha\rangle \\ &= \langle u_n^*, \beta\rangle + \langle x_n^* - u_n^*, \alpha\rangle - \langle x_n^*, \alpha\rangle\end{aligned}$$

$$\leq \psi(\beta) + \psi^*(u_n^*) + \phi(\alpha) + \phi^*(x_n^* - u_n^*) - \langle x_n^*, \alpha \rangle$$

$$\leq (\phi^* \nabla \psi^*)(x_n^*) + \varepsilon_n - \langle x_n^*, \alpha \rangle + \phi(\alpha) + \psi(\beta).$$

If $\liminf (\phi^* \nabla \psi^*)(x_n^*) = +\infty$, there is nothing to prove. Otherwise, by extraction of a subsequence, we can assume that

$$\sup_n (\phi^* \nabla \psi^*)(x_n^*) < +\infty.$$

From the above inequality it follows that the sequence $(u_n^*)_{n \in \mathbb{N}}$ is weakly bounded, and by the Banach-Steinhaus theorem, strongly bounded. Extracting a weakly converging subsequence

$$u_{n_k}^* \longrightarrow \underline{u}^* \text{ in } w\text{-}X^*$$

and going to the lower limit on (3.15), it follows

$$\liminf_n (\phi^* \nabla \psi^*)(x_n^*) \geq \phi^*(x^* - \underline{u}^*) + \psi^*(\underline{u}^*)$$

$$\geq \phi^* \nabla \psi^*(x^*). \quad \square$$

REMARK 3.5 A classical assumption which guarantees equality (3.14), $(\phi + \psi)^* = \phi^* \nabla \phi^*$, is that one of the two functions is continuous at one point of the domain of the other, $\text{dom}\, \phi \cap \text{Int}(\text{dom}\, \psi) = \emptyset$, for example. (Cf. Moreau [2], Rockafellar [2], Ekeland & Temam [1], Brezis [3]). Indeed, under this stronger assumption, conclusion (3.14) is valid in a general locally convex topological vector space: it just relies on Hahn-Banach theorem.

(3.13) is clearly a weaker assumption. On the other hand, the proof relies on the Banach-Steinhaus theorem and requires X to be a Banach space.

In this direction, Attouch & Brezis [1] have obtained that conclusion (3.14) holds in a general Banach space (not necessarily reflexive) under (3.13), or under the slightly weaker assumption: $\bigcup_{\lambda > 0} \lambda(\text{dom}\, \phi - \text{dom}\, \psi)$ is a closed subspace of X. $\square$

3.2.2 Continuity properties with respect to epi convergence of the Fenchel transformation

For simplicity of exposition, from now on, we assume X to be reflexive Banach space. The two topologies which play an important role are the strong topology, denoted by s-X or X_s, and the weak topology, denoted by w-X or X_w.

When working with the weak topology, a major difficulty is that this topology is not metrizable (when X is separable, bounded subsets are, indeed, metrizable). So, two different concepts of epi-convergence can be defined:

(a) The epi-convergence notions for the weak topology of X,

$$X_w\text{-}li_e F^n, \quad X_w\text{-}ls_e F^n, \quad X_w\text{-}lm_e F^n.$$

They are not easy to handle, since X_w is not metrizable.

(b) The corresponding sequential notions:

DEFINITION 3.6

$$\text{seq } X_w\text{-}li_e F^n(x) = \inf \{\liminf_n F^n(x_n)/x_n \xrightarrow{w} x\}$$

$$\text{seq } X_w\text{-}ls_e F^n(x) = \inf \{\limsup_n F^n(x_n)/x_n \xrightarrow{w} x\}.$$

When these last two quantities coincide, the sequence $(F^n)_{n\in\mathbb{N}}$ is said to be sequentially epi convergent (for the weak topology) at x, and this common value is denoted

$$\text{seq } X_w\text{-}lm_e F^n(x).$$

Comments

These two notions can be defined for any topology τ. When (X,τ) is first countable they coincide (refer to Theorem 1.13). Otherwise, they may differ as shown by the following example: Take $F^n = I_{K^n}$, the indicator function of the set $K^n = \{u_n\}$, where $u_n = n^\alpha e_n$ $(0 < \alpha < \frac{1}{2})$ and $\{e_n\}_{n\in\mathbb{N}}$ is an orthonormal base of a separable Hilbert space X.

Then one can verify that

$$\text{seq } X_w\text{-}li_e F^n \equiv +\infty$$

(since one cannot extract any bounded subsequence from the sequence $(u_n)_{n\in\mathbb{N}}$). But

$$X_w\text{-}li_e F^n = I_{\{0\}}$$

(cf. Sonntag [1] for example).

The notion which turns to be more interesting and easy to handle is the sequential one: from now on, we will deal only with this.

The following theorem is the basic tool in the study of the continuity properties of the Fenchel transformation. Its more general version can be found in Joly [1]. We give here a new direct proof (relying on the Moreau-Yosida approximation) in the reflexive Banach case. We say that a sequence $(F^n)_{n\in\mathbb{N}}$, $F^n:X \to]-\infty, +\infty]$ is *"uniformly proper"* if

There exists a bounded sequence $(x_{on})_{n\in\mathbb{N}}$ in X such that

$$\sup_{n\in\mathbb{N}} F^n(x_{on}) < + \infty. \tag{3.16}$$

THEOREM 3.7 Let X be a reflexive Banach space. For any uniformly proper sequence of closed convex functions $(F^n)_{n\in\mathbb{N}}$, $F^n:X \to]-\infty, +\infty]$, the following equality holds:

$$(\text{seq } X_w\text{-li}_e F^n)^* = X_s^*\text{-ls}_e F^{n^*}. \tag{3.17}$$

Proof of Theorem 3.7 In order to simplify the notation let us write

$$w\text{-li}_e F^n \text{ instead of seq } X_w\text{-li}_e F^n$$

and

$$s\text{-ls}_e (F^n)^* \text{ instead of } X_s^* - \text{ls}_e (F^n)^*.$$

(a) The inequality, which is easy to verify is:

$$(w\text{-li}_e F^n)^* \leq s\text{-ls}_e (F^n)^*$$

that is:

$$\forall x_n^* \xrightarrow{X_s^*} x^* \quad (w\text{-li}_e F^n)^*(x^*) \leq \limsup_n F^{n^*}(x_n^*).$$

By definition of $w\text{-li}_e F^n$, for every $\varepsilon > 0$, for every $x \in X$, there exists a sequence

$$x_n^\varepsilon \xrightarrow{X_w} x$$

such that

$$\sup \{w\text{-}li_e F^n(x) + \varepsilon; -\frac{1}{\varepsilon}\} \geq \lim_n \inf F^n(x_n^\varepsilon). \tag{3.18}$$

By definition of F^{n^*},

$$F^{n^*}(x_n^*) \geq \langle x_n^*, x_n^\varepsilon\rangle - F^n(x_n^\varepsilon),$$

where we choose x_n^ε given by (3.18). Taking the upper limit as $n \to +\infty$,

$$\lim_n \sup F^{n^*}(x_n^*) \geq \langle x^*, x\rangle - \lim_n \inf F^n(x_n^\varepsilon),$$

which, from (3.18), implies

$$\lim_n \sup F^{n^*}(x_n^*) \geq \langle x^*, x\rangle - \sup \{w\text{-}li_e F^n(x) + \varepsilon; -\frac{1}{\varepsilon}\} .$$

This being true for any $\varepsilon > 0$ and $x \in X$, we get

$$\lim_n \sup F^{n^*}(x_n^*) \geq \sup_x \{\langle x^*, x\rangle - w\text{-}li_e F^n(x)\}$$

$$\geq (w\text{-}li_e F^n)^*(x^*).$$

(b) The difficult point is to prove the opposite inequality

$$s\text{-}ls_e (F^n)^* \leq (w\text{-}li_e F^n)^*.$$

Let us denote $F = w\text{-}li_e F^n$. We have to show, for any $x^* \in X^*$, the existence of a sequence $(x_n^*)_{n\in\mathbb{N}}$ which is strongly convergent to x^* in X^* and satisfies

$$\lim_n \sup F^{n^*}(x_n^*) \leq F^*(x^*). \tag{3.19}$$

The idea of the proof is the following: if the functionals $(F^n)_{n\in\mathbb{N}}$ were "equicoercive", we could take $x_n^* = x^*$: The inequality (3.19), when taking $x_n^* = x^*$, can be reformulated

$$\lim_n \inf \inf_{x\in X} \{F^n(x) - \langle x^*, x\rangle\} \geq \inf_{x\in X} \{F(x) - \langle x^*, x\rangle\}.$$

When the $\{F^n\}_{n\in\mathbb{N}}$ are equicoercive, the infimum

$$\inf_{x\in X} \{F^n(x) - \langle x^*, x\rangle\}$$

is realized for each $n \in \mathbb{N}$ at a point x_n, and the sequence $(x_n)_{n\in\mathbb{N}}$ is bounded in X. The space X being reflexive, this sequence is weakly relatively compact in X. Using that $F = \text{w-li}_e F^n$, the inequality (3.19) follows immediately. (In fact one uses only the inequality $F \leqslant \text{w-li}_e F^n$).

So, the idea of the demonstration is to "coercify" the functions $(F^n)_{n\in\mathbb{N}}$ in order to reduce to the above situation. For every $\lambda > 0$, let us introduce

$$F^{n,\lambda}(x) = F^n(x) + \frac{\lambda}{2}\|x\|^2$$

$$F^{\lambda}(x) = F(x) + \frac{\lambda}{2}\|x\|^2 .$$

The above argument now works, noticing that

$$\begin{aligned} \text{w-li}_e F^{n,\lambda}(x) &\geqslant \text{w-li}_e F^n(x) + \frac{\lambda}{2}\|x\|^2 \\ &\geqslant F(x) + \frac{\lambda}{2}\|x\|^2 \\ &\geqslant F^{\lambda}(x) \end{aligned}$$

(we need only this inequality) and that the $F^{n,\lambda}$ are equi-coercive. In fact, in order to be sure that the minimizing sequence $(x_n)_{n\in\mathbb{N}}$ is bounded in X one needs, besides the uniform proper assumption (3.15), a uniform minorization property:

$$\text{There exists } r \geqslant 0 \text{ such that: } \forall n \in \mathbb{N} \quad \forall x \in X \quad F^n(x)+r(\|x\|+1) \geqslant 0. \tag{3.20}$$

We shall explain later how one can reduce to the above situation. So, assuming (3.20), we have

$$\forall\lambda > 0 \ \forall x^* \in X^* \ \limsup_n (F^{n,\lambda})^*(x^*) \leqslant (F^{\lambda})^*(x^*). \tag{3.21}$$

The function $x \to \frac{\lambda}{2}\|x\|^2$ being continuous on X, from Proposition 3.4, for every closed convex proper function G

$$\begin{aligned} (G + \frac{\lambda}{2}\|\cdot\|^2)^* &= G^* \nabla \frac{1}{2\lambda}\|\cdot\|_*^2 \\ &= (G^*)_{\lambda}. \end{aligned}$$

Applying the above formula with $G = F^n$, and returning to (3.21), we obtain

$$\limsup_n (F^{n^*})_\lambda (x^*) \leq (F^\lambda)^*(x^*).$$

This being true for any $\lambda > 0$,

$$\limsup_{\lambda \to 0} \limsup_{n \to +\infty} (F^{n^*})_\lambda (x^*) \leq \sup_{\lambda > 0} (F^\lambda)^*(x^*). \tag{3.22}$$

The right hand side of (3.22) is equal to

$$\sup_{\lambda>0} \sup_{x \in X} \{\langle x^*, x\rangle - F(x) - \frac{\lambda}{2} \|x\|^2\}$$

$$= \sup_{x \in X} \sup_{\lambda>0} \{\langle x^*, x\rangle - F(x) - \frac{\lambda}{2} \|x\|^2\}$$

$$= F^*(x).$$

So (3.22) can be rewritten

$$\limsup_{\lambda \to 0} \limsup_{n \to +\infty} (F^{n^*})_\lambda (x^*) \leq F^*(x^*).$$

Using the diagonalization Lemma 1.16, there exists a map $n \to \lambda(n)$ decreasing to zero as n increases to $+\infty$ such that

$$\limsup_n (F^{n^*})_{\lambda(n)} (x^*) \leq \limsup_{\lambda \to 0} \limsup_{n \to +\infty} (F^{n^*})_\lambda (x^*).$$

Thus,

$$\limsup_n (F^{n^*})_{\lambda(n)} (x^*) \leq F^*(x^*).$$

Let us examine the quantity

$$(F^{n^*})_{\lambda(n)} (x^*) = \inf_{u \in X^*} \{F^{n^*}(u^*) + \frac{1}{2\lambda(n)} \|x^*-u^*\|^2\}.$$

We minimize the convex lower semicontinuous coercive function

$$u^* \to F^{n^*}(u^*) + \frac{1}{2\lambda(n)} \|x^*-u^*\|^2$$

on the reflexive Banach space X^*. The infimum is achieved for some $x_n^* \in X^*$. Therefore

$$F^*(x^*) \geq \limsup_n \{F^{n^*}(x_n^*) + \frac{1}{2\lambda(n)} \|x^*-x_n^*\|^2\}. \qquad (3.23)$$

The quantity $\frac{1}{2\lambda(n)} \|x^*-x_n^*\|^2$ being positive, it follows that

$$F^*(x^*) \geq \limsup_n F^{n^*}(x_n^*).$$

The only thing we have to prove is that $x_n^* \xrightarrow{X_s^*} x^*$ as n goes to $+\infty$.

By the uniform proper assumption (3.16) on the $(F^n)_{n\in\mathbb{N}}$,

$$F^{n^*}(x_n^*) \geq \langle x_n^*, x_{on}\rangle - F^n(x_{on})$$

$$\geq -C(1 + \|x_n^*\|).$$

Returning to (3.23), for n sufficiently large

$$F^*(x^*) + 1 \geq -C(1 + \|x_n^*\|) + \frac{1}{2\lambda(n)} \|x_n^*-x^*\|^2.$$

If $F^*(x^*) = +\infty$, there is nothing to prove (in (3.19) take for example $x_n^* = x^*$). Otherwise, the above inequality clearly implies that $x_n^* \xrightarrow[(n\to+\infty)]{X_s^*} x^*$ (since $\lambda(n) \to 0$ as $n \to +\infty$).

So Theorem 3.7 has been proved under the additional uniform minorization assumption (3.20). With the help of the following lemma, we can extend the conclusion of Theorem 3.7 to the general situation.

LEMMA 3.8 Let $(F^n)_{n\in\mathbb{N}}$, $F^n: X \to]-\infty, +\infty]$, a sequence of closed convex proper functions which is uniformly proper.

Then, the following statements are equivalent:

(i) $\forall x \in X \quad \text{seq}X_w\text{-li}_e F^n(x) > -\infty$

(ii) $\exists r \geq 0$ such that: $\forall n \in \mathbb{N}, \forall x \in X \quad F^n(x) + r(\|x\| + 1) \geq 0$.

Proof of Lemma 3.8 Let us argue by contradiction and prove that (i) $\Rightarrow$ (ii). Denying (ii), we obtain the existence of a subsequence $(n(k))_{k\in\mathbb{N}}$ and $(x_k)_{k\in\mathbb{N}}$ such that

$$F^{n(k)}(x_k) + k(\|x_k\| + 1) < 0.$$

Without restriction, we can assume the map $k \to n(k)$ to be increasing to $+\infty$. There are two possible situations:

(a) The sequence $(x_k)_{k\in\mathbb{N}}$ is bounded. Let z be a weak limit point of this sequence (we still denote $x_k \xrightarrow{X_w} z$). By Definition 3.6 of $\text{seq}X_w - \text{li}_e F^n$

$$\text{seq}X_w\text{-li}_e F^n(z) \leq \liminf_k \, [-k \, \|x_k\| - k] = -\infty,$$

which is in contradiction with (i).

(b) The sequence $(x_k)_{k\in\mathbb{N}}$ is not bounded. Let us still denote by $(x_k)_{k\in\mathbb{N}}$ a subsequence such that $\|x_k\| \to +\infty$ as $k \to +\infty$. Let us define

$$z_k = t_k x_k + (1-t_k)x_{0k}$$

(x_{0k} given by (3.16), $x_{0k} \xrightarrow{X_w} \xi_0$) and choose $t_k \in (0,1)$ such that

$$z_k \xrightarrow{X_w} \xi_0 \text{ as } k \to +\infty:$$

$$\|z_k - x_{0k}\| = t_k \, \|x_k - x_{0k}\| .$$

Take $t_k = \dfrac{1}{\sqrt{k}\,\|x_k - x_{0k}\|}$ for example. We notice that $t_k \to 0$ as $k \to +\infty$ (because $\|x_k - x_{0k}\| \to +\infty$). By convexity of $F^{n(k)}$

$$F^{n(k)}(z_k) \leq t_k \, F^{n(k)}(x_k) + (1-t_k)F^{n(k)}(x_{0k})$$

$$\leq \frac{1}{\sqrt{k}.\,\|x_k - x_{0k}\|} \, [-k \, \|x_k\| - k] + (1-t_k)F^{n(k)}(x_{0k})$$

$$\leq -\sqrt{k} \, \frac{\|x_k\|}{\|x_k - x_{0k}\|} + C.$$

Thus,

$$\text{seq}X_w\text{-li}_e F^n(\xi_o) \leq \liminf_k \, F^{n(k)}(z_k) \leq -\infty$$

which again contradicts (i).

The other implication (ii) ⇒ (i) is clear: just use the fact that every weakly convergent sequence is bounded.

End of the proof of Theorem 3.7 Let $(F^n)_{n\in\mathbb{N}}$ be a sequence of closed convex functions which is uniformly proper. Let us assume that the sequence $(F^n)_{n\in\mathbb{N}}$ does not satisfy uniform minorization assumption (3.20). Then, from Lemma 3.8, there exists x_o belonging to X such that

$$w\text{-}li_e F^n(x_o) = -\infty.$$

Therefore, for any $x^* \in X^*$

$$(w\text{-}li_e F^n)^*(x^*) = \sup_{x\in X} \{\langle x^*,x\rangle - w\text{-}li_e F^n(x)\}$$

$$\geqslant \langle x^*,x_o\rangle - w\text{-}li_e F^n(x_o)$$

$$\geqslant +\infty.$$

On the other hand, since $w\text{-}li_e F^n(x_o) = -\infty$, there exists a sequence $(x_n)_{n\in\mathbb{N}}$ which weakly converges to x_o and satisfies

$$\liminf_n F^n(x_n) = -\infty.$$

It follows that, for any strongly converging sequence $x_n^* \to x^*$,

$$\limsup_n \sup_{x\in X} \{\langle x_n^*,x\rangle - F^n(x)\}$$

$$\geqslant \limsup_n \{\langle x_n^*,x_n\rangle - F^n(x_n)\}$$

$$\geqslant +\infty.$$

Thus,

$$X_s^*\text{-}ls_e F^{n^*} \equiv +\infty$$

and

$$(seqX_w\text{-}li_e F^n)^* = X_s^* - ls_e F^{n^*} \equiv +\infty.$$

So, equality (3.17) is always satisfied. □

The next theorem shows up the symmetric role played by the strong and the weak topologies in the above argument. The idea of the proof is the same as for Theorem 3.7, but more complicated from the technical point of view: we have to construct in an explicit way a weakly convergent sequence (which cannot be expected in general to be strongly converging!). Theorems 3.7 and 3.9 are the basic tools in the duality theory for epi-convergence.

THEOREM 3.9 Let X be a separable reflexive Banach space. Let $\{F^n; n \in \mathbb{N}\}$ be a sequence of closed convex functions from X into $]-\infty, +\infty]$ such that the $\{F^{n^*}; n \in \mathbb{N}\}$ satisfy the following "uniform coerciveness" property:

$$\text{for every sequence } \{x_n^*; n \in \mathbb{N}\} \text{ in } X^*, \ \sup_{n\in\mathbb{N}} F^{n^*}(x_n^*) < +\infty \Rightarrow \sup_{n\in\mathbb{N}} \|x_n^*\| < +\infty. \tag{3.24}$$

Then, the following equality holds:

$$(X_s\text{-li}_e F^n)^* = \text{seq } X_w^*\text{-ls}_e F^{n^*}.$$

Proof of Theorem 3.9 Let us use the notation $F = X_s\text{-li}_e F^n$. What is clear is that $F^* \leqslant \text{seq } X_w^*\text{-ls}_e F^{n^*}$ (it is the same argument as in Theorem 3.7, after exchanging strong and weak convergences). The difficult point is to prove that $F^* \geqslant \text{seq } X_w^*\text{-ls}_e F^{n^*}$, i.e.

$$\forall x^* \in X^*, \ \exists x_n^* \xrightarrow{X_w^*} x^* \text{ such that } F^*(x^*) \geqslant \limsup_n F^{n^*}(x_n^*).$$

The idea is, as in Theorem 3.7, to coercify the functions F^n, but now we need to make them inf-compact for the strong topology of X (and no longer, as in Theorem 3.7, for the weak topology of X: it was enough to add some $\lambda\|\cdot\|_X^2$ and let $\lambda \to 0$). So, let us introduce $\{x_1, x_2, \ldots, x_k, \ldots\}$, a dense denumerable subset of X, and for each $k \in \mathbb{N}$, $E_k = \text{span}\{x_1, \ldots, x_k\}$, the finite dimensional subspace of X generated by $x_1, \ldots, x_k$. Take $B_k = \{x \in E_k / \|x\|_X \leqslant k\}$ the closed ball of radius k in B_k and, for every $k, n \in \mathbb{N}$, let us define

$$F^{n,k} = F^n + I_{B_k}$$

$$F^k = F + I_{B_k}$$

that is

$$F^{n,k} = \begin{cases} F^n & \text{on } B_k \\ +\infty & \text{elsewhere} \end{cases} \qquad F^k = \begin{cases} F & \text{on } B_k \\ +\infty & \text{elsewhere} \end{cases}.$$

We first claim that:

$$\forall x^* \in X^* \quad F^{k^*}(x^*) \geq \limsup_n F^{n,k^*}(x^*) \tag{3.25}$$

Equivalently

$$\liminf_n \inf_{u \in B_k} \{F^n(u) - \langle x^*,u\rangle\} \geq \inf_{u \in B_k} \{F(u) - \langle x^*,u\rangle\}.$$

If the left member of this inequality is finite (otherwise (3.25) is clearly verified) one can find a subsequence $\{n_\nu ; \nu \in \mathbb{N}\}$ and $\{x_\nu ; x_\nu \in B_k$ for every $\nu \in \mathbb{N}\}$ such that

$$\liminf_n \inf_{u \in B_k} \{F^n(u) - \langle x^*,u\rangle\} = \lim_\nu \{F^{n_\nu}(x_\nu) - \langle x^*,x_\nu\rangle\}.$$

The set B_k being compact for the strong topology of X (it is bounded and of finite dimension), one can extract a strongly convergent subsequence from $\{x_\nu ; \nu \in \mathbb{N}\}$. Using that $F \leq X_s\text{-}li_e F^n$, one obtains (3.25).

From $F^k \geq F$ it follows that $F^* \geq (F^k)^*$ which, combined with (3.25), yields:

$$\forall x^* \in X^* \quad F^*(x^*) \geq \limsup_k \ \limsup_n F^{n,k^*}(x^*).$$

From the diagonalization Lemma 1.16, there exists a map $n \to k(n)$ strictly increasing to $+\infty$ with n such that

$$F^*(x^*) \geq \limsup_n F^{n,k(n)^*}(x^*). \tag{3.26}$$

Let us compute

$$\begin{aligned}(F^{n,k})^* &= (F^n + I_{B_k})^* \\ &= (F^{n^*} \nabla (I_{B_k})^*)^{**}.\end{aligned}$$

There is a slight difficulty (compare with Theorem 3.7) because we cannot assert that $F^{n^*} \nabla (I_{B_k})^*$ is closed. The sufficient condition (refer to Remark 3.5)

$$\bigcup_{\lambda>0} \lambda(\mathrm{dom}\, F^n - B_k) = X$$

is not satisfied in general.

Noticing that the function $F^{n^*} \nabla (I_{B_k})^*$ is convex,

$$(F^{n,k})^* = c\ell(F^{n^*} \nabla (I_{B_k})^*),$$

the closure being taken for the strong or the weak topology, it is equivalent because of the convexity of $F^{n^*} \nabla (I_{B_k})^*$. Thus, there exists a sequence $\{u^*_{k,n};\ k,\ n \in \mathbb{N}\}$ in X^* which satisfies

$$\begin{cases} (F^{n,k})^*(x^*) \geq [F^{n^*} \nabla (I_{B_k})^*](u^*_{k,n}) - \frac{1}{n} \\ \\ \|x^* - u^*_{k,n}\| < \frac{1}{n}. \end{cases} \tag{3.27}$$

Denoting $u_n = u_{k(n),n}$, from (3.26) and (3.27),

$$\begin{cases} F^*(x^*) \geq \limsup_n \, [F^{n^*} \nabla (I_{B_{k(n)}})^*](u^*_n) \\ \\ u^*_n \to x^* \text{ strongly in } X^* \text{ as } n \text{ goes to } + \infty. \end{cases} \tag{3.28}$$

By definition of the inf-convolution, this implies the existence of a sequence $\{x^*_n;\ n \in \mathbb{N}\}$ such that

$$F^*(x^*) \geq \limsup_n \, \{F^{n^*}(x^*_n) + I^*_{B_{k(n)}}(u^*_n - x^*_n)\}.$$

Thus,

$$F^*(x^*) \geq \limsup_n \, F^{n^*}(x^*_n). \tag{3.29}$$

The only point we need to verify is that

$$x^*_n \rightharpoonup x^* \text{ weakly in } X^* \text{ as } n \text{ goes to } + \infty.$$

If $F^*(x^*) = + \infty$ there is nothing to prove. So let us assume that $F^*(x^*)$ is finite. From the uniform coerciveness assumption on the $\{F^{n^*};\ n \in \mathbb{N}\}$ and inequality (3.29) we obtain that the sequence $\{x^*_n;\ n \in \mathbb{N}\}$ is bounded in X^*.

On the other hand,

$$F^*(x^*) \geq I^*_{B_{k(n)}}(u^*_n - x^*_n). \tag{3.30}$$

Let us compute $(I_{B_k})^* = (I_{E_k} + I_{\{\|x\| \leq k\}})^*$

$$= (I_{E_k})^* \nabla (I_{\{\|x\| \leq k\}})^*$$

$$= I_{(E_k{}^{\perp})} \nabla \; k \|\cdot\|_{X^*}$$

$$(I_{B_k})^*(x^*) = k.\inf_{u^* \in E_k^{\perp}} \|x^* - u^*\|$$

$$= k \text{ dist } (x^*, E_k^{\perp})$$

where $E_k^{\perp}$ denotes the orthogonal subspace of E_k.

Returning to (3.30), we obtain

$$\frac{1}{k(n)} F^*(x^*) \geq \text{dist}(x^*_n - u^*_n, E^{\perp}_{k(n)}).$$

Since $k(n) \to +\infty$ with n, we derive

$$\text{dist}(x^*_n - u^*_n, E^{\perp}_{k(n)}) \to 0 \text{ as n goes to } + \infty. \tag{3.31}$$

Let $D = \bigcup_{k \in \mathbb{N}} E_k$ which is a dense subset of X; for any $x \in D$, x belongs to $E_{k(n)}$ for n sufficiently large. Denoting $\xi^*_n = x^*_n - u^*_n$, we derive

$$\langle \xi^*_n, x \rangle = \langle \xi^*_n - \text{proj}_{E^{\perp}_{k(n)}} \xi^*_n, x \rangle.$$

Thus

$$|\langle \xi^*_n, x \rangle| \leq \text{dist}(x^*_n - u^*_n, E^{\perp}_{k(n)}). \|x\|$$

which from (3.31) implies

$$\langle x^*_n - u^*_n, x \rangle \to 0 \text{ as } n \to +\infty.$$

From the strong convergence of u^*_n to x^* (refer to (3.28)), the density of D in X^*, and the boundedness of the sequence $\{x^*_n; n \in \mathbb{N}\}$, we finally get:

$x_n^* \longrightarrow x^*$ weakly in X^* as n goes to $+\infty$. □

REMARK 3.10 The following example shows that the conclusion of Theorem 3.9 fails to be true if one drops the uniform coerciveness assumption on the $\{F^{n*}; n \in \mathbb{N}\}$. Let us assume the conclusion is true for any sequence $(F^n)_{n\in\mathbb{N}}$ of closed convex functions. By taking $F^{n*} = I_{K^n}$, K^n closed convex, noticing that the left member of Theorem 3.9 is closed convex, we would derive that "for any sequence of closed convex sets $\{K^n; n\in \mathbb{N}\}$, the limit set $\text{seq}X_w\text{-Li}K^n$ is closed and convex", where we have denoted

$$\text{seq}X_w\text{-Li}K^n = \{x \in X; \exists (x_n)_{n\in\mathbb{N}}\ x_n \in K^n \text{ for every } n \in \mathbb{N},\ x_n \xrightarrow{X_w} x\}.$$

This limit set seq $X_w\text{-Li}K^n$ is clearly convex. *But we claim that in general, it is not closed:* Take $H = \ell^2(\mathbb{N})$ and $\{e_1,e_2,\ldots,e_n,\ldots\}$ a Hilbertian base of H.

For every $n \in \mathbb{N}$, let us define

$$K^n = \{\lambda e_1 + \mu e_n / \mu \geq \frac{1}{\lambda},\ \lambda > 0\}$$

$$K = \{\lambda e_1 / \lambda > 0\}.$$

Then, $K = \text{seq}X_w\text{-Lm}K^n$. But clearly, K is not closed! One verifies that, if $x_n \in K^n$ for every $n \in \mathbb{N}$ i.e. $x_n = \lambda_n e_1 + \mu_n e_n$, $\mu_n \geq \frac{1}{\lambda_n}$ and $x_n \xrightarrow{X_w} x$, necessarily

$$+\infty > C \geq \|x_n\|^2 = \lambda_n^2 + \mu_n^2 \text{ and thus, } \{\mu_n; n \in \mathbb{N}\} \text{ is bounded.}$$

From the equalities

$$\langle x_n,e_1\rangle = \lambda_n \xrightarrow[(n\to+\infty)]{} \langle x,e_1\rangle$$

$$\langle x_n,e_p\rangle \xrightarrow[(n\to+\infty)]{} 0 = \langle x,e_p\rangle \quad (p > 1)$$

we derive that $x = \langle x,e_1\rangle e_1$ with $\langle x,e_1\rangle$ strictly positive: from $\mu_n \geq \frac{1}{\lambda_n}$ and μ_n bounded, λ_n can only converge to a strictly positive value. □

THEOREM 3.11 Let $(F^n)_{n\in\mathbb{N}}$, F be a sequence of closed convex proper functions from X into $]-\infty, +\infty]$, where X is a reflexive Banach space.

(a) The following implication holds:

(i) $F = \text{seq}X_w\text{-lm}_e F^n$

$\Downarrow$

(ii) $F^* = X_s^*\text{-lm}_e F^{n*}$

(b) When X is separable and the $\{F^n;\ n \in \mathbb{N}\}$ satisfy the following equi-coerciveness property:

$$\text{"For every sequence } \{x_n;\ n \in \mathbb{N}\} \text{ in } X,\ \sup_{n\in\mathbb{N}} F^n(x_n) < +\infty \Rightarrow \sup_{n\in\mathbb{N}} \|x_n\| < +\infty\text{"} \tag{3.32}$$

then, the two above statements are equivalent:

$$F = \text{seq}X_w\text{-lm}_e F^n \iff F^* = X_s^*\text{-lm}_e F^{n*}.$$

Proof of Theorem 3.11

(i) $\Rightarrow$ (ii) By assumption

$$w\text{-ls}_e F^n \leq F \leq w\text{-li}_e F^n.$$

Thus,

$$(w\text{-li}_e F^n)^* \leq F^* \leq (w\text{-ls}_e F^n)^*. \tag{3.33}$$

From Theorem 3.7, (since $F = w\text{-ls}_e F^n$ is proper, condition (3.16) is satisfied)

$$(w\text{-li}_e F^n)^* = s\text{-ls}_e F^{n*}.$$

Let us assume a moment that

$$(w\text{-ls}_e F^n)^* \leq s\text{-li}_e F^{n*}. \tag{3.34}$$

Then, returning to (3.33), we have

$$s\text{-ls}_e F^{n*} \leq F^* \leq s\text{-li}_e F^{n*}$$

that is:

$$F^* = s\text{-lm}_e F^{n*}.$$

In order to prove (3.34), we have to verify that for every $x^* \in X^*$, for every sequence $x_n^* \xrightarrow{X_s^*} x^*$, for every $x \in X$,

$$\liminf_n F^{n*}(x_n^*) \geq \langle x^*,x\rangle - (w\text{-}ls_e F^n)(x). \tag{3.35}$$

By definition of $w\text{-}ls_e F^n(x)$, there exists a sequence $x_n \xrightarrow{X_w} x$, such that

$$w\text{-}ls_e F^n(x) = \limsup_n F^n(x_n).$$

By definition of F^{n*}

$$F^{n*}(x_n^*) \geq \langle x_n^*,x_n\rangle - F^n(x_n).$$

Taking the lower limit, as n goes to $+\infty$, on the above inequality, we obtain

$$\begin{aligned}\liminf_n F^{n*}(x_n^*) &\geq \langle x^*,x\rangle - \limsup_n F^n(x_n)\\ &\geq \langle x^*,x\rangle - (w\text{-}ls_e F^n(x)),\end{aligned}$$

which is (3.35).

(ii) $\Rightarrow$ (i) in the case where the $\{F^n;\ n \in \mathbb{N}\}$ are equi-coercive. By assumption $F^* = X_s^*\text{-}lm_e F^{n*}$, that is,

$$s\text{-}ls_e F^{n*} \leq F^* \leq s\text{-}li_e F^{n*}.$$

Thus

$$(s\text{-}li_e F^{n*})^* \leq F \leq (s\text{-}ls_e F^{n*})^*.$$

From Theorem 3.7,

$$\begin{aligned}(s\text{-}ls_e F^{n*})^* &= (\text{seq}X_w\text{-}li_e F^n)^{**}\\ &\leq \text{seq}X_w\text{-}li_e F^n.\end{aligned}$$

From Theorem 3.9, and here we use the equicoerciveness assumption on the $\{F^n;\ n \in \mathbb{N}\}$,

$$(s\text{-}li_e F^{n*})^* = \text{seq}X_w\text{-}ls_e F^n.$$

Thus

$$\mathrm{seq}X_w\text{-}\mathrm{ls}_e F^n \leq F \leq \mathrm{seq}X_w\text{-}\mathrm{li}_e F^n$$

i.e.

$$F = \mathrm{seq}X_w\text{-}\mathrm{lm}_e F^n. \quad \square$$

The same example as in Remark 3.10 shows that if $X_s\text{-}\mathrm{lm}_e F^n = F$ exists, and if the $\{F^{n*}; n \in \mathbb{N}\}$ are only assumed to be closed and convex, then $\mathrm{seq}X_w\text{-}\mathrm{lm}_e F^{n*}$ (when it exists!) may fail to be closed and hence $F^* \neq \mathrm{seq}X_w\text{-}\mathrm{lm}_e F^{n*}$!

Let us give another proof of the implication (ii) $\Rightarrow$ (i) (in the uniform coercive case) which is of independent interest. The following lemma from Ambrosetti & Sbordone [1] throws light on how the equicoerciveness assumption allows us to overcome the lack of metrizability of the weak convergence.

LEMMA 3.12 Let X be a separable reflexive Banach space and $(F^n)_{n\in\mathbb{N}}$, a sequence of functions from X into $]-\infty, +\infty]$ which is equicoercive (in the sense of Theorem 3.11, (3.32)). Then, the following statements are equivalent:

(α) $F = X_w\text{-}\mathrm{lm}_e F^n$

(β) $F = \mathrm{seq}\ X_w\text{-}\mathrm{lm}_e F^n$

(γ) $F = \tau_d\text{-}\mathrm{lm}_e F^n$

where d is a metric on X

$$d(x,y) = \sum_{k\in\mathbb{N}} 2^{-k}\ |\langle x_k^*, x-y\rangle|$$

associated with a dense denumerable subset $\{x_k^*; k \in \mathbb{N}\}$ in the unit sphere of X^*.

Let us admit the conclusions of the above lemma and show how one can derive from the implication (i) $\Rightarrow$ (ii) of Theorem (3.11) the converse implication (ii) $\Rightarrow$ (i). Let us recall that, although epi-convergence is not in general attached to a topology, it enjoys the following topological property (refer to Proposition 2.72):

Let (X,τ) be a second countable space and $\{F^n; n \in \mathbb{N}\}$, $F^n: X \to \bar{\mathbb{R}}$ a sequence which has a unique τ-epi limit point F. Then, the sequence $\{F^n; n \in \mathbb{N}\}$ τ-epi converges to F. Let us notice that the relative compactness of the sequence

$\{F^n; n \in \mathbb{N}\}$ is automatically satisfied by Theorem 2.22).

Let us verify that the conditions of application of the above argument are satisfied. We first notice that because of the separability of X (and hence of X*) the topological space (X,τ_d) is metrizable and separable and hence second countable.

Let $\{F^{n_k}; k \in \mathbb{N}\}$ be an extracted subsequence of the sequence $\{F^n; n \in \mathbb{N}\}$ which τ_d-epi converges to some G:

$$G = \tau_d\text{-lm}_e F^{n_k}$$

from the implication (i) ⇒ (ii) of Theorem 3.11 and the above equality, we derive

$$G^* = X_s^*\text{-lm}_e \; F^{n_k *}.$$

Since by assumption $F^* = X_s^*\text{-lm}_e F^{n^*}$, this implies

$$G^* = F^*.$$

Noticing that $G = \tau_d\text{-lm}_e F^{n_k} = X_w\text{-lm}_e F^{n_k}$ (from Lemma 3.12) is still closed convex (refer to Proposition 3.1) we obtain

$$G = F.$$

So the whole sequence $\{F^n; n \in \mathbb{N}\}$ τ_d-epi-converges to F, that is

$$F = \text{seq}X_w\text{-lm}_e F^n. \quad \square$$

<u>COROLLARY 3.13</u> Let X be a separable reflexive Banach space and $\{F^n; n \in \mathbb{N}\}$ a sequence of closed convex functions from X into $]-\infty, +\infty]$ satisfying the equicoerciveness assumption:

$$F^n(x) \geq C(\|x\|) \text{ with } \lim_{r\to+\infty} \frac{C(r)}{r} = +\infty.$$

Then, the following statements are equivalent:

(i) $F = \text{seq}\, X_w\text{-lm}_e F^n$

(ii) $F^* = X_s^*\text{-lm}_e F^{n*}$

(iii) $\forall f \in X^* \quad F^*(f) = \lim_{n\to+\infty} F^{n^*}(f)$

Proof of Corollary 3.13

Just notice that under the above coerciveness assumption, the family $\{F^{n*}; n \in \mathbb{N}\}$ is uniformly locally Lipschitz. Thus, X_s^*-epi convergence turns to be equivalent to pointwise convergence. Then apply the conclusions of Theorem 3.11. □

3.2.3 Examples of computation of epi-limits by duality

The equivalence $F = \text{seq}X_w\text{-lm}_e F^n \Leftrightarrow F^* = X_s^*\text{-lm}_e F^{n*}$ (for a sequence of equicoercive closed convex functions F^n) provides a flexible tool in the study of epi-convergence by the duality method. One can distinguish two types of situations:

(a) The convergence result $F = \text{seq}X_w\text{-lm}_e F^n$ is known. One then derives a new convergence result, $F^* = X_s^*\text{-lm}_e F^{n*}$, by computing the conjugate functions of F^{n*} and F^*.

(b) In order to study the "primal" epi-convergence problem, it may be simpler in some situations to study the "dual" epi-convergence problem

$$F^* = X_s^*\text{-lm}_e F^{n*}$$

and then return to the primal problem by computing $(X_s^*\text{-lm}_e F^{n*})^*$.

$$\begin{array}{ccc} F^n & \dashrightarrow & F \\ {}^*\downarrow & & \uparrow^* \\ F^{n*} & \xrightarrow{e} & F^* \end{array}$$

Let us illustrate these two different aspects on the following examples.

PROPOSITION 3.14 "Dual formulation of homogenization results"

Let

$$\phi : \mathbb{R}^N \times \mathbb{R}^N \to \mathbb{R}^+$$
$$(y,z) \to \phi(y,z)$$

be Y-periodic in y, convex continuous in z and satisfying

$$\lambda_0(|z|^2-1) \leq \phi(y,z) \leq \Lambda_0(|z|^2+1) \qquad 0 < \lambda_0 \leq \Lambda_0 < +\infty.$$

Then, for every $f \in H^{-1}(\Omega)$,

$$\min_{\{-\mathrm{div}v=f\}}\int_\Omega \phi(\frac{x}{\varepsilon},v(x))dx \xrightarrow[(\varepsilon\to 0)]{} \min_{\{-\mathrm{div}v=f\}}\int_\Omega \phi_{hom}(v(x))dx$$

where

$$\phi_{hom}(z) = \min_{\begin{cases}\mathrm{div}w=0\\ \int_Y w=0\\ w\cdot n \text{ takes opposite values}\\ \quad\text{on opposite faces of } Y\end{cases}} \int_Y \phi(y,w(y)+z)dy. \qquad (3.36)$$

Proof of Proposition 3.14 Let us denote $j = \phi^*$ and

$$F^\varepsilon(u) = \int_\Omega j(\frac{x}{\varepsilon},\mathrm{grad}u(x))dx.$$

From Theorem 1.20, $\mathrm{seq}X_w\text{-}\mathrm{lm}_e F^\varepsilon = F^{hom}$ exists, with $X = H_0^1(\Omega)$,

$$F^{hom}(u) = \int_\Omega j^{hom}(\mathrm{grad}u(x))dx,$$

and

$$j^{hom}(z) = \min_{\{w\ Y\text{-periodic}\}}\int_Y j(y,\mathrm{grad}w(y)+z)dy.$$

The dual formulation of this epi-convergence result is, by Corollary 3.13 (let us notice that the functionals F^ε are equicoercive): for every $f\in H^{-1}(\Omega)$,

$$\lim_{\varepsilon\to 0}(F^\varepsilon)^*(f) = F^{hom*}(f).$$

A classical computation relying on Proposition 3.4 (cf. Attouch & Damlamian [2]) yields

$$F^{\varepsilon*}(f) = \min_{\{-\mathrm{div}\ v=f\}}\int_\Omega j^*(\frac{x}{\varepsilon}, v(x))dx$$

$$F^{hom^*}(f) = \min_{\{-\mathrm{div}\ v=f\}}\int_\Omega j^{hom^*}(v(x))dx.$$

In order to complete the proof of Proposition 3.14 we just have to compute j^{hom^*}:

$$j^{hom*}(z) = \sup_{\xi\in\mathbb{R}^N} \{\langle z,\xi\rangle - \min_{\{w\ Y\text{-periodic}\}} \int_Y j(y,\text{grad}w(y)+\xi)dy\}$$

$$= \sup_{\{\substack{\xi\in\mathbb{R}^N \\ w\ Y\text{-periodic}}} \{\langle z,\xi\rangle - \int_Y j(y,\text{grad}w(y) + \xi)dy\}.$$

Noticing that for any Y-periodic function w and any $z \in \mathbb{R}^N$,

$$\int_Y \langle z,\text{grad}w(y)\rangle dy = 0$$

$$j^{hom*}(z) = \sup_{\eta\in\mathbb{E}} \{\langle z,\eta\rangle - \int_Y j(y,\eta(y))dy\},$$

where $\mathbb{E}$ is the subspace of functions of the form ξ + grad$w(y)$, $\xi \in \mathbb{R}^N$, w Y-periodic. Applying again Proposition 3.4, it follows that

$$(j^{hom})^*(z) = \min_{\{\eta_1+\eta_2=z\}} \{ \int_Y j^*(y,\eta_1(y))dy + I_{\mathbb{E}^\perp}(\eta_2)\},$$

where $\mathbb{E}^\perp$ is the orthogonal subspace of $\mathbb{E}$, that is

$$\mathbb{E}^\perp = \{\vec{w}/ \int_Y \vec{w} = 0,\ \text{div}\ \vec{w} = 0,\ \vec{w}.\vec{n} \text{ takes opposite values on opposite face of } Y\}.$$

We finally obtain

$$(j^{hom})^*(z) = \min_{\begin{cases} \text{div } w = 0 \\ \int_Y w = 0 \\ w.n \text{ takes opposite values} \\ \quad \text{on opposite face of } Y \end{cases}} \int_Y \phi(y,w(y)+z)dy.$$

This dual approach to homogenization is close to the one developed by P. Suquet [1]: it can be systematically extended, and its interpretation is of physical interest, to systems of continum mechanics, elasticity (refer to Attouch [10], Aze [1]).

PROPOSITION 3.15 Let $\{F^\varepsilon:H_o^1(0,1) \to \bar{\mathbb{R}}^+;\ \varepsilon \to 0\}$ be the sequence of functionals

$$F^\varepsilon(u) = \int_0^1 a_\varepsilon(x)\dot{u}^2(x)dx + I_{K_\rho}(u)$$

where the convex of constraints K_ρ is equal to

$$K_\rho = \{u \in H_0^1(0,1) / \int_0^1 \dot{u}^2(x)dx \leq \rho\}, \quad \rho > 0$$

and, as in model homogenization problems, the "conductivity" coefficients oscillate more and more rapidly between two values α and β, $0 < \alpha \leq \beta < +\infty$:

$$a_\varepsilon(x) = \frac{\alpha+\beta}{2} + \frac{\beta-\alpha}{2} \; \frac{\sin(\frac{\pi x}{\varepsilon})}{|\sin(\frac{\pi x}{\varepsilon})|} .$$

Then, $X_w\text{-lm}_e F^\varepsilon = F^{hom}$ where, for every $u \in H_0^1(\Omega)$,

$$F^{hom}(u) = h_{\alpha,\beta} \left(\int_0^1 \dot{u}^2(x)dx\right) \tag{3.37}$$

and

$$h_{\alpha,\beta}(t) = \sup_{\lambda > 0} \; \{-\lambda\rho + \frac{2(\alpha+\lambda)(\beta+\lambda)}{\alpha+\beta+2\lambda} t\} \tag{3.38}$$

is a convex increasing function satisfying, $h_{\alpha,\beta}(0) = 0$, $h_{\alpha,\beta}(t) < +\infty \Leftrightarrow t \leq \rho$, and $\alpha t \leq h_{\alpha,\beta}(t) \leq \beta t \quad \forall t \in [0,\rho]$.

<u>Proof of Proposition 3.15</u> The following proof illustrates how by a duality method, one can reduce a limit analysis constrained problem to an unconstrained one (refer to Attouch & Sbordone [1]).

The functions $\{F^\varepsilon; \varepsilon > 0\}$ being equicoercive, the computation of $X_w\text{-lm}_e F^n$ is equivalent to the study of the convergence, for every f belonging to $X^* = H^{-1}(0,1)$, of the sequence of real numbers $\{m_\varepsilon(f); \varepsilon \to 0\}$, where

$$m_\varepsilon(f) = \min_{u \in X} \{F^\varepsilon(u) - \langle f,u\rangle\}.$$

We stress the fact that, thanks to the convexity of the functionals F^ε, it is enough, in order to compute the epi-limit of the sequence $\{F^\varepsilon; \varepsilon \to 0\}$, to consider the convergence of such minima for *linear continuous perturbations*.

Let us introduce the following penalized approximation of the constraints K_ρ:

$$m_\varepsilon(f) = \min_{u \in X} \; \sup_{\lambda \geq 0} \{F^\varepsilon(u) + \lambda \left(\int_0^1 \dot{u}^2(x)dx - \rho\right) - \langle f,u\rangle\}.$$

By a classical min-max theorem (cf. Aubin [1], for example)

$$m_\varepsilon(f) = \sup_{\lambda\geq 0} \min_{u\in X} \{- \lambda\rho + \int_0^1 (a_\varepsilon(x) + \lambda)\dot{u}^2(x)dx - \langle f,u\rangle\}$$

$$= \sup_{\lambda\geq 0} \phi_\varepsilon(\lambda)$$

where we denote

$$\phi_\varepsilon(\lambda) = - \lambda\rho + \min_{u\in X} \{\int_0^1 (a_\varepsilon(x) + \lambda)\dot{u}^2(x)dx - \langle f,u\rangle\}. \tag{3.39}$$

Now assuming $\lambda \geq 0$ fixed, from Theorem 1.1, case N = 1, we obtain that, for every $\lambda \geq 0$,

$$\lim_{\varepsilon\to 0} \phi_\varepsilon(\lambda) = \phi(\lambda)$$

exists, where

$$\phi(\lambda) = - \lambda\rho + \min_{u\in X} \{\int_0^1 \frac{1}{\text{w-lim}(\frac{1}{a_\varepsilon+\lambda})} \dot{u}^2(x)dx - \langle f,u\rangle\}. \tag{3.40}$$

In order to derive from the pointwise convergence of the $(\phi_\varepsilon)_{\varepsilon>0}$, the convergence of the sequence $(m_\varepsilon)_{\varepsilon>0}$, $m_\varepsilon = \sup_{\lambda>0} \phi_\varepsilon(\lambda)$, we use the following geometrical observations (Fig. 3.1):

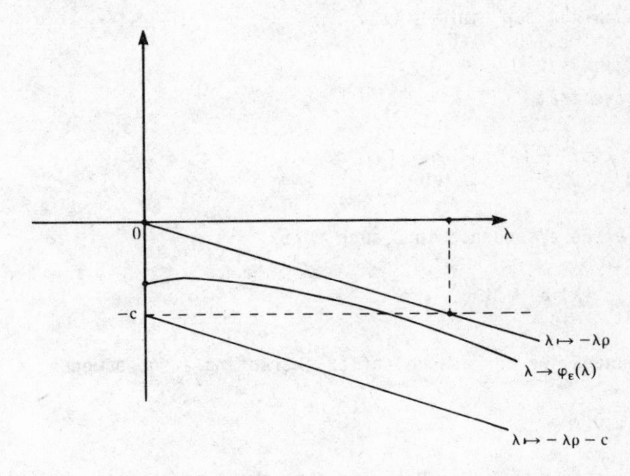

Figure 3.1

(a) the functions $\lambda \to \phi_\varepsilon(\lambda)$ are concave and continuous on $[0, +\infty[$.

(b) $\forall \lambda \geq 0 \quad \phi_\varepsilon(\lambda) \leq -\lambda\rho$ (take $u = 0$ in (3.39).

(c) $\forall \lambda \geq 0 \quad \phi_\varepsilon(\lambda) \geq -\lambda\rho + \min_{u \in X} \{\alpha \int_0^1 \dot{u}^2(x)dx - \langle f,u \rangle\}$

$$\geq -\lambda\rho - c.$$

Clearly, the supremum of the functions $(\phi_\varepsilon)_{\varepsilon>0}$ is attained on the bounded interval $[0, \frac{c}{\rho}]$, which is independent of ε.

We conclude to the convergence of the supremum $(m_\varepsilon)_{\varepsilon>0}$ of the $(\phi_\varepsilon)_{\varepsilon>0}$ with the help of the following result.

LEMMA 3.16 Let D be a bounded convex open set in $\mathbb{R}^N$ and $(\phi^\varepsilon)_{\varepsilon>0}$ a sequence of convex continuous functions on D such that

$$\forall \lambda \in D \quad \lim_{\varepsilon \to 0} \phi^\varepsilon(\lambda) = \phi(\lambda) \text{ exists (and is finite).}$$

Then,

$$\lim_{\varepsilon \to 0} \inf_{\lambda \in D} \phi^\varepsilon(\lambda) = \inf_{\lambda \in D} \phi(\lambda).$$

Proof of Lemma 3.16 Clearly

$$\inf_{\lambda \in D} \phi(\lambda) \geq \limsup_{\varepsilon \to 0} \inf_{\lambda \in D} \phi^\varepsilon(\lambda).$$

So, let us prove that

$$\liminf_{\varepsilon \to 0} \inf_{\lambda \in D} \phi^\varepsilon(\lambda) \geq \inf_{\lambda \in D} \phi(\lambda).$$

Let $\{\lambda_\varepsilon ; \varepsilon \to 0\}$ be a sequence in D such that

$$\varepsilon + \inf_{\lambda \in D} \phi^\varepsilon(\lambda) \geq \phi^\varepsilon(\lambda_\varepsilon).$$

Since D is bounded, we can assume (after extracting a subsequence) that

$$\lambda_\varepsilon \to \bar{\lambda} \in \bar{D}.$$

Let us introduce some $\lambda_0 \in D$. Noticing that the sequence $\{\phi^\varepsilon ; \varepsilon \to 0\}$ converges uniformly to ϕ on compact subsets of D, we obtain that, for every $t \in]0,1[$,

$$\inf_{\lambda \in D} \phi(\lambda) \leq \phi(t\bar{\lambda} + (1-t)\lambda_o)$$

$$\leq \lim_{\varepsilon \to 0} \phi^\varepsilon(t\lambda_\varepsilon + (1-t)\lambda_o)$$

$$\leq t \liminf_{\varepsilon \to 0} \phi^\varepsilon(\lambda_\varepsilon) + (1-t) \lim_{\varepsilon \to 0} \phi^\varepsilon(\lambda_o)$$

$$\leq t \liminf_{\varepsilon \to 0} \inf_{\lambda \in D} \phi^\varepsilon(\lambda) + (1-t)\phi(\lambda_o).$$

This being true for any $t \in]0,1[$, letting t go to one, it follows that

$$\inf_{\lambda \in D} \phi(\lambda) \leq \liminf_{\varepsilon \to 0} \inf_{\lambda \in D} \phi^\varepsilon(\lambda). \quad \square$$

End of Proof of Proposition 3.15 From the above lemma and (3.40) it follows that

$$\lim_{\varepsilon \to 0} m_\varepsilon(f) = \sup_{\lambda > 0} \phi(\lambda)$$

$$= \sup_{\lambda > 0} \min_{u \in X} \{- \lambda\rho + \int_o^1 \frac{1}{\text{w-}\lim_{\varepsilon \to 0} (\frac{1}{a_\varepsilon + \lambda})} \dot{u}^2(x)dx - \langle f,u\rangle\}.$$

Applying once more a min-max argument,

$$\lim_{\varepsilon \to 0} m_\varepsilon(f) = \min_{u \in X} \{F(u) - \langle f,u\rangle\}$$

where

$$F(u) = \sup_{\lambda > 0} \{- \lambda\rho + \int_o^1 \frac{1}{\text{w-}\lim_{\varepsilon \to 0}(\frac{1}{a_\varepsilon + \lambda})} \dot{u}^2(x)dx\}.$$

So, from Theorem 3.11, F is equal to the epi-limit, for the weak topology of X, of the sequence $\{F^\varepsilon; \varepsilon \to 0\}$. Making $\frac{1}{\text{w-}\lim(\frac{1}{a_\varepsilon + \lambda})}$ explicit, we finally obtain

$$F(u) = h_{\alpha,\beta}(\int_o^1 \dot{u}^2(x)dx),$$

where

$$h_{\alpha,\beta}(t) = \sup_{\lambda>0} \{ - \lambda\rho + \frac{2(\alpha+\lambda)(\beta+\lambda)}{\alpha+\beta+2\lambda} t\}$$

and $h_{\alpha,\beta}$ is a convex, increasing function of t, satisfying $h_{\alpha,\beta}(t) < +\infty \Leftrightarrow t \leq \rho$, and

$$\forall t \in [0,\rho] \quad \alpha t \leq h_{\alpha,\beta}(t) \leq \beta t$$

(Fig. 3.2).

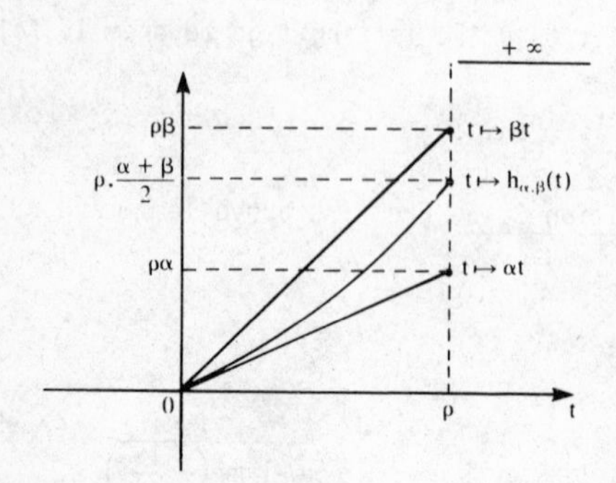

Figure 3.2

3.3 MOSCO CONVERGENCE. BICONTINUITY OF THE FENCHEL TRANSFORMATION

3.3.1 Introduction to Mosco-convergence via the bicontinuity of the Fenchel transformation

There are many ways to introduce Mosco convergence: one of them is to try to define a notion of convergence for convex functions which makes the Fenchel transformation $F \to F^*$ bicontinuous.

Let us return to Theorem 3.11 and notice that (at least in the equicoercive case)

$$F = \text{seq}X_w - \text{lm}_e F^n \Leftrightarrow F^* = X_s^* - \text{lm}_e F^{n*}$$

$$F = X_s\text{-lm}_e F^n \Leftrightarrow F^* = \text{seq}X_w^*\text{-lm}_e F^{n*}.$$

The two topologies, weak and strong, are exchanged by conjugation. Thus, it is natural to introduce a convergence for *both* for the weak and the strong topology: this is Mosco convergence:

DEFINITION 3.17 Let X be a reflexive Banach space and $(F^n)_{n\in\mathbb{N}}$, F a sequence of functions from X into $]-\infty, +\infty]$. The sequence $(F^n)_{n\in\mathbb{N}}$ is said to be Mosco-convergent to F (and we then write $F^n \xrightarrow{M} F$) if:

$$F = \mathrm{seq}X_w\text{-}\mathrm{lm}_e F^n = X_s\text{-}\mathrm{lm}_e F^n. \tag{3.41}$$

Before studying in detail this notion of convergence, let us formulate the bicontinuity property for Mosco convergence of the Fenchel transformation.

THEOREM 3.18 Let X be a reflexive Banach space and $(F^n)_{n\in\mathbb{N}}$, F a sequence of closed convex proper functions from X into $]-\infty, +\infty]$. The following statements are equivalent:

$$\begin{array}{ll} F^n \xrightarrow{M} F & \text{(on } X) \\ \Updownarrow & \\ F^{n*} \xrightarrow{M} F^* & \text{(on } X^*). \end{array} \tag{3.42}$$

In other words the Fenchel transformation is bicontinuous for the Mosco-convergence from $\Gamma(X)$ onto $\Gamma(X^*)$.

Proof of Theorem 3.18 The argument which naturally led us to introduce the Mosco convergence notion yields the bicontinuity of the Fenchel duality transformation only under some restrictive equicoerciveness assumption. The conclusion in the general case can be easily derived from Theorem 3.7, which states that:

$$(\mathrm{seq}\ X_w\text{-}\mathrm{li}_e F^n)^* = X_s^* - \mathrm{ls}_e F^{n*}.$$

Let us assume that $F^n \xrightarrow{M} F$. Equivalently,

$$\mathrm{seq}X_w\text{-}\mathrm{li}_e F^n = \mathrm{seq}X_w\text{-}\mathrm{ls}_e F^n = X_s\text{-}\mathrm{li}_e F^n = X_s\text{-}\mathrm{ls}_e F^n.$$

This sequence of equalities is clearly equivalent to the equality of the smallest and the largest of these four functions, which are (refer to

Proposition 2.53) respectively $\text{seq}X_w\text{-li}_e F^n$ and $X_s\text{-ls}_e F^n$. So, the following equivalence holds:

$$F^n \xrightarrow{M} F$$

$$\Updownarrow \qquad (3.43)$$

$$X_s\text{-ls}_e F^n \leq F \leq \text{seq}X_w\text{-li}_e F^n.$$

By conjugation, the inequalities are reversed:

$$(\text{seq}X_w\text{-li}_e F^n)^* \leq F^* \leq (X_s\text{-ls}_e F^n)^*.$$

From Theorem 3.7, and the formula obtained by conjugation,

$$X_s^*\text{-ls}_e F^{n^*} \leq F^* \leq (\text{seq}X_w^*\text{-li}_e F^{n^*})^{**}$$

(we use the fact that $F^n = (F^{n^*})^*$. Noticing that for any function G the inequality

$$G^{**} \leq G$$

holds, we obtain

$$X_s^*\text{-ls}_e F^{n^*} \leq F^* \leq \text{seq}X_w^*\text{-li}_e F^{n^*}.$$

Finally, we have proved (using again formulation (3.43) of the Mosco convergence) that

$$F^n \xrightarrow{M} F$$

$$\Downarrow$$

$$F^{n^*} \xrightarrow{M} F^*.$$

Using once more that the functions $(F^n)_{n\in\mathbb{N}}$ are closed convex, $(F^{n^*})^* = F^n$, we obtain the equivalence (3.42). □

Let us now give a series of equivalent formulations of Mosco-convergence, some, such as (3.44) being interesting from a practical point of view, some such as (3.45), demonstrating clearly the symmetric role played by F^n and F^{n^*} in this notion of convergence.

PROPOSITION 3.19 Let $(F^n)_{n \in \mathbb{N}}$, F be a sequence of closed convex proper functions from X, a reflexive Banach space, into $]-\infty, +\infty]$. The following statements are equivalent

(i) $F^n \xrightarrow{M} F$

(ii) $X_s\text{-}\mathrm{ls}_e F^n \leqslant F \leqslant \mathrm{seq}X_w\text{-}\mathrm{li}_e F^n$

(iii)
$$\begin{cases} \forall x \in X, \exists x_n \xrightarrow{X_s} x \text{ such that } F^n(x_n) \to F(x). \\ \forall x \in X, \forall x_n \xrightarrow{X_w} x, \quad F(x) \leqslant \liminf_n F^n(x_n) \end{cases} \tag{3.44}$$

(iv)
$$\begin{cases} \forall x \in X, \exists x_n \xrightarrow{X_s} x \text{ such that } F^n(x_n) \to F(x) \\ \forall x^* \in X^*, \exists x_n^* \xrightarrow{X_s^*} x^* \text{ such that } F^{n^*}(x_n^*) \to F^*(x^*). \end{cases} \tag{3.45}$$

Proof of Proposition 3.19 When proving (3.43), we obtained the equivalence (i) $\Leftrightarrow$ (ii). The formulation (iii) follows directly from (ii) and the sequential definition (cf. Theorem 1.13) of $X_s\text{-}\mathrm{ls}_e F^n$ and $\mathrm{seq}X_w\text{-}\mathrm{li}_e F^n$. From the equivalence (3.42), we obtain implication (iii) $\Rightarrow$ (iv). The only thing we have to prove is (iv) $\Rightarrow$ (iii):

Let $x_n \xrightarrow{X_w} x$; by assumption, for every $x^* \in X^*$, there exists a sequence $x_n^* \xrightarrow{X_s^*} x^*$ such that $F^{n^*}(x_n^*) \to F^*(x^*)$. From $F^n = (F^{n^*})^*$, for every $n \in \mathbb{N}$,

$$F^n(x_n) \geqslant \langle x_n, x_n^* \rangle - F^{n^*}(x_n^*).$$

Going to the lower limit, as n goes to $+\infty$, on the above inequality,

$$\liminf_n F^n(x_n) \geqslant \langle x, x^* \rangle - F^*(x^*).$$

This being true for every $x^* \in X^*$, noticing that $F = (F^*)^*$, it follows that

$$\liminf_n F^n(x_n) \geqslant F(x). \quad \square$$

We stressed in Section 2.5 that the monotone convergence is a variational convergence. When applied to sequences of closed convex functions this result takes a striking form:

THEOREM 3.20 Let $(F^n)_{n\in\mathbb{N}}$ be a sequence of closed convex proper functions from X a reflexive Banach space into $]-\infty, +\infty]$. Then the following properties hold:

(i) If the sequence $(F^n)_{n\in\mathbb{N}}$ increases, then it Mosco-converges to $F = \sup_{n\in\mathbb{N}} F^n$.

(ii) If the sequence $(F^n)_{n\in\mathbb{N}}$ decreases, then it Mosco-converges to

$F = c\ell(\inf_{n\in\mathbb{N}} F^n)$.

These two results can be viewed as dual one from the other, via the Fenchel duality transformation, and the bicontinuity of this transformation for Mosco-convergence.

Proof of Theorem 3.20 From Theorem 2.40, the sequence $(F^n)_{n\in\mathbb{N}}$, being monotonically increasing, is τ-epi convergent for any topology τ, with $\tau\text{-lm}_e F^n = \sup_{n\in\mathbb{N}} (c\ell_\tau F^n)$.

Taking τ = strong topology of X, then τ = weak topology of X and noticing that, because of their convexity, the functions $(F^n)_{n\in\mathbb{N}}$ are closed for these two topologies we obtain

$$\sup_{n\in\mathbb{N}} F^n = X_s\text{-lm}_e F^n = X_w\text{-lm}_e F^n,$$

which expresses the Mosco-convergence of the sequence $(F^n)_{n\in\mathbb{N}}$ to F. Similarly, using Theorem 2.46, and noticing that $\inf F^n$, as a decreasing limit of convex functions, is still convex, and thus has the same closure for the strong and the weak topology of X, we obtain conclusion (ii). Finally, we notice that, via the transformation $F \to F^*$, any monotonically increasing sequence is transformed into a decreasing one. □

3.3.2 Mosco-convergence of convex sets

In Proposition 1.40 we established that the correspondence

$$K \to I_K$$

which associates with every closed set K its indicator function I_K (we recall that $I_K = 0$ on K, $+\infty$ elsewhere) is one to one, bicontinuous (when the family of the closed sets is equipped with set convergence and the family of closed functions with epi-convergence). Thus, it is natural to introduce the following notion of convergence, also called Mosco convergence, for sequences of sets in a reflexive Banach space:

DEFINITION-PROPOSITION 3.21 Let X be a reflexive Banach space and $(K^n)_{n\in\mathbb{N}}$, K a sequence of subsets of X. The following properties are equivalent:

(i) $I_{K^n} \to I_K$ in Mosco sense, as n goes to $+\infty$.

(ii) $K = X_s\text{-}LmK^n = \text{seq}X_w\text{-}LmK^n$

(iii) $\begin{cases} \forall x \in K\ \exists (x_n)_{n\in\mathbb{N}},\ x_n \in K^n \text{ for every } n \in \mathbb{N} \text{ such that } x_n \xrightarrow[(n\to+\infty)]{X_s} x. \\ \forall (n_k)_{k\in\mathbb{N}},\ \forall (x_k)_{k\in\mathbb{N}},\ x_k \in K^{n_k} \text{ for every } k \in \mathbb{N},\ (x_k \xrightarrow{X_w} x) \Rightarrow x \in K. \end{cases}$

when one of these equivalent properties is satisfied, the sequence $(K^n)_{n\in\mathbb{N}}$ is said to be Mosco-convergent to K.

Transcribing the preceding results (Theorems 3.20 and 3.18) we obtain

PROPOSITION 3.22 Let X be a reflexive Banach space and $(K^n)_{n\in\mathbb{N}}$ a sequence of closed convex sets in X;

(i) If the sequence $(K^n)_{n\in\mathbb{N}}$ is increasing, i.e. $K^n \subset K^{n+1}$ for every $n \in \mathbb{N}$, then $K^n \xrightarrow{M} K$ in the Mosco sense, where $K = c\ell(\bigcup_{n\in\mathbb{N}} K^n)$.

(ii) If the sequence $(K^n)_{n\in\mathbb{N}}$ is decreasing, i.e. $K^{n+1} \subset K^n$ for every $n \in \mathbb{N}$, then $K^n \xrightarrow{M} K$ in the Mosco sense, where $K = \bigcap_{n\in\mathbb{N}} K^n$.

PROPOSITION 3.23 Let X be a reflexive Banach space and $(K^n)_{n\in\mathbb{N}}$ a sequence of closed convex subsets of X. The following statements are equivalent:

(i) $K^n \xrightarrow{M} K$ (i.e. the sequence $(K^n)_{n\in\mathbb{N}}$ Mosco-converges to K)

(ii) $s(K^n) \xrightarrow{M} s(K)$ (i.e. the sequence $(s(K^n))_{n\in\mathbb{N}}$ Mosco-converges to s(K))

where, for any closed convex set C, s(C) denotes the support function of C:

$$\forall x^* \in X^* \quad s(C)(x^*) = \sup_{x\in C} \langle x^*,x\rangle.$$

The above proposition gives us another way to connect Mosco convergence of convex sets with Mosco convergence of convex functions. In the following sections, we give a geometrical interpretation of the Mosco convergence of convex sets, then illustrate its importance in the study of approximation or perturbation problems in optimization theory. For a detailed study of this notion of convergence and further examples, refer to Mosco [1-4], Joly [1], Sonntag [1], Zolezzi [1-4], Attouch [2-4].

3.4 MOSCO-CONVERGENCE AND CONVERGENCE OF MOREAU-YOSIDA APPROXIMATES

3.4.1 Differentiability properties of Moreau-Yosida approximates

Given X a reflexive Banach space and $F:X \to]-\infty, +\infty]$ a closed convex proper function, the *subdifferential* ∂F of F is the (possibly) multivalued operator from X into X* whose graph is equal to

$$\partial F = \{(x,f) \in X \times X^*/\forall u \in X \quad F(u) \geq F(x) + \langle f,u-x\rangle\}$$

$$= \{(x,f) \in X \times X^*/F(x) + F^*(f) = \langle f,x\rangle\}. \tag{3.46}$$

The operator ∂F is a maximal monotone operator from X into X* (cf. Rockafellar [1,2], Brezis [1] for further details).

The subdifferential of the convex continuous function $x \to \frac{1}{2}\|x\|_X^2$ is called the *duality map* from X into X*. We denote it by H. One can easily verify (it is just an application of the Hahn-Banach theorem) that, for every $x \in X$, H(x) is non-void and equal to

$$H(x) = \{f \in X^*/ \|f\|_{X^*} = \|x\| \text{ and } \langle f,x\rangle = \|x\|^2\}. \tag{3.47}$$

By renorming, we can assume without restriction that $\|\cdot\|_X$ and $\|\cdot\|_{X^*}$ are *strictly convex*. We recall that a function $\phi:X \to]-\infty, +\infty]$ is strictly convex, if for every u and v in X, $u \neq v$, for every $0 < \lambda < 1$ the convexity inequality

$$\phi(\lambda u + (1-\lambda)v) < \lambda\phi(u) + (1-\lambda)\phi(v)$$

is strict. The duality map H is, consequently, a one to one mapping from X onto X*.

THEOREM 3.24 Let X be a reflexive Banach space, strictly convex along with its dual. Let $F:X \to]-\infty, +\infty]$ be a closed, convex, proper function. Then, for every $\lambda > 0$, for every $x \in X$,

$$F_\lambda(x) = \min_{u\in X} \{F(u) + \frac{1}{2\lambda}\|u-x\|^2\}$$
$$= F(J_\lambda^F x) + \frac{1}{2\lambda}\|x - J_\lambda^F x\|^2, \tag{3.48}$$

where $J_\lambda^F x$ denotes the unique point of X where this infimum is achieved. It satisfies the extremality relation

$$\frac{1}{\lambda} H(x - J_\lambda^F x) \in \partial F(J_\lambda^F x). \tag{3.49}$$

For every x belonging to the domain of F,

$$J_\lambda^F x \to x \text{ strongly in } X, \text{ as } \lambda \text{ goes to zero.} \tag{3.50}$$

$$F(J_\lambda^F x) \to F(x) \text{ as } \lambda \text{ goes to zero.} \tag{3.51}$$

When, in addition, the norms of X *and* X^* satisfy the following smoothness property (R)

(R) "The weak convergence and the norm convergence imply the strong convergence" (3.52)

then, for every $\lambda > 0$, F_λ is a $\mathcal{C}^1$ function whose Frechet derivative ∇F_λ is equal to the Yosida approximation of the maximal monotone operator ∂F:

$$\forall x \in X \quad \nabla F_\lambda(x) = (\partial F)_\lambda(x)$$
$$= \frac{1}{\lambda} H(x - J_\lambda^F x) \tag{3.53}$$

The mapping $x \to J_\lambda^F x$ is continuous from X_s into X_s and the mapping $x \to F(J_\lambda^F x)$ is continuous from X_s into $\mathbb{R}$. (3.54)

Proof of Theorem 3.24 The function $u \to F(u) + \frac{1}{2\lambda}\|u-x\|^2$ being lower semi-continuous, proper and coercive on a reflexive Banach space X reaches its minimum at a unique point (uniqueness follows from the strict convexity of the norm of X). From classical theorems (cf. Rockafellar [2]) about additivity of subdifferentials, $J_\lambda^F x$ satisfies the extremality relation (3.49).

Introducing some minorizing affine function, $F(x) + r\,(\|x\| + 1) \geq 0$, from (3.48),

$$F(x) \geq \sup_{\lambda>0} F_\lambda(x)$$

$$\geq -r\,\|J_\lambda^F x\| - r + \frac{1}{2\lambda}\|x - J_\lambda^F x\|^2 .$$

Assuming that x belongs to dom(F), i.e. $F(x) < +\infty$, we first derive (3.50). Using the inequality

$$F(x) \geq F(J_\lambda^F x)$$

the convergence of $J_\lambda^F x$ to x (as λ goes to zero), and the lower semicontinuity of F we obtain

$$F(x) \geq \limsup_{\lambda \to 0} F(J_\lambda^F x) \geq \liminf_{\lambda \to 0} F(J_\lambda^F x) \geq F(x),$$

which ensures the convergence of $F(J_\lambda^F x)$ to F(x) as λ goes to zero, (3.51).

Let us now prove under the norm regularity assumptions (R), the Frechet differentiability of the Moreau-Yosida approximates F_λ: for every $x, y \in X$,

$$F_\lambda(y) - F_\lambda(x) - \frac{1}{\lambda}\langle H(x-J_\lambda^F x),\, y-x\rangle = \|y-x\|\,.\varepsilon(\|y-x\|). \qquad (3.55)$$

From the definition (3.48) of J_λ (we omit the subscript F for J_λ^F since there is no ambiguity)

$$F_\lambda(y) - F_\lambda(x) = F(J_\lambda y) - F(J_\lambda x) + \frac{1}{2\lambda}\{\|y-J_\lambda y\|^2 - \|x-J_\lambda x\|^2\} \qquad (3.56)$$

$$\geq \langle \partial F(J_\lambda x), J_\lambda y - J_\lambda x\rangle + \frac{1}{\lambda}\langle H(x-J_\lambda x), (y-J_\lambda y)-(x-J_\lambda x)\rangle$$

where we use (which is rather convenient) the notation $\partial F(J_\lambda x)$ to denote any element belonging to the set $\partial F(J_\lambda x)$. From the extremality relation (3.49)

$$F_\lambda(y) - F_\lambda(x) \geq \frac{1}{\lambda}\langle H(x-J_\lambda x),\, y-x\rangle,$$

which expresses that the first member of (3.55) is positive. So let us estimate it from above. From the following subdifferential inequalities

$$F(J_\lambda x) \geq F(J_\lambda y) + \frac{1}{\lambda} \langle H(y-J_\lambda y),\ J_\lambda x - J_\lambda y\rangle$$

$$\frac{1}{2\lambda} \|x-J_\lambda x\|^2 \geq \frac{1}{2\lambda} \|y-J_\lambda y\|^2 + \frac{1}{\lambda} \langle H(y-J_\lambda y),\ (x-J_\lambda x) - (y-J_\lambda y)\rangle,$$

on returning to (3.56), we obtain

$$F_\lambda(y) - F_\lambda(x) - \frac{1}{\lambda} \langle H(x-J_\lambda x) y-x\rangle$$

$$\leq \frac{1}{\lambda} \langle H(y-J_\lambda y),\ J_\lambda y - J_\lambda x\rangle - \frac{1}{\lambda} \langle H(y-J_\lambda y), (x-J_\lambda x)-(y-J_\lambda y)\rangle$$

$$- \frac{1}{\lambda} \langle H(x-J_\lambda x), y-x\rangle$$

$$\leq \frac{1}{\lambda} \langle H(y-J_\lambda y) - H(x-J_\lambda x), y-x\rangle.$$

Thus,

$$0 \leq F_\lambda(y)-F_\lambda(x) - \frac{1}{\lambda} \langle H(x-J_\lambda x), y-x\rangle \leq \frac{1}{\lambda} \|H(y-J_\lambda y)-H(x-J_\lambda x)\|_{X^*} \|y-x\|_X .$$

In order to complete the proof we just need to prove the continuity of H and J_λ:

Let us first verify that under assumption (R), H is continuous from X_s onto X_s^*: given a sequence $(x_n)_{n\in\mathbb{N}}$ strongly converging in X, the sequence $(H(x_n))_{n\in\mathbb{N}}$ is bounded in the reflexive Banach space X* and thus weakly relatively compact. Extracting some weakly converging subsequence

$$H(x_{n_k}) \xrightarrow{X_w^*} \eta \quad \text{as } k \text{ goes to } + \infty,$$

and going to the limit on the subdifferential inequality

$$\frac{1}{2} \|u\|^2 \geq \frac{1}{2} \|x_{n_k}\|^2 + \langle H(x_{n_k}),\ u-x_{n_k}\rangle,$$

one obtains:

$$\forall u \in X \quad \frac{1}{2} \|u\|^2 \geq \frac{1}{2} \|x\|^2 + \langle \eta, u-x\rangle$$

that is, $\eta = H(x)$. So, the whole sequence $(H(x_n))_{n\in\mathbb{N}}$ converges weakly to H(x). This, combined with the equality

$$\|H(x_n)\| = \|x_n\|$$

and the (R) *property of* $\|\cdot\|_{X^*}$ implies the strong convergence of the sequence $(H(x_n))_{n\in\mathbb{N}}$ to $H(x)$.

Similarly, given a strongly converging sequence $(x_n)_{n\in\mathbb{N}}$, $x_n \xrightarrow{X_s} x$, we first verify that the sequence $(J_\lambda x_n)_{n\in\mathbb{N}}$ remains bounded in X. Extracting some weakly converging subsequence

$$J_\lambda x_{n_k} \xrightarrow{X_w} \xi \quad \text{as k goes to } + \infty,$$

and going to the limit on the equality

$$F_\lambda(x_n) = F(J_\lambda x_n) + \frac{1}{2\lambda} \|x_n - J_\lambda x_n\|^2$$

we obtain (we recall that F_λ is locally Lipschitz and thus continuous on X)

$$\begin{aligned} F_\lambda(x) &\geq \liminf_n F(J_\lambda x_n) + \limsup_n \frac{1}{2\lambda} \|x_n - J_\lambda x_n\|^2 \qquad (3.57) \\ &\geq F(\xi) + \limsup \frac{1}{2\lambda} \|x_n - J_\lambda x_n\|^2 \\ &\geq F(\xi) + \frac{1}{2\lambda} \|x-\xi\|^2 . \end{aligned}$$

Thus, $\xi = J_\lambda x$ and the sequence $(J_\lambda x_n)_{n\in\mathbb{N}}$ converges weakly to $J_\lambda x$ as n goes to $+ \infty$. Returning to (3.57) it follows that

$$\|x-J_\lambda x\|^2 \geq \limsup_n \|x_n-J_\lambda x_n\|^2$$

which, from *the* (R) *property of* $\|\cdot\|_X$, implies the strong convergence of the sequence $(J_\lambda x_n)_{n\in\mathbb{N}}$ to $J_\lambda x$. □

We have been naturally led to introduce the following class C of normed spaces.

<u>DEFINITION 3.25</u> C = {X/X is a reflexive Banach space, whose norm, along with its dual, is strictly convex and satisfies (R) (that is, weak-convergence and convergence of the norms imply norm-convergence)}.

Comments: This class contains all Banach spaces uniformly convex with their duals. For examples $L^p(\Omega,\mathcal{T},\mu)$, $(1 < p < +\infty)$, the Sobolev spaces $W^{k,p}(\Omega)$, $(k \in \mathbb{N}, 1 < p < +\infty)$ belong to this class. We refer to Sonntag [1] and its detailed bibliography) for following properties of this class of Banach spaces.

As far as one is only concerned with topological properties, it is not a restrictive assumption to assume for a reflexive Banach space to belong to the class C: from a renorming theorem of S. Trojanski and E. Asplund, every reflexive Banach space can be renormed so as to belong to the class C.

Let us finally notice that C is really the larger class in which one can expect the Moreau-Yosida approximates to be differentiable, since one can prove that a Banach space belongs to C iff its norm and its dual norm are Frechet differentiable (except at the origin!).

3.4.2 Equivalence between Mosco-convergence and pointwise convergence of the Moreau-Yosida approximates

THEOREM 3.26 Let X be a reflexive Banach space belonging to the class C. Let $(F^n)_{n\in\mathbb{N}}$, F be a sequence of closed convex proper functions from X into $]-\infty, +\infty]$. The following equivalences hold: (i) $\Leftrightarrow$ (ii) $\Leftrightarrow$ (iii)

(i) $F^n \xrightarrow{M} F$, i.e. the sequence $(F^n)_{n\in\mathbb{N}}$ Mosco converges to F.

(ii) $\begin{cases} \forall\lambda > 0, \quad \forall x \in X \quad J_\lambda^{F^n} x \xrightarrow[(n\to+\infty)]{} J_\lambda^F x \text{ strongly in } X. \\ \exists(u,v) \in \partial F, \exists(u_n,v_n) \in \partial F^n \text{ such that } u_n \xrightarrow{X_s} u,\ v_n \xrightarrow{X_s^*} v,\ F^n(u_n) \to F(u). \end{cases}$

(iii) $\forall\lambda > 0 \quad \forall x \in X \quad F_\lambda^n(x) \to F_\lambda(x)$ as n goes to $+\infty$.

Proof of Theorem 3.26 We are going to prove that (i) $\Rightarrow$ (ii) $\Rightarrow$ (iii) $\Rightarrow$ (i):

(i) $\Rightarrow$ (ii) Let us first prove that the sequence $(u_n)_{n\in\mathbb{N}}$, where we write $u_n = J_\lambda^{F^n}x$, is bounded in X. Taking some $x_0 \in \text{dom } F$, since $F^n \xrightarrow{M} F$ there exists a sequence $(x_{0n})_{n\in\mathbb{N}}$ converging to x_0 in X_s such that

$$F^n(x_{0n}) \longrightarrow F(x_0) \text{ as n goes to } +\infty.$$

Let us return to the definition of $J_\lambda^{F^n}(x) = u_n$

$$\frac{1}{\lambda} H(x-u_n) \in \partial F^n(u_n),$$

and write the subdifferential inequality

$$\begin{aligned} F^n(x_{on}) &\geq F^n(u_n) + \langle \tfrac{1}{\lambda} H(x-u_n), x_{on}-u_n \rangle \\ &\geq F^n(u_n) + \tfrac{1}{\lambda} \langle H(x-u_n), x_{on}-x \rangle + \tfrac{1}{\lambda} \|x-u_n\|^2 . \end{aligned} \tag{3.58}$$

Since $F^n \xrightarrow{M} F$, and F is proper, it follows from Lemma 3.8 that there exists $r \in \mathbb{R}^+$ such that

$$\forall n \in \mathbb{N} \quad \forall x \in X \quad F^n(x) + r(\|x\| + 1) \geq 0. \tag{3.59}$$

From (3.58) and (3.59),

$$F^n(x_{on}) + r(\|u_n\| + 1) + \frac{1}{\lambda} \|x-u_n\| \cdot \|x_{on}-x\| \geq \frac{1}{\lambda} \|x-u_n\|^2 ,$$

which clearly implies that the sequence $(u_n)_{n\in\mathbb{N}}$ is bounded. The space X being reflexive, the sequence $(u_n)_{n\in\mathbb{N}}$ is weakly relatively compact in X. Let us extract a converging subsequence

$$u_{n_k} \xrightarrow{X_w} u \text{ as } k \text{ goes to } +\infty$$

and prove that $u = J_\lambda^F x$.

To that end, let us use the variational formulation of $u_n = J_\lambda^{F^n} x_n$:

$$u_n = \arg\min_{u\in X} \{F^n(u) + \frac{1}{2\lambda} \|x-u\|^2 \}. \tag{3.60}$$

Since $F = X_s\text{-}\mathrm{lm}_e F^n$, for every $\xi \in X$, there exists a strongly converging sequence $(\xi^n)_{n\in\mathbb{N}}$, $\xi^n \xrightarrow{X_s} \xi$ as n goes to $+\infty$, such that

$$F^n(\xi^n) \to F(\xi) \text{ as n goes to } +\infty.$$

From (3.60),

$$F^{n_k}(u_{n_k}) + \frac{1}{2\lambda} \|x-u_{n_k}\|^2 \leq F^{n_k}(\xi^{n_k}) + \frac{1}{2\lambda} \|x-\xi^{n_k}\|^2 .$$

Therefore

$$\liminf_k F^{n_k}(u_{n_k}) + \liminf_k \frac{1}{2\lambda} \|x-u_{n_k}\|^2 \leqslant F(\xi) + \frac{1}{2\lambda} \|x-\xi\|^2 .$$

From $F = \text{seq}X_w\text{-lm}_e F^n$ and lower semicontinuity of the norm for the weak topology, it follows

$$\forall \xi \in X \quad F(u) + \frac{1}{2\lambda} \|x-u\|^2 \leqslant F(\xi) + \frac{1}{2\lambda} \|x-\xi\|^2$$

that is,

$$u = J_\lambda^F x.$$

(We use the strict convexity of the norm which implies that, for every $x \in X$, $J_\lambda^F x$ is reduced to a single point).

Thus, the whole sequence $(u_n)_{n\in\mathbb{N}}$ weakly converges to u. In order to obtain strong convergence of the sequence $(u_n)_{n\in\mathbb{N}}$, from the (R) property of the norm of X, we prove the convergence of the norms, $\|u_n\| \xrightarrow[n\to+\infty]{} \|u\|$, which is equivalent.

From

$$F^n(u_n) + \frac{1}{2\lambda} \|x-u_n\|^2 \leqslant F^n(\xi^n) + \frac{1}{2\lambda} \|x-\xi^n\|^2,$$

taking $(\xi^n)_{n\in\mathbb{N}}$ such that

$$\begin{cases} \xi^n \xrightarrow[n\to+\infty]{} u \text{ strongly in } X \\ F^n(\xi^n) \xrightarrow[n\to+\infty]{} F(u) \end{cases}$$

it follows that

$$\begin{aligned} \limsup_n \frac{1}{2\lambda} \|x-u_n\|^2 &\leqslant \limsup_n (-F^n(u_n)) + \limsup_n F^n(\xi^n) + \limsup_n \frac{1}{2\lambda}\|x-\xi^n\|^2 \\ &\leqslant -\liminf F^n(u_n) + F(u) + \frac{1}{2\lambda} \|x-u\|^2 . \end{aligned}$$

Since

$$F(u) \leqslant \liminf_n F^n(u_n) \quad \text{(we recall that } u_n \xrightarrow{X_w} u\text{)},$$

we finally obtain

$$\limsup_n \|x-u_n\|^2 \leq \|x-u\|^2$$

and the strong convergence of the sequence $(u_n)_{n\in\mathbb{N}}$ to u.

As a consequence of the above argument, we obtain that $F^n(u_n) \to F(u)$. Taking

$$v_n = \frac{1}{\lambda} H(x-u_n) \in \partial F^n(u_n), \quad v = \frac{1}{\lambda} H(x-u) \in \partial F(u),$$

we have

$$u_n \to u \text{ strongly in } X$$

$$v_n \to v \text{ strongly in } X^*$$

$$F^n(u_n) \to F(u)$$

which completes the proof of (i) $\Rightarrow$ (ii).

(ii) $\Rightarrow$ (iii) We want to derive from the pointwise convergence of the resolvents $(J^n_\lambda)_{n\in\mathbb{N}}$ (for the strong topology of X) the pointwise convergence of the Moreau-Yosida approximates $(F^n_\lambda)_{n\in\mathbb{N}}$. These two quantities are tied by the derivation formula (3.53)

$$\nabla F_\lambda(x) = \frac{1}{\lambda} H(x-J^F_\lambda x).$$

Thus, in order to express F_λ in terms of J^F_λ we need to use an integration formula: noticing that

$$\frac{d}{d\tau} F_\lambda[x_o + \tau(x-x_o)] = \langle \nabla F_\lambda(x_o + \tau(x-x_o)), x-x_o \rangle$$

and that the second member of this equality is a continuous function of τ (from (3.54) and continuity of H), integrating from $\tau = 0$ to $\tau = 1$, we obtain

$$F^n_\lambda(x) = F^n_\lambda(x_o) + \int_o^1 \langle \nabla F^n_\lambda(x_o + \tau(x-x_o)), x-x_o \rangle d\tau. \tag{3.61}$$

Take $x_o = u + \lambda H^{-1}(v)$, given by the normalization condition (ii); we claim that

$$F^n_\lambda(x_o) \to F_\lambda(x_o) \text{ as } n \to +\infty. \tag{3.62}$$

Let us introduce $x_n = u_n + \lambda H^{-1}(v_n)$. By (ii), $v_n \in \partial F^n(u_n)$; thus

$$\frac{1}{\lambda} H(x_n - u_n) \in \partial F^n(u_n),$$

equivalently $u_n = J^n_\lambda x_n$.

It follows that

$$F^n_\lambda(x_n) = F^n(u_n) + \frac{1}{2\lambda} \|x_n - u_n\|^2 .$$

Since

$$x_n \xrightarrow{X_s} x_o = u + \lambda H^{-1}(v) \text{ as n goes to } +\infty,$$

$$F^n_\lambda(x_n) \to F(u) + \frac{1}{2\lambda} \|x_o - u\|^2 = F_\lambda(x_o).$$

This last equality follows from the relations

$$x_o = u + \lambda H^{-1}(v) \text{ and } v \in \partial F(u), \text{ i.e. } u = J^F_\lambda(x_o).$$

We now use that $x_n \xrightarrow{X_s} x_o$ and that the $(F^n_\lambda)_{n\in\mathbb{N}}$ are equi locally Lipschitz to derive (from $F^n_\lambda(x_n) \to F_\lambda(x_o)$) that

$$F^n_\lambda(x_o) \to F_\lambda(x_o) \text{ as n goes to } +\infty.$$

(The equi-Lipschitz property follows from Theorem 2.64, noticing that $\sup_n F^n(u_n) < +\infty$, and $F^n(x) \geq F^n(u_n) + \langle u_n, x-u_n\rangle \geq r(1 + \|x\|)$ for r large enough!).

Using the strong convergence of the resolvents, $J^n_\lambda x \to J_\lambda x$ and the continuity of H, from (3.53) we obtain

$$\forall \tau \in [0,1] \; \langle \nabla F^n_\lambda(x_o + \tau(x-x_o)), x-x_o\rangle \xrightarrow[n\to+\infty]{} \langle \nabla F_\lambda(x_o+\tau(x-x_o)), x-x_o\rangle.$$

In order to proceed in (3.61), to the convergence of the integrals, let us apply the Lebesgue dominated convergence theorem: to that end, we notice that, on every bounded subset K of X,

$$\sup_{\substack{x\in K \\ n\in\mathbb{N}}} \|\nabla F^n_\lambda(x)\|_{X^*} \leqslant C(K) < +\infty. \tag{3.63}$$

This is a consequence of the (convex) subdifferential inequality

$$\forall n \in \mathbb{N} \quad \forall\lambda > 0 \quad \forall x,y \in X \quad F^n_\lambda(y) \geqslant F^n_\lambda(x) + \langle\nabla F^n_\lambda(x), y-x\rangle \tag{3.64}$$

and of the equi-locally Lipschitz property of the $(F^n_\lambda)_{n\in\mathbb{N}}$. So, we can go to the limit on (3.61) and finally obtain

for every $\lambda > 0$, for every $x \in X$,

$$\lim_{n\to+\infty} F^n_\lambda(x) = F_\lambda(x_o) + \int_o^1 \langle\nabla F_\lambda(x_o+\tau(x-x_o)), x-x_o\rangle d\tau$$

$$= F_\lambda(x). \quad \square$$

(iii) ⇒ (i) This is the most difficult point. The proof we present here is an extension to the Banach space (with some technical modifications) of the argument developed by the author in the Hilbertian case (cf. Attouch [4]). Before entering into the details, we would like to stress the fact that we have to use the fact that *all* the Moreau-Yosida approximates $\{(F^n_\lambda)_{n\in\mathbb{N}}; \lambda > 0\}$ converge. The proof relies on a precise study of *the dependence with respect to* $\lambda > 0$ of all the elements of the Moreau-Yosida approximation

$$\lambda \to F_\lambda(x), \quad \lambda \to J_\lambda(x), \quad \lambda \to F(J^F_\lambda x).$$

Step one: Let us prove that

$$\forall\lambda > 0 \quad \forall x \in X \qquad A^n_\lambda x \xrightarrow[w]{X^*} A_\lambda x \text{ as n goes to } +\infty, \tag{3.65}$$

where we denote

$$A^n_\lambda x = \frac{1}{\lambda} H(x-J^n_\lambda x) = \nabla F^n_\lambda(x). \tag{3.66}$$

We first notice that, from subdifferential inequality (3.64), and assumption (iii),

for every $\xi \in X$, the sequence $\langle A^n_\lambda x,\xi\rangle$ is bounded.

From the Banach-Steinhaus theorem, it follows that

for every $x \in X$, for every $\lambda > 0$, the sequence $(A_\lambda^n x)_{n\in\mathbb{N}}$ is bounded in X^*.

The space X, and hence X*, being reflexive, the sequence $(A_\lambda^n x)_{n\in\mathbb{N}}$ is weakly relatively compact in X*. For every weakly converging subsequence

$$A_\lambda^{n_k} x \xrightarrow{X^*_w} z \quad \text{(as k goes to } + \infty),$$

by going to the limit on subdifferential inequality (3.64) and using hypothesis (iii) we obtain

$$\forall y \in X \quad F_\lambda(y) \geq F_\lambda(x) + \langle z, y-x\rangle;$$

that is, $z = \nabla F_\lambda(x) = A_\lambda x$. Therefore, the whole sequence $(A_\lambda^n x)_{n\in\mathbb{N}}$ weakly converges to $A_\lambda x$.

Step two: Let us prove that

$$\forall x \in X, \quad \forall \lambda > 0, \quad A_\lambda^n x \xrightarrow[(n \to + \infty)]{X^*_s} A_\lambda x \tag{3.67}$$

Using the (R) property of the norm of X*, it is equivalent to prove that

$$\forall x \in X, \quad \forall \lambda > 0, \quad \|A_\lambda^n x\|_{X^*} \xrightarrow[(n \to +\infty)]{} \|A_\lambda x\|_{X^*}.$$

Noticing that

$$F_\lambda^n(x) = F^n(J_\lambda^n x) + \frac{\lambda}{2} \|A_\lambda^n x\|_{X^*}^2$$

and using again hypothesis (iii), what we have to prove is reduced to:

$$\forall x \in X, \quad \forall \lambda > 0 \quad F^n(J_\lambda^n x) \xrightarrow[n \to + \infty]{} F(J_\lambda x). \tag{3.68}$$

The proof relies on the following lemma, which links the quantities $F_\lambda(x)$ and $F(J_\lambda x)$ via a derivation formula (with respect to λ).

LEMMA 3.27 Let $F: X \to]-\infty, +\infty]$ a closed convex proper function from X, a reflexive C Banach space, into the extended reals. For every $x \in X$, the function $\lambda \to \lambda F_\lambda(x)$ is a concave, $\mathbb{C}^1$ function from $]0, +\infty[$ into $\mathbb{R}$. Its first derivative is equal to:

$$\frac{d}{d\lambda} \lambda F_\lambda(x) = F(J_\lambda x) \tag{3.69}$$

Proof of Lemma 3.27 From the equality

$$\lambda F_\lambda(x) = \inf_{u \in X} \{\lambda F(u) + \frac{1}{2} \|x-u\|^2\}$$

it follows that $\lambda \to \lambda F_\lambda(x)$ is a concave s.c.s. (and in fact continuous cf. (2.168) function.

Given $\lambda_1, \lambda_2 > 0$, by definition of F_λ

$$\begin{aligned}\lambda_1 F_{\lambda_1}(x) - \lambda_2 F_{\lambda_2}(x) &\leq [\lambda_1 F(J_{\lambda_2} x) + \tfrac{1}{2} \|x - J_{\lambda_2} x\|^2] - [\lambda_2 F(J_{\lambda_2} x) + \tfrac{1}{2} \|x - J_{\lambda_2} x\|^2] \\ &\leq (\lambda_1 - \lambda_2) F(J_{\lambda_2} x).\end{aligned}$$

Exchanging λ_1 and λ_2, we obtain

$$(\lambda_1 - \lambda_2) F(J_{\lambda_1} x) \leq \lambda_1 F_{\lambda_1}(x) - \lambda_2 F_{\lambda_2}(x) \leq (\lambda_1 - \lambda_2) F(J_{\lambda_2} x). \tag{3.70}$$

Let us now prove that

$$\lambda \to F(J_\lambda x) \text{ is continuous.} \tag{3.71}$$

Given $\{\lambda_n; n \in \mathbb{N}\}$ a converging sequence, $\lambda_n \to \lambda$ as $n \to +\infty$, $\lambda > 0$, we write

$$G^n(u) = F(u) + \frac{1}{2\lambda_n} \|x-u\|^2, \quad G(u) = F(u) + \frac{1}{2\lambda} \|x-u\|^2.$$

We clearly have

$$G = X_w\text{-}\mathrm{lm}_e\, G^n.$$

Moreover Argmin $G^n = \{J^F_{\lambda_n} x\}$ is bounded in X and hence weakly relatively compact. Since Argmin $G = \{J^F_\lambda x\}$, it follows from variational Theorem 1.10 that

$$J^F_{\lambda_n} x \xrightarrow{X_w} J^F_\lambda x. \tag{3.72}$$

Moreover,

$$\inf_{u \in X} G^n(u) \to \inf_{u \in X} G(u),$$

that is

$$F(J_{\lambda_n} x) + \frac{1}{2\lambda_n} \|x - J_{\lambda_n} x\|^2 \xrightarrow[(n \to + \infty)]{} F(J_\lambda x) + \frac{1}{2\lambda} \|x - J_\lambda x\|^2 \tag{3.73}$$

From (3.72) and the lower semicontinuity for the weak topology of X of the convex functions F and $\|\cdot\|_X$,

$$F(J_\lambda x) \leq \liminf_n F(J_{\lambda_n} x) \tag{3.74}$$

$$\frac{1}{2\lambda} \|x - J_\lambda x\|^2 \leq \liminf_n \frac{1}{2\lambda_n} \|x - J_{\lambda_n} x\|^2. \tag{3.75}$$

Combining (3.73), (3.74) and (3.75), it follows that

$$F(J_{\lambda_n} x) \to F(J_\lambda x),$$

which is (3.71). Moreover, $\|x - J_{\lambda_n} x\| \to \|x - J_\lambda x\|$; combined with (3.72) and the (R) property of the norm $\|\cdot\|_X$, we obtain in addition

$$\forall x \in X \quad \lambda \to J_\lambda^F x \text{ is continuous from }]0, +\infty[\text{ into } X_s. \tag{3.76}$$

Returning to (3.70), dividing by $\lambda_1 - \lambda_2$ (first with $\lambda_1 > \lambda_2$, then with $\lambda_1 < \lambda_2$) and letting λ_2 tend to λ_1, we obtain, by using (3.71), the derivation formula (3.69). □

End of the proof of Step two Let us return to the proof of point (3.68). To that end, let us introduce the two sequences of real valued functions $(h_n)_{n \in \mathbb{N}}$ and $(f_n)_{n \in \mathbb{N}}$:

$$\forall \lambda > 0 \quad h_n(\lambda) \doteq \lambda F_\lambda^n(x)$$

$$\forall \lambda > 0 \quad f_n(\lambda) \doteq F^n(J_\lambda^n x).$$

From Lemma 3.27,

$$\forall n \in \mathbb{N} \quad \forall \lambda > 0 \quad \frac{d}{d\lambda} h_n(\lambda) = f_n(\lambda). \tag{3.77}$$

By assumption (iii),

$$\forall \lambda > 0 \quad h_n(\lambda) \xrightarrow[(n \to +\infty)]{} h(\lambda) \doteq \lambda F_\lambda(x).$$

What we want to do is to derive from the pointwise convergence of the sequence $(h_n)_{n \in \mathbb{N}}$, *the pointwise convergence of the sequence* $(h'_n) = (f_n)_{n \in \mathbb{N}}$ *to* $h' = f$. In the Hilbertian case, we did that by proving (via the Ascoli theorem) that the sequence $(f_n)_{n \in \mathbb{N}}$ is uniformly locally Lipschitz on $]0,+\infty[$. In general Banach spaces, we cannot directly apply this type of argument, because we have not enough information on the modules of continuity of the functions $\lambda \to F^n(J^n_\lambda x)$. The proof relies on the following lemma, which uses essentially the concavity (resp. monotonicity) of the functions $\lambda \to h_n(\lambda) = \lambda F^n_\lambda(x)$ (resp. $\lambda \to h'_n(\lambda) = F^n(J^n_\lambda x)$).

<u>LEMMA 3.28</u> Let $(h_n)_{n \in \mathbb{N}}$, $h_n :]0,+\infty[\to \mathbb{R}$ be a sequence of concave, $\mathcal{C}^1$ functions which is pointwise converging to a concave $\mathcal{C}^1$ function h. Then,

$$\forall \lambda > 0 \quad h'_n(\lambda) \to h'(\lambda)$$

and in fact, this convergence is uniform on every compact subset of $]0, +\infty[$.

<u>Proof of Lemma 3.28</u> Let us first verify that the sequence $(h'_n)_{n \in \mathbb{N}}$ is relatively compact in $L^1_{loc}(0, +\infty)$. Take any $0 < \lambda_0 < \lambda_1 < +\infty$; then, by monotonicity of $-h'_n$, it follows that its variation on (λ_0, λ_1) is equal to

$$\mathrm{Var}_{[\lambda_0,\lambda_1]}(h'_n) = h'_n(\lambda_0) - h'_n(\lambda_1).$$

By concavity of the functions $\lambda \to h_n(\lambda)$,

$$\forall \lambda, \mu \in]0, +\infty[\quad h_n(\mu) \leq h_n(\lambda) + \langle h'_n(\lambda), \mu - \lambda \rangle.$$

Taking successively $\mu = \lambda \pm \varepsilon$ ($\varepsilon < |\lambda|$), and using the fact that for every $\mu > 0$, the sequence $(h_n(\mu))_{n \in \mathbb{N}}$ is bounded, it follows that

$$\forall \lambda > 0 \quad \sup_{n \in \mathbb{N}} |h'_n(\lambda)| < +\infty.$$

Thus

$$\forall 0 < \lambda_0 < \lambda_1 < +\infty \qquad \sup_{n\in\mathbb{N}} \mathrm{Var}_{[\lambda_0,\lambda_1]}(h_n') < +\infty,$$

which from the compact injection of $BV(\lambda_0,\lambda_1)$ into $L^1(\lambda_0,\lambda_1)$ implies that the sequence (h_n') is relatively compact in $L^1_{loc}(0,+\infty)$. Since h_n converges to h (pointwise and in fact uniformly on every compact subset of $(0,+\infty)$) this implies:

$$h_n' \to h' \text{ in } L^1_{loc}(0,+\infty), \text{ as } n \text{ goes to } +\infty. \tag{3.78}$$

In order to derive from the above result the uniform convergence of h_n' to h' on every compact subset of $(0,+\infty)$, let us argue by contradiction and assume that (we denote $f_n = h_n'$, $f = h'$):

there exists $\varepsilon > 0$, a subsequence $(f_{n_k})_{k\in\mathbb{N}}$, a converging

sequence $\lambda_k \xrightarrow[(k\to+\infty)]{} \lambda_0$ such that: $\forall k \in \mathbb{N}\quad |f_{n_k}(\lambda_k)-f(\lambda_k)| \geq \varepsilon_0$.

Considering the indexes k such that

$$\begin{aligned} &f_{n_k}(\lambda_k) \leq f(\lambda_k) - \varepsilon_0 \\ &f_{n_k}(\lambda_k) \geq f(\lambda_k) + \varepsilon_0 \quad \text{(resp.)} \end{aligned} \tag{3.79}$$

one of these two sets contains an infinity of elements k. Let us assume for example we are in the first situation (the argument is similar in the second one) and denote by $(f_k)_{k\in\mathbb{N}}$ the corresponding subsequence.

The function $f = h'$ being continuous (we use in an essential way the fact that $h_n \to h$ with $h \in \mathcal{C}^1$!), there exists some $\eta > 0$ such that

$$\forall \lambda,\mu \in\,]\lambda_0-\eta,\ \lambda_0+\eta[\quad |f(\lambda) - f(\mu)| \leq \frac{\varepsilon_0}{2}. \tag{3.80}$$

The functions $(f_k)_{k\in\mathbb{N}}$ being decreasing,

$$\forall \lambda \in [\lambda_k, \lambda_0 + \eta[\quad f_k(\lambda) \leq f_k(\lambda_k). \tag{3.81}$$

From (3.79), (3.80), (3.81) it follows that

$$\forall \lambda \in [\lambda_k, \lambda_0 + \eta[\quad f_k(\lambda) < f_k(\lambda_k) < f(\lambda_k) - \varepsilon_0 < f(\lambda) - \frac{\varepsilon_o}{2}.$$

Noticing that for k sufficiently large $[\lambda_k, \lambda_0 + \eta] \supset [\lambda_0 + \frac{\eta}{2}, \lambda_o + \eta]$ (since $\lambda_k \to \lambda_0$) we obtain

$$\int_{\lambda_0 + \frac{\eta}{2}}^{\lambda_0 + \eta} |f_k(\lambda) - f(\lambda)| d\lambda > \frac{\varepsilon_o \eta}{4}$$

which contradicts (3.78). This ends the proof of step two.

Let us summarize the situation. We know that

$$\begin{aligned} &\forall \lambda > 0 \quad \forall x \in X \quad F_\lambda^n(x) \to F_\lambda(x) \quad (\text{as } n \to +\infty) \\ &\forall \lambda > 0 \quad \forall x \in X \quad A_\lambda^n(x) \to A_\lambda(x) \quad \text{strongly in } X \ (\text{as } n \text{ goes to } +\infty) \end{aligned} \tag{3.82}$$

It is now relatively easy to derive from (3.82) the Mosco-convergence of the sequence $(F^n)_{n\in\mathbb{N}}$:

Step three

$$F^n \xrightarrow{M} F.$$

Let us verify directly on the definition the Mosco-convergence of the sequence $(F^n)_{n\in\mathbb{N}}$ to F. To that end, let us first notice that, by (3.82), the sequence $(F^n)_{n\in\mathbb{N}}$ satisfies a uniform minorization property:

$$\exists r > 0 \text{ such that } \forall n \in \mathbb{N} \quad \forall x \in X \quad F^n(x) + r(\|x\| + 1) > 0. \tag{3.83}$$

From (3.82) and the equality

$$F_\lambda^n(x) = F(J_\lambda^n x) + \frac{\lambda}{2} \|A_\lambda^n x\|^2$$

it follows that, for every $x \in X$, for every $\lambda > 0$, the sequences $(F^n(J_\lambda^n x))_{n\in\mathbb{N}}$ and $(A_\lambda^n x)_{n\in\mathbb{N}}$ are bounded. Picking up some $x_o \in X$, $\lambda_0 > 0$, from the sub-differential inequality

$$\forall x \in X \quad \forall n \in \mathbb{N} \quad F^n(x) > F^n(J_{\lambda_o}^n x_o) + \langle A_{\lambda_o}^n x_o, x - x_o \rangle$$

(which follows from $A_\lambda^n x \in \partial F^n(J_\lambda^n x)$ - refer to (3.49)), we obtain (3.83).

(a) It follows from Theorem 2.64 (which states that F_λ increases to F as λ decreases to zero) and assumption (iii) (convergence of $F^n_\lambda(x)$ to $F_\lambda(x)$) that, for every $x \in X$

$$F(x) = \lim_{\lambda\to 0} \lim_{n\to+\infty} F^n_\lambda(x)$$

$$= \lim_{n\to+\infty} F^n_{\lambda(n)}(x) \quad \text{(where we use the diagonalization Lemma 1.18)}$$

$$= \lim_{n\to+\infty} \{F^n(J^n_{\lambda(n)}x) + \frac{1}{2\lambda(n)} ||x-J^n_{\lambda(n)}x||^2\}.$$

Taking $x_n = J^n_{\lambda(n)}x$, we obtain

$$F(x) \geq \limsup_n F^n(x_n). \tag{3.84}$$

On the other hand, using (3.83) and the fact that $\lambda(n) \to +\infty$ as n goes to $+\infty$, we obtain that, whenever $F(x) < +\infty$,

$$x_n \to x \text{ strongly in } X \text{ as n goes to } +\infty. \tag{3.85}$$

(When $F(x) = +\infty$, there is nothing to prove.)

(b) Let $x_n \xrightarrow{X_w} x$ as n goes to $+\infty$. From the sequence of inequalities

$$F^n(x_n) \geq F^n_\lambda(x_n)$$

$$\geq F^n_\lambda(x) + \langle A^n_\lambda(x), x_n-x\rangle$$

and from convergence properties (3.82), it follows that

$$\liminf_n F^n(x_n) \geq F_\lambda(x).$$

This being true for any $\lambda > 0$, taking the supremum with $\lambda > 0$, we finally obtain

$$\liminf_n F^n(x_n) \geq F(x). \quad \square$$

Let us now comment on Theorem 3.26. The equivalence between properties (ii) and (iii) takes on a striking form when written as follows:

PROPOSITION 3.29 Under the hypothesis of Theorem 3.26, the following equivalences hold: (i) $\Leftrightarrow$ (ii) $\Leftrightarrow$ (iii)

(i) $F^n \xrightarrow{M} F$

(ii) $\forall\lambda > 0 \quad \forall x \in X \quad A_\lambda^n x \to A_\lambda x$ strongly in X^*, as $n \to +\infty$.

(+ Normalization condition)

(iii) $\forall\lambda > 0 \quad \forall x \in X \quad A_\lambda^n x \rightharpoonup A_\lambda x$ weakly in X^*, as $n \to +\infty$.

(+ Normalization condition)

where, $A_\lambda^n = \nabla F_\lambda^n$ (resp. $A_\lambda = \nabla F_\lambda$) is the Yosida approximation of the maximal monotone operator $A^n = \partial F^n$ (resp. $A = \partial F$), and the normalization condition, as given by Theorem 3.26, allows us to rebuild the functionals F^n from their subdifferentials ∂F^n:

$$\exists(u,v) \in \partial F \quad \exists(u_n,v_n) \in \partial F^n \text{ such that } u_n \xrightarrow{X_s} u,\ v_n \xrightarrow{X_s^*} v,\ F^n(u_n) \to F(u).$$

Proof of Proposition 3.29 Since $A_\lambda^n x = \frac{1}{\lambda} H(x - J_\lambda^n x)$, and H is continuous from X_s into X_s^*, implication (i) $\Rightarrow$ (ii) is a direct consequence of Theorem 3.26 (i) $\Rightarrow$ (ii). The implication (ii) $\Rightarrow$ (iii) being clear, in order to prove that (iii) $\Rightarrow$ (i) it is enough to verify (by using implication (iii) $\Rightarrow$ (i) of Theorem 3.26) that

$$\forall\lambda > 0 \quad \forall x \in X \quad A_\lambda^n x \rightharpoonup A_\lambda x \text{ weakly in } X^*, \text{ as } n \to +\infty$$

implies

$$\forall\lambda > 0 \quad \forall x \in X \quad F_\lambda^n x \to F_\lambda x, \quad \text{as } n \to +\infty.$$

This follows from the equality

$$F_\lambda^n(x) = F_\lambda^n(x_o) + \int_0^1 \langle A_\lambda^n(x_o + \tau(x - x_o)), x - x_o\rangle d\tau$$

and the convergence, $F_\lambda^n(x_o) \to F_\lambda(x_o)$, where $x_o = u + \lambda H^{-1}(v)$ is given by the normalization condition (refer to the proof of Theorem 3.26 (ii) $\Rightarrow$ (iii)). □

REMARK 3.30 (a) We stress the fact that the equivalence between strong and weak convergence of the Yosida approximates is correct in such a general setting (X reflexive, $X \in C$) only when expressed in terms of the A_λ! In terms of J_λ, *the following equivalence*

$$\begin{array}{l} \forall\lambda > 0 \quad \forall x \in X \quad J^n_\lambda x \longrightarrow J_\lambda x \text{ strongly in } X \\ \Updownarrow \\ \forall\lambda > 0 \quad \forall x \in X \quad J^n_\lambda x \rightharpoonup J_\lambda x \text{ weakly in } X \end{array} \tag{3.86}$$

(+ normalization condition) *is true in a Hilbert space* (H being linear, closed, weakly continuous) but not in general in such Banach spaces C.

The following example, from Sonntag [1] illustrates this situation: take X a reflexive Banach space belonging to C, for which the duality map $H:X \to X^*$, is not weakly continuous. For example, $X = L^p(0,1)$ $(1 < p < +\infty)$. Then, one can find a sequence $(x_n)_{n\in\mathbb{N}}$ in X satisfying:

$$\begin{cases} \|x_n\| = 1 \text{ for every } n \in \mathbb{N},\ x_n \rightharpoonup 0 \text{ weakly in } X. \\ H(x_n) \text{ does not converge weakly to } 0 \text{ in } X^*. \end{cases}$$

Take $F^n = I_{K^n}$, $F = I_X \equiv 0$ where $K^n = \{x \in X / \langle x,H(x_n)\rangle = 0\}$ is the "orthogonal" of the subspace generated by $H(x_n)$. Then, for every $\lambda > 0$, $x \in X$

$$J^n_\lambda x = \operatorname{proj}_{K^n} x = x - \langle x,H(x_n)\rangle x_n.$$

For every $x \in X$, $J^n_\lambda x \xrightarrow[(n\to+\infty)]{} x = J_\lambda x$ weakly in X (since $x_n \rightharpoonup 0$ and $|\langle x,H(x_n)\rangle| \leq \|x\|$). On the other hand, the normalization condition is satisfied at the origin. But

$$\|J^n_\lambda x - J_\lambda x\| = |\langle x,H(x_n)\rangle|$$

fails to converge to zero for at least some $x_o \in X$ (since by hypothesis, $H(x_n)$ does not weakly converge to zero).

(b) The equivalence between the strong and the weak convergence of the Yosida approximates looks like a paradox! In fact, it has to be understood in the following way:

If the Yosida approximates $\{A^n_\lambda\}_{\substack{n\in\mathbb{N}\\ \lambda>0}}$ are such that, for every $x \in X$ and $\lambda > 0$, the sequence $\{A_\lambda^n x\}_{n\in\mathbb{N}}$ weakly converges in X^* to some $f(\lambda,x)$ *and* if the $(f(\lambda,\cdot))_{\lambda>0}$ are the Yosida approximates of some maximal monotone operator $A = \partial F$, then the Yosida approximates do converge strongly in X.

We have already mentioned the intimate relation between the epi-convergence of a sequence of functions in a metrizable space and the pointwise convergence of the Moreau-Yosida approximates. We stressed the fact that (refer to Remark 2.71) in general, the limit family, when it exists, is no more the Moreau-Yosida approximation of a given function F. Theorem 3.26 yields the interpretation of this result when X is a reflexive Banach space: the limit family is the Moreau-Yosida approximation of a given function F iff the sequence $(F^n)_{n\in\mathbb{N}}$ epi-converges both for the strong and the weak topology of X, i.e. Mosco-converges.

So it is interesting to know, given a family $(A^\lambda)_{\lambda>0}$ of operators (resp. (F^λ) a family of functions) when it is the Yosida (resp. Moreau-Yosida) approximation of a given operator A (resp. function F): the answer is given by the *resolvent equation*. Let us explain it here for functions. We shall do the same for operators in Section 3.4.3.

PROPOSITION 3.31 Let $(F^\lambda)_{\lambda>0}$ be a family of closed convex functions from X, a reflexive Banach space, into $\mathbb{R}$. Then, the $(F^\lambda)_{\lambda>0}$ are the Moreau-Yosida approximates of a given closed convex function F iff they satisfy the resolvent equation: $\forall\lambda,\mu > 0\ (F^\lambda)_\mu = F^{\lambda+\mu}$.

Proof of Proposition 3.31 From Proposition 2.68, the condition is necessary. Conversely, take $F = \sup_{\lambda>0} F^\lambda$. As a supremum of closed convex functions, F is closed convex. From the inequality

$$\forall\lambda,\ \mu > 0 \quad F^\lambda \geq (F^\lambda)_\mu = F^{\lambda+\mu}$$

it follows that the sequence $(F^\lambda)_{\lambda>0}$ converges increasingly to F (as λ decreases to zero). From Theorem 3.20, it Mosco-converges to F, and thus from Theorem 3.26, (i) ⇒ (iii), we obtain

$$\forall\mu > 0 \quad \forall x \in X \quad F_\mu(x) = \lim_{\lambda\to 0} (F^\lambda)_\mu(x).$$

Using the resolvent equation

$$(F^\lambda)_\mu = F^{\lambda+\mu} = (F^\mu)_\lambda$$

we obtain

$$F_\mu(x) = \lim_{\lambda \to 0} (F^\mu)_\lambda(x)$$
$$= F^\mu(x)$$

(this last property following from the convergence, for any closed function G, of $G_\lambda(x)$ to $G(x)$ as λ goes to zero - refer to Theorem 2.64). □

REMARK 3.32 From Lemma 3.27, we can derive the following formula:

$$\forall x \in X \quad \forall \lambda > 0 \quad \frac{d}{d\lambda} F_\lambda(x) = -\frac{1}{2} \|(\partial F)_\lambda(x)\|^2 . \tag{3.87}$$

To that end, let us write

$$F_\lambda(x) = \frac{1}{\lambda} \cdot \lambda F_\lambda(x)$$

and differentiate this last expression as a product. Using formula (3.69), we obtain

$$\frac{d}{d\lambda} F_\lambda(x) = \frac{1}{\lambda} F(J_\lambda x) - \frac{1}{\lambda^2} \lambda F_\lambda(x)$$
$$= -\frac{1}{\lambda} [F_\lambda(x) - F(J_\lambda x)]$$
$$= -\frac{1}{2\lambda^2} \|x - J_\lambda x\|^2$$
$$= -\frac{1}{2} \|A_\lambda x\|^2 ,$$

which is (3.87).

When x belongs to dom (A), $A = \partial F$, noticing that (refer to 3.49 and 3.50),

$$A_\lambda x \xrightarrow[(\lambda \to 0^+)]{} (\partial F)^o(x),$$

the element of minimal norm of ∂F,

$$\frac{d}{d\lambda} F_\lambda(x)\Big|_{\lambda = 0^+} = -\frac{1}{2} \|(\partial F)^o(x)\|^2 .$$

This suggests that we approximate the operator $x \to \frac{1}{2} \|\nabla F(x)\|^2$ by $\frac{1}{\lambda}(F-F_\lambda)$ (as λ goes to 0^+): in [1], Barbu and Da Prato use this device in order to approximate some Hamilton-Jacobi equations.

3.4.3 Geometric interpretation of Mosco-convergence of convex sets

Let us transcribe the results of the preceding section when F^n is equal to I_{K^n}, the indicator function of a closed convex set K^n. Noticing that

$$F = I_K$$

$$F_\lambda(x) = \frac{1}{2\lambda} \operatorname{dist}^2(x,K) \qquad \text{(distance from } x \text{ to } K)$$

$$J_\lambda^F(x) = \operatorname{proj}_K(x) \qquad \text{(projection of } x \text{ on } K)$$

$$A_\lambda^F(x) = \frac{1}{\lambda} H(x-\operatorname{proj}_K x).$$

Theorem 3.26 can be translated in the following way:

THEOREM 3.33 Let $(K^n)_{n\in\mathbb{N}}$, K a sequence of non empty closed convex sets in a reflexive Banach space X (the space X belongs to the class (C) Definition 3.25). The following statements are equivalent:

(i) $K^n \xrightarrow{M} K$ i.e. the sequence $(K^n)_{n\in\mathbb{N}}$ Mosco-converges to K.

(ii) $\forall x \in X$ $\operatorname{proj}_{K^n} x \to \operatorname{proj}_K x$ strongly in X, as $n \to +\infty$.

(iii) $\forall x \in X$ $\operatorname{dist}(x,K^n) \to \operatorname{dist}(x,K)$.

Let us notice that, in the situation $F^n = I_{K^n}$, the above property (ii) automatically implies the normalization condition of Theorem 3.26: take any $u \in K$ and $v = 0$; clearly $v \in \partial I_K(u)$, $I_K(u) = 0$. Taking $u_n = \operatorname{proj}_{K^n} u$, $v_n = 0$ we have

$$u_n \xrightarrow{X_s} u, \quad v_n \xrightarrow{X_s^*} v, \quad I_{K^n}(u_n) \to I_K(u).$$

PROPOSITION 3.34 Under the hypothesis of Theorem 3.33, the following statements are equivalent to the Mosco convergence of the sequence of closed convex sets $(K^n)_{n\in\mathbb{N}}$ to K:

(i) $\forall x \in X \quad \mathrm{proj}_{K^n} x \to \mathrm{proj}_K x$ strongly in X

(ii) $\forall x \in X \quad H(x-\mathrm{proj}_{K^n} x) \to H(x-\mathrm{proj}_K x)$ strongly in X*

(iii) $\begin{cases} \forall x \in X \quad H(x-\mathrm{proj}_{K^n} x) \rightharpoonup H(x-\mathrm{proj}_K x) \text{ weakly in } X^* \\ \exists x_o \in K\ \exists (x_{on})_{n\in\mathbb{N}} \text{ such that } x_{on} \in K^n \text{ for every } n \in \mathbb{N},\ x_{on} \xrightarrow{X_s} x_o. \end{cases}$

When X is a Hilbert space these properties are equivalent to

(iv) $\begin{cases} \forall x \in X \quad \mathrm{proj}_{K^n} x \longrightarrow \mathrm{proj}_K x \text{ weakly in } X \\ + \text{ normalization condition.} \end{cases}$

3.5 TOPOLOGY OF MOSCO-CONVERGENCE

In this section, we show that on $\Gamma(X)$, the convex cone of closed convex proper functions from X into $]-\infty,+\infty]$, Mosco-convergence is attached to a topology. As in the whole chapter, X denotes a reflexive Banach space. By renorming, we can assume, following Definition 3.25, that X is a C-normed space. We then denote $F_\lambda = F \nabla \frac{1}{2\lambda} \|\cdot\|^2$ the Moreau-Yosida approximation of index $\lambda > 0$ of $F \in \Gamma(X)$. From Theorem 3.26, Mosco-convergence $F^n \xrightarrow[(n\to+\infty)]{} F$ of functions of $\Gamma(X)$ is equivalent to pointwise convergence of all the Moreau-Yosida approximates $F^n_\lambda \to F_\lambda$. Relying on the equi-Lipschitz property of $\{F^n_\lambda, n \in \mathbb{N}\}$, λ fixed, we are naturally led to introduce a uniform structure inducing Mosco-convergence on $\Gamma(X)$.

3.5.1 Definition and properties of Mosco-convergence topology

DEFINITION 3.35 Let X be a reflexive Banach space. On $\Gamma(X)$, let us consider the uniform structure associated with the family of "pseudo-metrics" $\{e_{\lambda,x};\ \lambda > 0,\ x \in X\}$:

$$\text{for every } F,G \in \Gamma(X) \quad e_{\lambda,x}(F,G) = |F_\lambda(x) - G_\lambda(x)|. \tag{3.88}$$

We denote $\Gamma_M(X)$ the convex cone $\Gamma(X)$ equipped with the topology associated with this family of "pseudo-metrics". This topology is called the Mosco-convergence topology.

Let us study the properties of this topology and justify the above terminology:

THEOREM 3.36 Let us assume that X is a reflexive *separable* Banach space. Then the topological space $\Gamma_M(X)$ is a Polish space, that is: it is metrizable, separable, complete for a metric inducing the topology. Moreover for any sequence $(F^n)_{n\in\mathbb{N}}$ in $\Gamma(X)$ the following statements are equivalent:

$$F^n \xrightarrow{\Gamma_M(X)} F \Leftrightarrow \text{the sequence } (F^n)_{n\in\mathbb{N}} \text{ Mosco-converges to } F. \tag{3.89}$$

Proof of Theorem 3.36 By definition, the topology Γ_M is the weakest topology which makes continuous all the applications $\{F \to F_\lambda(x)/\lambda > 0,\ x \in X\}$ from $\Gamma(X)$ into $\mathbb{R}$.

Consequently, a filtered sequence $\{F^\nu;\ \nu \in \mathcal{H}\}$ is Γ_M-convergent to F iff

$$\forall\lambda > 0,\quad \forall x \in X \quad F_\lambda^\nu(x) \to F_\lambda(x).$$

When considering a sequence $(F^n;\ n \in \mathbb{N})$ by equivalence (i)$\Leftrightarrow$(iii) of Theorem 3.26, Γ_M convergence turns out to be equivalent to the Mosco-convergence of the sequence $(F^n)_{n\in\mathbb{N}}$ to F.

Let us prove that, *when* X *is separable, this topology is metrizable:* To that end, let us introduce a dense denumerable subset $(x_{op})_{p\in\mathbb{N}}$ in X and $(\lambda_k)_{k\in\mathbb{N}}$ a sequence in $\mathbb{R}^+$ strictly decreasing to zero, and prove the implication (and hence the equivalence) (3.91) $\Rightarrow$ (3.90):

$$\forall\lambda > 0,\quad \forall x \in X \quad F_\lambda^\nu(x) \to F_\lambda(x) \tag{3.90}$$

$$\forall(k,p) \in \mathbb{N}\times\mathbb{N} \quad F_{\lambda_k}^\nu(x_{op}) \to F_{\lambda_k}(x_{op}). \tag{3.91}$$

Picking up some $x_{op} \doteq x_o$ and $\lambda_{k_o} \doteq \lambda_o$, from (3.91) there exist some $c_1 \in \mathbb{R}^+$ and $H_1 \in \mathcal{H}$:

$$\forall\nu \in H_1 \quad \inf_{u\in X} \{F^\nu(u) + \frac{1}{2\lambda_o}\|u-x_o\|^2\} \geq C_1 > -\infty.$$

Thus the filtered sequence $\{F^\nu;\ \nu \in \mathcal{H}\}$ satisfies:

$$\forall\nu \in H_1,\ \forall u \in X \quad F^\nu(u) + r(\|u\|^2 + 1) \geq 0 \tag{3.92}$$

for some r sufficiently large.

On the other hand, from (3.91) too, there exists some $c_2 \in \mathbb{R}^+$ and $H_2 \in \mathbb{H}$:

$$\forall \nu \in H_2 \quad c_2 \geq \inf_{u \in X} \{F^\nu(u) + \frac{1}{2\lambda_o} \|u-x_o\|^2\} \geq F^\nu(J^\nu_{\lambda_o} x_o) + \frac{1}{2\lambda_o} \|J^\nu_{\lambda_o} x_o - x_o\|^2 . \tag{3.93}$$

Using (3.92), it follows that the filtered sequences $\{x_\nu = J^\nu_{\lambda_o} x_o; \nu \in H_2 \cap H_1\}$ $\{F^\nu(x_\nu); \nu \in H_2 \cap H_1\}$ are bounded.

Taking $H = H_2 \cap H_1$, we observe that the conditions (2.152) are satisfied in a uniform way by the family $\{F^\nu; \nu \in H\}$. From Theorem 2.64, for every $\lambda > 0$ the family $\{F^\nu_\lambda; \nu \in \mathbb{H}\}$ is locally equi-Lipschitz on X. Thus the convergence on a dense subset of X of the family $\{F^\nu_{\lambda_k}; \nu \in \mathbb{H}\}$ implies the convergence of the family on the whole of X:

For every $k \in \mathbb{N}$, for every $x \in X$ $F^\nu_{\lambda_k}(x) \to F_{\lambda_k}(x)$.

Noticing that the functions F_λ depend monotonically and continuously on λ, conclusion (3.90) follows.

Consequently, the topological space $\Gamma_M(X)$ is metrizable: take, for example,

for every $F,G \in \Gamma(X)$

$$d(F,G) = \sum_{(p,k) \in \mathbb{N} \times \mathbb{N}} \frac{1}{2^{p+k}} \frac{|F_{\lambda_k}(x_{op}) - G_{\lambda_k}(x_{op})|}{1 + |F_{\lambda_k}(x_{op}) - G_{\lambda_k}(x_{op})|} \tag{3.94}$$

Let us now prove that one can define *a metric on* $\Gamma(X)$ *which induces the Mosco-convergence topology and which is complete:* we first notice that the above metric (associated with the family of pseudo-metrics $\{e_{\lambda,x}\}$) is not complete. This is related to the fact that if

$$\forall \lambda > 0 \quad \forall x \in X \quad F^n_\lambda(x) \to F^\lambda(x)$$

$\{F^\lambda; \lambda > 0\}$ is not in general the Moreau-Yosida approximation of a given function F of $\Gamma(X)$ (refer to Remarks 2.71 and 3.30).

So, let us introduce the uniform structure associated to the pseudo-metrics $\{e_{\lambda_k,x_{op}} / (k,p) \in \mathbb{N}^2\}$ *and* $\{h_{\lambda_k,x_{op}} / (k,p) \in \mathbb{N}^2\}$ where $e_{\lambda,x}$ is given by (3.88) and

$$\text{for every } F,G \in \Gamma(X) \quad h_{\lambda,x}(F,G) = \|A_\lambda^F(x) - A_\lambda^G(x)\|_{X^*}. \tag{3.95}$$

As before, $\{x_{op}; p \in \mathbb{N}\}$ is a dense subset of X, $\{\lambda_k; k \in \mathbb{N}\}$ is a strictly decreasing sequence converging to zero. The same argument as we used to prove (3.91) ⇒ (3.90) and the implications (iii) ⇒ (i) and (i) ⇒ (ii) of Theorem 3.26, imply that this uniform structure (which is a metric) induces topology $\Gamma_M(X)$.

Let us verify that it is a complete metric and take $(F^n)_{n\in\mathbb{N}}$ a Cauchy sequence. The spaces X^* and $\mathbb{R}$ being complete and the functions $\{F_\lambda^n; n \in \mathbb{N}\}$ equi-locally Lipschitz, it follows that

$$\forall x \in X \quad \forall \lambda_k > 0 \quad F_{\lambda_k}^n(x) \xrightarrow[(n\to+\infty)]{} f(\lambda_k, x) \tag{3.96}$$

$$\forall p \in \mathbb{N} \quad \forall \lambda_k > 0 \quad A_{\lambda_k}^n(x_{op}) \xrightarrow[(n\to+\infty)]{} A(\lambda_k, x_{op}) \text{ strongly in } X^*. \tag{3.97}$$

Let us define

$$\text{for every } x \in X \quad F(x) \doteq \sup_{k\in\mathbb{N}} f(\lambda_k, x) \tag{3.98}$$

and verify that

$F^n \to F$ in the Mosco sense (i.e. the sequence $(F^n)_{n\in\mathbb{N}}$

converges to F in $\Gamma_M(X)$).

(i) From (3.96) and (3.97)

$$\forall x \in X \quad F(x) \geq \limsup_{k\to+\infty} \limsup_{n\to+\infty} F_{\lambda_k}^n(x).$$

From the diagonalization Lemma (1.16) there exists a strictly increasing mapping $n \to k(n)$ such that

$$F(x) \geq \limsup_n F_{\lambda_{k(n)}}^n(x).$$

From definition of F_λ^n, using the notation $x_n = J_{\lambda_{k(n)}}^n(x)$ and noticing that the $\{F^n; n \in \mathbb{N}\}$ are uniformly minorized (refer to (3.92) we obtain that

$$F(x) \geq \limsup_n F^n(x_n)$$

$$x_n \to x \quad \text{strongly in } X \quad (\text{when } x \in \text{dom } F).$$

(ii) Let $(x_n)_{n \in \mathbb{N}}$ be a weakly converging sequence in X, $x_n \xrightarrow[w]{X} x$. Then,

$$\forall (k,p) \in \mathbb{N} \quad \liminf_n F^n(x_n) \geq \liminf_n F^n_{\lambda_k}(x_n)$$

$$\geq \liminf_n F^n_{\lambda_k}(x_{op}) + \liminf_n \langle A^n_{\lambda_k} x_{op}, x_n - x_{op} \rangle$$

From (3.96) and (3.97),

$$\liminf_n F^n(x_n) \geq f(\lambda_k, x_{op}) + \langle A(\lambda_k, x_{op}), x - x_{op} \rangle. \tag{3.99}$$

This being true for any $p \in \mathbb{N}$, by density of the sequence $\{x_{op};\ p \in \mathbb{N}\}$ in X, we can extract a subsequence (which we still denote by x_{op}) strongly converging to x. The local Lipschitz property of the $\{F^n_{\lambda_k};\ n \in \mathbb{N}\}$, being uniform with respect to $n \in \mathbb{N}$, is preserved by the limit process (3.96). Thus, $f(\lambda_k, \cdot)$ is a continuous function and

$$f(\lambda_k, x_{op}) \to f(\lambda_k, x) \text{ as } p \to +\infty. \tag{3.100}$$

On the other hand, from the subdifferential inequality for every $x \in X$, $(n,k,p) \in \mathbb{N}^3$,

$$F^n_{\lambda_k}(x + x_{op}) \geq F^n_{\lambda_k}(x_{op}) + \langle A^n_{\lambda_k}(x_{op}), x \rangle$$

it follows that (we recall that $x_{op} \xrightarrow[s]{X} x$ as $p \to +\infty$), for every $k \in \mathbb{N}$

$$\sup_{(n,p) \in \mathbb{N} \times \mathbb{N}} \| A^n_{\lambda_k}(x_{op}) \|_{X^*} < +\infty.$$

From (3.97), it follows that the sequence $\{A(\lambda_k, x_{op});\ p \in \mathbb{N}\}$ is bounded. Thus,

$$\lim_{p \to +\infty} \langle A(\lambda_k, x_{op}), x - x_{op} \rangle = 0. \tag{3.101}$$

Returning to (3.99), using (3.100) and (3.101), we obtain

$$\lim_n \inf F^n(x_n) \geq f(\lambda_k, x).$$

This being true for any $k \in \mathbb{N}$, from the definition (3.98) of F, we finally obtain

$$\forall x_n \xrightarrow{X_w} x \quad \lim_n \inf F^n(x_n) \geq F(x),$$

which completes the proof of Mosco-convergence of the sequence $(F^n)_{n\in\mathbb{N}}$ to F.

Let us complete the proof of the Theorem 3.36 by proving that $\Gamma_M(X)$ *is separable*. This is a consequence of the following approximation result, the proof of which is given in the next section:

Let X be a reflexive and separable Banach space. For any F belonging to $\Gamma(X)$ there exists a sequence $(F^n)_{n\in\mathbb{N}}$ of polyhedral convex functions such that the sequence $(F^n)_{n\in\mathbb{N}}$ Mosco-converges to F (in fact, as we shall see one can take the sequence $(F^n)_{n\in\mathbb{N}}$ to be increasing). By polyhedral, we mean that each of the functions F^n can be written as a supremum of a finite number of continuous affine functions:

$$\text{for every } x \in X \quad F^n(x) = \sup_{1 \leq i \leq \ell(n)} \{\alpha_{i,n} + \langle x^*_{i,n}, x\rangle\}. \tag{3.102}$$

The space X being reflexive and separable, so is the space X*. Take $\{x^*_{on}; n \in \mathbb{N}\}$ a dense subset of X* and $\{\alpha_n; n\in\mathbb{N}\}$ a dense subset of $\mathbb{R}$. Then, notice that the polyhedral functions which can be written as a supremum of a finite number of affine functions with slopes in $\{x^*_{on}; n \in \mathbb{N}\}$ and values at the origin in $\{\alpha_n; n \in \mathbb{N}\}$ form a denumerable family which is dense for the uniform convergence on bounded subsets (and hence for Mosco-convergence) in the convex polyhedral functions. From the density of the convex polyhedral functions in $\Gamma(X)$, one derives the separability of $\Gamma_M(X)$. □

3.5.2 Dense subsets of $\Gamma_M(X)$ and approximation of closed convex functions

Let us describe two approximation schemes which, both with their duals, turn out to be useful in convex analysis.

(a) The first, which has been used in an intensive way throughout this chapter, is the Moreau-Yosida approximation: given $F \in \Gamma(X)$, denoting $F_\lambda = F \triangledown \frac{1}{2\lambda} \|\cdot\|_X^2$ (renorming X with a C-norm, refer to Definition 3.25) we know (refer to Theorem 3.24) that

$$\begin{cases} \text{the sequence } \{F_\lambda ; \lambda > 0\} \text{ increases to } F \text{ as } \lambda \text{ decreases to zero} \\ \text{the functions } \{F_\lambda ; \lambda > 0\} \text{ are } \mathbb{C}^1 \text{ convex functions.} \end{cases}$$

Since monotone convergence implies Mosco-convergence (refer to Theorem 3.20) we can translate the above results in a topological way:

PROPOSITION 3.37 Let X be a reflexive Banach space. Then, the subset of $\Gamma(X)$ of convex functions of class $\mathbb{C}^1$ is dense in $\Gamma(X)$ for the Mosco-topology.

This approximation scheme may be useful in some optimization problems, for example minimization of a non-differentiable convex function. By regularization by inf-convolution one obtains approximating convex $\mathbb{C}^1$ functions for which one can easily write the optimality conditions.

By duality, one obtains the approximation scheme called the "viscosity method": The sequence $\{F^* + \frac{\lambda}{2} \|\cdot\|^2_{X^*} ; \lambda > 0\}$ decreases to F^* as λ decreases to zero and hence (refer again to Theorem 3.20) Mosco-converges to F^*. In other words, the subset of $\Gamma(X)$ formed by convex coercive functions is dense in $\Gamma(X)$ for the Mosco-topology.

(b) Let us now describe a finite dimensional approximation scheme for general closed convex functions. Its dual formulation yields an increasing approximation by convex polyhedral functions (refer to Attouch [8]).

THEOREM 3.38 Let X be a reflexive, separable Banach space and $F \in \Gamma(X)$, i.e. F is a convex lower semicontinuous proper function from X into $]-\infty,+\infty]$.

Given $\{\xi_i ; i \in \mathbb{N}\}$ a dense denumerable subset of X, let us define:

$$\text{for every } n \in \mathbb{N}, \quad X_n = \overline{\text{conv}}\, \{u_i = J_1^F(\xi_i); \quad i = 1,2,\dots,n\} \tag{3.103}$$

i.e. X_n is the closed convex hull of $J_1^F(\xi_1),\dots,J_1^F(\xi_n)$ (refer to (3.48) for definition of proximal approximation J_λ^F).

Considering for every integer n

$$F^n = F + I_{X^n} \quad (\text{i.e. } F^n = F \text{ on } X_n, +\infty \text{ elsewhere}) \tag{3.104}$$

the sequence $\{F^n ; n \in \mathbb{N}\}$ is decreasing and F is equal to the closure of $\inf_{n \in \mathbb{N}} F^n$:

$$F = c\ell(\inf F^n). \tag{3.105}$$

In other words, the sequence $\{F^n; n \in \mathbb{N}\}$ Mosco-converges to F.

Proof of Theorem 3.38 Clearly, the sequence $\{F^n; n \in \mathbb{N}\}$ decreases to the function G equal to

$$G = F \text{ on } \bigcup_{n \in \mathbb{N}} X_n, \quad G = +\infty \text{ elsewhere.}$$

From Theorem 3.20, the sequence $\{F^n; n \in \mathbb{N}\}$ Mosco-converges to $c\ell(G)$. So we have to prove that $F = c\ell G$.

Since for every $n \in \mathbb{N}$, $F \leqslant F^n$ and F is closed,

$$F \leqslant c\ell G.$$

We are finally reduced to proving that

$$F \geqslant c\ell G$$

which will follow from the approximation result:

for every $x \in \text{Dom } F$, there exists a sequence $\{x_n; n \in \mathbb{N}\}$ satisfying

$$x_n \in X_n \text{ for every } n \in \mathbb{N}, \; x_n \to x \text{ strongly in } X \text{ as } n \to +\infty \tag{3.106}$$

and

$$F(x) = \lim_{n\to+\infty} F^n(x_n).$$

Let us first prove (3.106) when x belongs to Dom (∂F); then there exists $x^* \in \partial F(x)$. Denoting $z = x + H^{-1}(x^*)$ (we recall that H is the duality map from X onto X^*, refer to (3.47)), we obtain

$$0 \in \partial F(x) + H(x-z), \text{ i.e.}$$

$$x = J_1^F z.$$

By density of the sequence $\{\xi_i; i \in \mathbb{N}\}$ in X, there exists a sequence $\{z_n; n \in \mathbb{N}\}$ such that

$$z_n \in \{\xi_1, \xi_2, \ldots, \xi_n\} \text{ for every } n \in \mathbb{N}$$

$$z_n \to z \text{ strongly in } X, \text{ as } n \to +\infty.$$

After renorming X as a C-normed space, we know from Theorem 3.24, (3.54) that for every $\lambda > 0$, the mapping $x \to J_\lambda^F x$ is continuous from X_s into X_s and the mapping $x \to F(J_\lambda^F x)$ is continuous from X_s into $\mathbb{R}$. Denoting $x_n = J_1^F z_n$, we thus have

$$x_n \in \overline{\text{conv}}\,\{J_1^F\,\xi_1,\ldots,J_1^F\xi_n\} = X_n$$

$$x_n = J_1^F z_n \to J_1^F z = x \text{ strongly in } X, \text{ as } n \to +\infty$$

$$F^n(x_n) = F(J_1^F z_n) \to F(J_1^F z) = F(x) \quad \text{as } n \to +\infty.$$

Now, if $x \in \text{Dom}\, F$, we reduce to the preceding situation by introducing an approximating sequence $\{x^k;\ k \in \mathbb{N}\}$ satisfying:

$$x^k \in \text{Dom}\,(\partial F) \text{ for every } k \in \mathbb{N}$$

$$x^k \to x \text{ strongly in } X \text{ as } k \to +\infty$$

$$F(x^k) \to F(x) \text{ as } k \to +\infty.$$

Take for example (refer to (3.50) and (3.51))

$$x^k = J_{\lambda_k}^F\, x \text{ with } \lambda_k \to 0 \text{ as } k \to +\infty.$$

We achieve the proof thanks to a diagonalization argument: for every $k \in \mathbb{N}$, since $x^k \in \text{Dom}\,\partial F$ by the preceding argument there exists a sequence $\{x_n^k;\ n \in \mathbb{N}\}$ such that $x_n^k \in X_n$ for every $n \in \mathbb{N}$ and

$$\begin{array}{ccc} (x_n^k,\ F(x_n^k)) & \xrightarrow[(n\to+\infty)]{X_s\times\mathbb{R}} & (x^k,\ F(x^k)) \\ & & \Big\downarrow {\scriptstyle (k\to+\infty)}\ \ X_s\times\mathbb{R} \\ & & (x,\ \ F(x)). \end{array}$$

From the diagonalization result Corollary 1.18, there exists a sequence $n \to k(n)$ such that

$$x_n^{k(n)} \longrightarrow x \text{ strongly in } X \text{ as } n \to +\infty\,, \quad x_n^{k(n)} \in X_n,$$
$$F(x_n^{k(n)}) \longrightarrow F(x) \text{ as } n \to +\infty.$$

The sequence $x_n = x_n^{k(n)}$ satisfies (3.106).

REMARK 3.39 Because of the lack of continuity of the function F (which is only assumed to be lower semicontinuous) the finite dimensional approximation subspaces very much depend on F and are constructed in a variational way from F. □

Let us now examine the dual statement (refer to Attouch [8] for justification of such a "dual" qualification).

THEOREM 3.40 Let X be a reflexive separable Banach space and $F \in \Gamma(X)$. Given $\{\xi_i; i \in \mathbb{N}\}$ a dense denumerable subset of X, let us consider for every $i \in \mathbb{N}$ and $x \in X$ the affine function G^i

$$G^i(x) = F(J_1^F \xi_i) + \langle H(\xi_i - J_1^F \xi_i), x - J_1^F \xi_i \rangle_{(X^*,X)}. \tag{3.107}$$

Then,

$$F = \sup_{i \in \mathbb{N}} G^i = \sup_{n \in \mathbb{N}} F^n \tag{3.108}$$

where for every $n \in \mathbb{N}$ we denote $F^n = \sup_{1 \leq i \leq n} G^i$.

The sequence $\{F^n; n \in \mathbb{N}\}$ is an increasing sequence of convex, continuous, polyhedral functions converging (pointwise and in the Mosco sense) to F.

Proof of Theorem 3.40 By definition of J_λ^F, (refer to Theorem 3.24)

$$H(\xi_i - J_1^F \xi_i) \in \partial F(J_1^F \xi_i)$$

which, by the subdifferential inequality

$$F(x) \geq F(J_1^F \xi_i) + \langle H(\xi_i - J_1^F \xi_i), x - J_1^F \xi_i \rangle$$

and by definition of G^i, implies that

$$F \geq \sup_{i \in \mathbb{N}} G^i. \tag{3.109}$$

In order to obtain the opposite inequality, let us prove that

$$\text{for every } i \in \mathbb{N} \quad F_1(\xi_i) = (G^i)_1(\xi_i). \tag{3.110}$$

Let us assume this equality for the present and see how to conclude: From (3.110)

$$F_1(\xi_j) \leqslant (\sup_{i\in\mathbb{N}} G^i)_1(\xi_j).$$

From the density of the sequence $\{\xi_j; j \in \mathbb{N}\}$ and the continuity property of Moreau-Yosida approximation, it follows that

$$F_1 \leqslant (\sup G^i)_1.$$

From (3.109),

$$F_1 \geqslant (\sup G^i)_1$$

thus

$$F_1 = (\sup G^i)_1 \tag{3.111}$$

which clearly implies $F = \sup_{i\in\mathbb{N}} G^i$: Take for example the conjugate of the two functions in (3.111). One obtains $F^* + \|\cdot\|^2_{X^*} = (\sup G^i)^* + \|\cdot\|^2_{X^*}$ and hence $F^* = (\sup G^i)^*$. The two functions F and $\sup G^i$ being closed convex, this implies

$$F = \sup_{i\in\mathbb{N}} G^i.$$

So let us verify (3.110):

$$(G^i)_1(\xi_i) = \min_{u\in X} \{G^i(u) + \frac{1}{2}\|\xi_i - u\|^2\}$$

$$= F(J_1^F\xi_i) + \min_{u\in X} \{\langle H(\xi_i - J_1^F\xi_i), u - J_1^F\xi_i\rangle + \frac{1}{2}\|u - \xi_i\|^2\}.$$

This minimum is achieved at the point u_i such that

$$H(\xi_i - J_1^F\xi_i) = H(\xi_i - u_i).$$

The duality map H being one to one, $u_i = J_1^F\xi_i$ and

$$(G^i)_1(\xi_i) = F(J_1^F\xi_i) + \frac{1}{2}\|\xi_i - J_1^F\xi_i\|^2$$

$$= F_1(\xi_i). \quad \square$$

REMARK 3.41 In Theorem 2.30 (Step 3) we mentioned a nice application of the above approximation result to obtain an integral representation of general unilateral constraint functionals in the calculus of variations. (refer to Attouch and Picard [4]).

COROLLARY 3.42 The convex continuous polyhedral functions (i.e. the supremum of a finite number of continuous affine functions) form a dense subset of $\Gamma(X)$ for the Mosco-topology (X is still assumed reflexive separable). □

Let us now apply Theorems 3.38 and 3.40 to the case where $F = I_K$ is the indicator function of a closed convex set K in X. Their geometric interpretation takes a striking form:

THEOREM 3.43 Let X be a reflexive separable Banach space and K a closed convex subset of X.

Given $\{\xi_i; i \in \mathbb{N}\}$ a dense denumerable subset of X, let us denote for every $n \in \mathbb{N}$

$$K_n = \overline{\text{conv}}\,\{\text{proj}_K \xi_1, \ldots, \text{proj}_K \xi_n\}$$

$$K^n = \{x \in X / \langle x - \text{proj}_K \xi_i, H(\xi_i - \text{proj}_K \xi_i)\rangle \leq 0 \quad \forall i = 1,2,\ldots,n\}.$$

We call K_n (resp. K^n) the internal (resp. external) approximation of K. Then,

$$K = \overline{\bigcup_{n \in \mathbb{N}} K_n}$$

$$K = \bigcap_{n \in \mathbb{N}} K^n.$$

Consequently, the sequences $\{K_n; n \in \mathbb{N}\}$ and $\{K^n; n \in \mathbb{N}\}$ Mosco-converge to K.

Figure 3.3 shows this in pictorial form.

Let us now give some applications of the above approximation theorem:

Let K be a convex compact subset of a reflexive Banach space X, and S be a continuous mapping from K into K. In this context, the Schauder fixed point theorem can easily be reduced from the Brouwer fixed point theorem via the following finite dimensional approximation: (3.112)

for every $n \in \mathbb{N}$, let us consider the continuous map $S_n : K_n \to K_n$

$$S_n(x) = \text{proj}_{K_n} S(x). \tag{3.113}$$

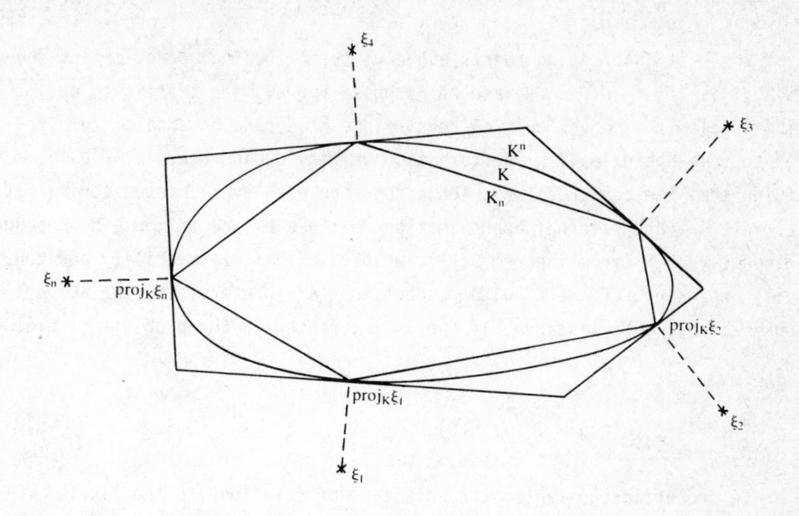

Figure 3.3

Then S_n is continuous from K_n into K_n; from the Brouwer theorem there exists $x_n \in K_n$ such that

$$S_n(x_n) = x_n.$$

The sequence $\{x_n;\ n \in \mathbb{N}\}$ being contained in the compact set K, extracting some converging subsequence $x_{n_k} \to x$, and using the continuity of S and the Mosco convergence of K_n to K (which implies the convergence of the projections - refer to Theorem 3.33) we obtain

$$S(x) = x$$

i.e. the existence of a fixed point for S. □

In a quite similar way, one can reduce the proof of the Kakutani fixed point theorem (for a multivalued u.s.c mapping S with closed convex image from a convex compact set K into itself) to the finite dimensional case. Take, for example,

$$S_n : K_n \to K_n \text{ given by } S_n(x) = \overline{\text{conv}} \{\text{proj}_{K_n} S(x)\}. \tag{3.114}$$

One of the most interesting points above of the preceding approximation Theorem 3.43 is its *constructive* aspect: suppose for example that in an optimization problem the convex sets of constraints K^ν depend on a parameter (related to some approximation, perturbation or time dependence...), $\nu \in Y$. Noticing that the construction which associated with every closed convex set K its internal and external approximation $\{K_n; n \in \mathbb{N}\}$ and $\{K^n; n \in \mathbb{N}\}$ depends continuously on K (from Theorem 3.33), we obtain that measurability and continuity dependence of the K^ν with respect to ν are preserved by the above approximations. For instance, if one is interested in the problems of evolution,

$$\frac{du}{dt} + \partial I_{K(t)}(u(t)) \ni f(t),\ u(0) = u_o$$

the above properties guarantee the existence of solutions to the finite dimensional approximation problems

$$\frac{du_n}{dt} + \partial I_{K_n(t)}(u_n(t)) \ni f_n(t); u_n(0) = u_{on}.$$

($\partial I_K(x)$ is the subdifferential of the indicator function of K at x, that is the outward normal cone to K at x). □

3.6 APPLICATIONS OF MOSCO-CONVERGENCE

In this section we do not aim for generality. We illustrate by means of two examples, one from singular perturbation theory in optimal control, the other dealing with varying obstacles in calculus of variations, how Mosco-convergence naturally arises in perturbation or approximation of optimization problems. Its main feature is to imply strong stability properties.

3.6.1 Convergence problems in optimal control

In Theorem 1.50, Remark 1.51, we met an example of Mosco convergence in a singular perturbation problem in optimal control. Let us analyse a similar type problem (cf. J.-L. Lions [1]) which is a little simpler because of the linearity of the state equation. For every $\varepsilon > 0$, the state equation and cost function are given respectively by

$$\begin{cases} -\varepsilon\Delta y + y = v \text{ on } \Omega \\ \dfrac{\partial y}{\partial n} = 0 \qquad \text{on } \Gamma = \partial\Omega \\ J_\varepsilon(v) = \int_\Gamma |\gamma(y_\varepsilon(v)) - z_d|^2 d\sigma + N \int_\Omega |v|^2 dx \end{cases} \tag{3.116}$$

where $y_\varepsilon(v)$ denotes the state associated with the control v via (3.116), $z_d \in L^2(\Gamma)$. Introducing $X = L^2(\Omega) \times L^2(\Gamma)$ with the product norm

$$\|(v,y)\|_X^2 = N\cdot \|v\|^2_{L^2(\Omega)} + \|y\|^2_{L^2(\Gamma)}$$

and

$$K_\varepsilon = \{(v,\gamma(y)) \in L^2(\Omega) \times L^2(\Gamma)/v \text{ and } y \text{ are linked by (3.116)}, v \in U_{ad}\},$$

γ denotes the trace operator from $H^1(\Omega)$ onto $H^{1/2}(\Gamma)$, U_{ad} the convex subset of admissible constraints), we have

$$\min_{v\in U_{ad}} J_\varepsilon(v) = \text{dist}^2((0,z_d); K_\varepsilon),$$

the distance being computed for the Hilbertian norm of X.

From Theorem 3.33, this is equivalent to requiring, for every $\xi \in X$,

$$\text{dist}(\xi, K_\varepsilon) \xrightarrow[\varepsilon\to 0]{} \text{dist}(\xi, K)$$

or to requiring

$\{K_\varepsilon; \varepsilon \to 0\}$ converges in the Mosco sense to K in X.

The analysis of the convergence of the sequence $\{K_\varepsilon; \varepsilon \to 0\}$ depends very much indeed, on the convex subset of constraints U_{ad}. Let us describe two typical cases:

1. Let us first assume U_{ad} to be a *closed convex bounded subset of* $H^1(\Omega)$. (3.117)
Then we claim that:

$$K_\varepsilon \xrightarrow[(\varepsilon\to 0)]{} K = \{(v,\gamma(v)) \in H^1(\Omega) \times H^{1/2}(\Gamma)/v \in U_{ad}\} \textit{ in the Mosco sense in } X$$

To that end, we observe that, if $v \in H^1(\Omega)$ and $y_\varepsilon(v)$ is the corresponding solution, multiplying (3.116) by $-\Delta y_\varepsilon(v)$, one obtains

$$\varepsilon|\Delta y_\varepsilon(v)|^2_{L^2(\Omega)} + |\text{grad } y_\varepsilon(v)|^2_{L^2(\Omega)} \leq |\text{grad} v|_{L^2(\Omega)} \; |\text{grad} y_\varepsilon(v)|_{L^2(\Omega)}.$$

Hence

$$|\text{grad} y_\varepsilon(v)|_{L^2(\Omega)} \leq |\text{grad} v|_{L^2(\Omega)}.$$

On the other hand, multiplying (3.116) by $y_\varepsilon(v)$ in $L^2(\Omega)$, one obtains

$$|y_\varepsilon(v)|_{L^2(\Omega)} \leq |v|_{L^2(\Omega)}.$$

Combining the two last inequalities,

$$\|y_\varepsilon(v)\|_{H^1(\Omega)} \leq \|v\|_{H^1(\Omega)}. \tag{3.118}$$

(a) Given $(v,\gamma(v)) \in K$, taking $v_\varepsilon = v$ and $y_\varepsilon = y_\varepsilon(v)$, by definition of K_ε, $(v_\varepsilon,\gamma(y_\varepsilon)) \in K_\varepsilon$. On the other hand,

$$v_\varepsilon \to v \text{ in } L^2(\Omega) \text{ as } \varepsilon \to 0,$$

$$\gamma(y_\varepsilon) \to \gamma(v) \text{ in } L^2(\Gamma) \text{ as } \varepsilon \to 0.$$

This last point follows from inequality (3.118): since y_ε remains bounded in $H^1(\Omega)$, going to the limit on (3.116) and using the compact embedding from $H^1(\Omega)$ into $L^2(\Omega)$ we obtain

$$y_\varepsilon \to v \text{ in } L^2(\Omega) \text{ and } w\text{-}H^1(\Omega).$$

From the compact embedding $H^1(\Omega) \hookrightarrow L^2(\Gamma)$ it follows that

$$\gamma(y_\varepsilon) \to \gamma(v) \text{ in } L^2(\Gamma).$$

(b) Let us consider a weakly converging sequence in X

$$(v_\varepsilon,\gamma(y_\varepsilon)) \xrightarrow[(\varepsilon\to 0)]{} (v,z)$$

with

$$(v_\varepsilon,\gamma(y_\varepsilon)) \in K_\varepsilon.$$

The sequence v_ε being contained in U_{ad} is consequently bounded in $H^1(\Omega)$.

By estimation (3.118) so is the sequence $y_\varepsilon = y_\varepsilon(v_\varepsilon)$. The same argument as above yields

$$y_\varepsilon \longrightarrow v \text{ weakly in } H^1(\Omega),\ v \in U_{ad},$$

and hence

$$\gamma(y_\varepsilon) \longrightarrow z = \gamma(v) \text{ in } L^2(\Gamma).$$

Thus, $(v,z) = (v,\gamma(v))$ belongs to K. □

2. Let us now take $U_{ad} = L^2(\Omega)$.

Then, we claim that

$$K_\varepsilon \to K = L^2(\Omega) \times L^2(\Gamma) \tag{3.119}$$

in the Mosco sense in $X = L^2(\Omega) \times L^2(\Gamma)$, i.e. the set K_ε fills all the space X as $\varepsilon \to 0$.

The only point we have to verify is that

$$\forall (v,z) \in X, \text{ there exists a sequence } \{v_\varepsilon;\ \varepsilon \to 0\} \text{ such that} \tag{3.120}$$

$$v_\varepsilon \to v \text{ in } L^2(\Omega) \text{ as } \varepsilon \to 0$$

and

$$\gamma(y_\varepsilon(v_\varepsilon)) \to z \text{ in } L^2(\Gamma).$$

When taking $(v,z) = (v,\gamma(v))$ a couple in $H^1(\Omega) \times H^{1/2}(\Gamma)$, the same argument as before works when taking $v_\varepsilon = v$. (3.121)

Then, one can notice that, for any couple $(v,y) \in H^1(\Omega) \times H^{1/2}(\Gamma)$ there exists a sequence $\{v_k;\ k \in \mathbb{N}\}$ in $H^1(\Omega)$ such that

$$(v_k,\gamma(v_k)) \to (v,y) \text{ in } X, \text{ as } k \to +\infty. \tag{3.122}$$

It is sufficient to take $v_k = \theta_k w + (1-\theta_k)v$ where $w \in H^1(\Omega)$ is such that $\gamma(w) = y$ (this is possible since $y \in H^{1/2}(\Gamma)$) and $\theta_k \in W^{1,\infty}(\Omega)$, $0 \leqslant \theta_k \leqslant 1$ satisfies

$$\theta_k(x) = \begin{cases} 1 & \text{if dist } (x,\Gamma) \leqslant \frac{1}{k} \\ 0 & \text{if dist } (x,\Gamma) \geqslant \frac{2}{k}. \end{cases}$$

The property (3.121) can be interpreted as

$$\forall v \in H^1(\Omega) \quad (v,\gamma(v)) \in X_s\text{-Li}K_\varepsilon.$$

The set X_s-LiK_ε being closed in X_s (Proposition 1.33) it follows that

$$\forall (v,y) \in H^1(\Omega) \times H^{1/2}(\Gamma) \quad (v,y) \in X_s\text{-Li}K_\varepsilon.$$

Noticing that $H^1(\Omega) \times H^{1/2}(\Gamma)$ is dense in $X = L^2(\Omega) \times L^2(\Gamma)$, using once more that X_s-LiK_ε is closed, we obtain $X \subset X_s$-LiK_ε, which completes the proof of (3.120). As a consequence, one obtains that when $U_{ad} = L^2(\Omega)$,

$$\inf_{v \in L^2(\Omega)} J_\varepsilon(v) \to 0 \text{ as } \varepsilon \to 0.$$

One can prove more precisely (cf. J.-L. Lions [1]) that

$$c_2 \varepsilon^{1/2} \int_\Gamma |z_d|^2 d\sigma \leq \inf_{v \in L^2(\Omega)} J_\varepsilon(v) \leq c_1 \varepsilon^{1/2} \int_\Gamma |z_d|^2 d\sigma. \quad \square$$

3. Let us now give an answer to the initial problem, sufficiently general in order to include the two above situations. To that end, as in Theorem 1.50, let us formulate the problem as follows:

$$\begin{aligned}\min_{v \in U_{ad}} J_\varepsilon(v) &= \min_{\left\{\begin{array}{l}-\varepsilon\Delta y + y \in U_{ad} \\ \frac{\partial y}{\partial n} = 0\end{array}\right.} \left\{ \int_\Gamma |\gamma(y) - z_d|^2 d\sigma + N \int_\Omega (-\varepsilon\Delta y + y)^2 dx \right\} \\ &= \min_{y \in L^2(\Omega)} \left\{ \int_\Gamma |\gamma(y) - z_d|^2 d\sigma + N \int_\Omega (-\varepsilon\Delta y + y)^2 dx \right. \\ &\qquad \left. + I_{\{-\varepsilon\Delta y + y \in U_{ad};\ \frac{\partial y}{\partial n} = 0\}} \right\}.\end{aligned}$$

So the problem turns to be equivalent to the study of the epi-convergence and, in fact, Mosco-convergence in $L^2(\Omega)$, of the sequence $\{F^\varepsilon; \varepsilon \to 0\}$

$$F^\varepsilon(y) = \int_\Gamma |\gamma(y) - z_d|^2 d\sigma + N \int_\Omega (-\varepsilon\Delta y + y)^2 dx + I_{\{-\varepsilon\Delta w + w \in U_{ad};\ \frac{\partial w}{\partial n} = 0\}}(y).$$

(a) The same argument as in (3.117) relying on inequality (3.118) yields that:

for every $y \in U_{ad} \cap H^1(\Omega)$, y_ε defined by

$$\begin{cases} -\varepsilon\Delta y_\varepsilon + y_\varepsilon = y \\ \dfrac{\partial y_\varepsilon}{\partial n} = 0 \end{cases}$$

satisfies

$$-\varepsilon\Delta y_\varepsilon + y_\varepsilon \in U_{ad}, \ \frac{\partial y_\varepsilon}{\partial n} = 0$$

and

$$y_\varepsilon \to y \text{ in } L^2(\Omega), \ \gamma(y_\varepsilon) \to \gamma(y) \text{ in } L^2(\Gamma) \text{ as } \varepsilon \to 0.$$

Therefore,

$$F^\varepsilon(y_\varepsilon) = \int_\Gamma |\gamma(y_\varepsilon)-z_d|^2 d\sigma + N\int_\Omega y^2 dx \underset{(\varepsilon\to 0)}{\longrightarrow} \int_\Gamma |\gamma(y)-z_d|^2 d\sigma + N\int_\Omega y^2 dx.$$

Equivalently

$$\forall y \in L^2(\Omega)$$

$$s\text{-}L^2(\Omega)\text{-}ls_e F^\varepsilon(y) \leq \int_\Gamma |\gamma(y) - z_d|^2 d\sigma + N\int_\Omega y^2 dx + I_{U_{ad}\cap H^1}(y).$$

The function $s\text{-}L^2(\Omega)\text{-}ls_e F^\varepsilon$ being $s\text{-}L^2(\Omega)$ lower semicontinuous,

$$s\text{-}L^2(\Omega)\text{-}ls_e F^\varepsilon \leq cl\ \{\int_\Gamma |\gamma(y)-z_d|^2 d\sigma + N\int_\Omega y^2 dx + I_{U_{ad}\cap H^1}(y)\} \qquad (3.124)$$

where $c\ell$ denotes the closure operation in $L^2(\Omega)$ (for a convex function weak and strong closure are equal).

(b) Let us now consider a sequence y_ε weakly converging to some y in $L^2(\Omega)$, and try to minorize $\liminf_{\varepsilon\to 0} F^\varepsilon(y_\varepsilon)$ (when this quantity is finite).

$$F^\varepsilon(y_\varepsilon) = \int_\Gamma |\gamma(y_\varepsilon) - z_d|^2 d\sigma + N\int_\Omega (-\varepsilon\Delta y_\varepsilon + y_\varepsilon)^2 dx$$

$$= \int_\Gamma |\gamma(y_\varepsilon) - z_d|^2 d\sigma + N\int_\Omega \varepsilon^2(\Delta y_\varepsilon)^2 dx + 2\int_\Omega \varepsilon N|\text{grad} y_\varepsilon|^2 dx + N\int_\Omega y_\varepsilon^2$$

$$F^\varepsilon(y_\varepsilon) \geq \int_\Gamma |\gamma(y_\varepsilon)-z_d|^2 d\sigma + N\int_\Omega y_\varepsilon^2 dx + I_{\{-\varepsilon\Delta w_\varepsilon + w_\varepsilon \in U_{ad};\ \sup_{\varepsilon\to 0}|-\varepsilon\Delta w_\varepsilon + w_\varepsilon|_{L^2} < +\infty\}}(y_\varepsilon).$$

Let us suppose that given such a sequence y_ε, we are able to build another sequence $\{z_\varepsilon ; \varepsilon > 0\}$ such that

$$|z_\varepsilon - y_\varepsilon|_{L^2(\Omega)} \to 0$$

$$|\gamma(z_\varepsilon - y_\varepsilon)|_{L^2(\Gamma)} \to 0$$

and

$$z_\varepsilon \in U_{ad} \cap H^1.$$

Then

$$\liminf_{\varepsilon \to 0} F^\varepsilon(y_\varepsilon) \geq \liminf_{\varepsilon \to 0} \{\int_\Gamma |\gamma(z_\varepsilon) - z_d|^2 d\sigma + N \int_\Omega z_\varepsilon^2 dx + I_{U_{ad} \cap H^1}(z_\varepsilon)\} \tag{3.125}$$

$$\geq c\ell \{\int_\Gamma |\gamma(z) - z_d|^2 d\sigma + N \int_\Omega z^2 dx + I_{U_{ad} \cap H^1}(z)\}(y).$$

Let us summarize the preceding results:

PROPOSITION 3.44 Let us assume that U_{ad} is a closed convex subset of $L^2(\Omega)$ satisfying the following approximation property: $U_{ad} \cap H^1 \neq \emptyset$ and, for every sequence $\{y_\varepsilon ; \varepsilon \to 0\}$ weakly converging in $L^2(\Omega)$ to some y and satisfying

$$\begin{cases} -\varepsilon \Delta y_\varepsilon + y_\varepsilon \in U_{ad} \\ \dfrac{\partial y_\varepsilon}{\partial n} = 0 \text{ on } \Gamma \end{cases} \tag{3.126}$$

there exists a sequence $\{z_\varepsilon ; \varepsilon \to 0\}$ satisfying

$$\begin{cases} z_\varepsilon \in U_{ad} \cap H^1(\Omega) \text{ for every } \varepsilon > 0 \\ |\gamma(z_\varepsilon) - \gamma(y_\varepsilon)|_{L^2(\Gamma)} \to 0 \quad (\text{as } \varepsilon \to 0) \\ |z_\varepsilon - y_\varepsilon|_{L^2(\Omega)} \to 0 \quad (\text{as } \varepsilon \to 0). \end{cases}$$

Then the following convergence result holds in the *Mosco sense in* $L^2(\Omega)$

$$\underset{\varepsilon \to 0}{\text{M-lim}} \left\{ \int_\Gamma |\gamma(y)-z_d|^2 d\sigma + N \int_\Omega (-\varepsilon\Delta y + y)^2 dx \right.$$

$$\left. + I_{\{-\varepsilon\Delta w + w \in U_{ad};\ \frac{\partial w}{\partial n} = 0\}}(y) \right\} \tag{3.127}$$

$$= c\ell \left\{ \int_\Gamma |\gamma(y) - z_d|^2 d\sigma + N \int_\Omega y^2 dx + I_{U_{ad} \cap H^1}(y) \right\},$$

the closure $c\ell$ being taken for the $L^2(\Omega)$ topology.

Comments. The above result covers all classical situations for U_{ad}:

(a) $U_{ad} = L^2(\Omega)$; take $z_\varepsilon = y_\varepsilon$.

(b) U_{ad} = closed convex bounded set of $H^1(\Omega)$; take $z_\varepsilon = -\varepsilon\Delta y_\varepsilon + y_\varepsilon$.

Noticing that $z_\varepsilon \in U_{ad}$ and hence is bounded in $H^1(\Omega)$, the same argument as in (3.117), (3.118) yields that

$$|z_\varepsilon - y_\varepsilon|_{L^2(\Omega)} \to 0, \quad |\gamma(z_\varepsilon) - \gamma(y_\varepsilon)|_{L^2(\Gamma)} \to 0$$

as ε goes to zero.

(c) $U_{ad} = \{u \in L^2(\Omega),\ u \geqslant 0\}$; take $z_\varepsilon = y_\varepsilon$ and notice that

$$\begin{cases} -\varepsilon\Delta y_\varepsilon + y_\varepsilon \geqslant 0 \\ \dfrac{\partial y_\varepsilon}{\partial n} = 0 \end{cases}$$

implies, from maximum principle that y_ε is positive.

(d) $U_{ad} = \{u \in L^2(\Omega),\ \int_\Omega u(x)dx \geqslant 0\}$; take $z_\varepsilon = y_\varepsilon$ and notice that since

$$-\varepsilon\Delta y_\varepsilon + y_\varepsilon = u_\varepsilon, \quad \int_\Omega y_\varepsilon = \int_\Omega u_\varepsilon dx;$$

hence $y \in U_{ad}$. □

Noticing that a functional and its closure have the same infimum on any open set (Proposition 2.6) we obtain the following corollary from Proposition 3.44 and the above considerations.

COROLLARY 3.45 For any U_{ad} satisfying the approximation property (3.126), (take for example $U_{ad} = L^2(\Omega)$, U_{ad} a closed bounded convex subset of $H^1(\Omega)$, $U_{ad} = \{u \in L^2(\Omega), u \geqslant 0\}$, $U_{ad} = \{u \in L^2(\Omega), \int u \geqslant 0\}$) the following convergence result holds:

$$\min_{v \in U_{ad}} \{\int_\Gamma |\gamma y_\varepsilon(v) - z_d|^2 d\sigma + N\int_\Omega v^2 dx\} \xrightarrow[(\varepsilon \to 0)]{} \inf_{v \in U_{ad} \cap H^1} \{\int_\Gamma |\gamma(v) - z_d|^2 d\sigma + N\int_\Omega v^2 dx\}. \tag{3.128}$$

We are now going to prove that when one does not assume U_{ad} to satisfy an approximation property by $U_{ad} \cap H^1$ (like (3.126)) this convergence result may fail to be true.

A quite pathological situation is when

$$U_{ad} = \{v_o\} \quad v_o \in L^2(\Omega) \setminus H^1(\Omega).$$

Take $N = 1$, $\Omega = (0,1)$ and prove that for such a v_o suitably chosen, the sequence $\{m_\varepsilon, \varepsilon > 0\}$ defined by

$$m_\varepsilon = \min_{v \in U_{ad}} \{\int_\Gamma |\gamma y_\varepsilon(v) - z_d|^2 d\sigma + N\int_\Omega v^2 dx\} = |y_\varepsilon(v_o)(1) - z_d(1)|^2 + |y_\varepsilon(v_o)(0) - z_d(0)|^2 + N\int_\Omega v_o^2 dx$$

may not be convergent.

Equivalently, we have to prove that there exists $v_o \in L^2(\Omega)$ such that the sequence

$$\{y_\varepsilon(v_o)(0); \varepsilon \to 0\} \text{ does not converge.} \tag{3.129}$$

Let us compute $y_\varepsilon(v_o)(0)$:

$$\begin{cases} -\varepsilon y'' + y = v_o \text{ on } (0,1) \\ y'(0) = y'(1) = 0. \end{cases}$$

Taking

$$y(x) = C_1(x)\, e^{(1/\sqrt{\varepsilon})x} + C_2(x)\, e^{-(1/\sqrt{\varepsilon})x},$$

C_1, C_2 have to satisfy:

$$\begin{cases} -e^{(1/\sqrt{\varepsilon})x}\; C_1'(x) + e^{-(1\sqrt{\varepsilon})x}\; C_2'(x) = \dfrac{v_o}{v_\varepsilon} \\ \\ e^{(1/\sqrt{\varepsilon})x}\; C_1'(x) + e^{-(1\sqrt{\varepsilon})x}\; C_2'(x) = 0. \end{cases}$$

Hence,

$$C_2'(x) = \frac{1}{2}\, e^{(1/\sqrt{\varepsilon})x}\; \frac{v_o}{\sqrt{\varepsilon}}$$

$$C_1'(x) = -\frac{1}{2}\, e^{-(1/\sqrt{\varepsilon})x}\; \frac{v_o}{\sqrt{\varepsilon}}\,.$$

Expressing that

$$y'(0) = \frac{C_1(0) - C_2(0)}{\sqrt{\varepsilon}} = 0, \text{ i.e. } C_1(0) = C_2(0)$$

and

$$y'(1) = \frac{C_1(1)}{\sqrt{\varepsilon}}\; e^{(1/\sqrt{\varepsilon})} - \frac{C_2(1)}{\sqrt{\varepsilon}}\; e^{-1(\sqrt{\varepsilon})} = 0$$

we obtain (write $C_i(1) - C_i(0) = \int_0^1 C_i'(\tau)d\tau \quad i = 1,2$)

$$\begin{cases} C_1(0) = C_2(0) \\ \\ e^{(1\sqrt{\varepsilon})}[C_1(0) - \dfrac{1}{2\sqrt{\varepsilon}}\displaystyle\int_0^1 e^{-(1\;/\sqrt{\varepsilon})\tau} v_o(\tau)d\tau] - e^{-(1/\sqrt{\varepsilon})}[\,C_2(0) \\ \qquad\qquad + \dfrac{1}{2\sqrt{\varepsilon}}\displaystyle\int_0^1 e^{(1/\sqrt{\varepsilon})\tau} v_o(\tau)d\tau] = 0. \end{cases}$$

Therefore,

$$C_1(0)\;(e^{(1/\sqrt{\varepsilon})} - e^{-(1/\sqrt{\varepsilon})}) = \frac{1}{2\sqrt{\varepsilon}}\;[e^{(1/\sqrt{\varepsilon})}\int_0^1 e^{-(1/\sqrt{\varepsilon})\tau} v_o(\tau)d\tau$$

$$+ e^{-(1/\sqrt{\varepsilon})}\int_0^1 e^{(1/\sqrt{\varepsilon})\tau} v_o(\tau)d\tau].$$

Since

$$\begin{aligned} y_\varepsilon(v_o)(0) &= C_1(0) + C_2(0) \\ &= 2C_1(0) \end{aligned}$$

we finally obtain

$$y_\varepsilon(v_0)(0) = \frac{1}{(e^{(1/\sqrt{\varepsilon})}-e^{-(1/\sqrt{\varepsilon})})\sqrt{\varepsilon}}[e^{(1/\sqrt{\varepsilon})}\int_0^1 e^{-(1/\sqrt{\varepsilon})\tau}v_0(\tau)d\tau$$

$$+ e^{-(1/\sqrt{\varepsilon})}\int_0^1 e^{(1/\sqrt{\varepsilon})\tau}v_0(\tau)d\tau]$$

$$\sim \frac{1}{\sqrt{\varepsilon}}\int_0^1 e^{-(1/\sqrt{\varepsilon})} v_0(\tau)d\tau + \frac{1}{\sqrt{\varepsilon}\ e^{(2/\sqrt{\varepsilon})}}\int_0^1 e^{(1/\sqrt{\varepsilon})\tau}v_0(\tau)d\tau.$$

The last term can be majorized, using the Hölder inequality

$$\frac{1}{\sqrt{\varepsilon}\ e^{(2/\sqrt{\varepsilon})}}\int_0^1 e^{(1/\sqrt{\varepsilon})\tau}|v_0(\tau)|d\tau \leqslant \frac{1}{\sqrt{\varepsilon}\ e^{(2/\sqrt{\varepsilon})}}\left(\int_0^1 e^{(2/\sqrt{\varepsilon})\tau}d\tau\right)^{\frac{1}{2}}.\|v_0\|_{L^2(\Omega)}$$

$$\leqslant \frac{1}{\sqrt{\varepsilon}\ e^{(2/\sqrt{\varepsilon})}}\left[\frac{\sqrt{\varepsilon}}{2}(e^{(2/\sqrt{\varepsilon})}-1)\right]^{\frac{1}{2}}\|v_0\|_{L^2(\Omega)}$$

$$\leqslant \frac{1}{\sqrt{2}\varepsilon^{1/4}.e^{(2/\sqrt{\varepsilon})}}(e^{(1/\sqrt{\varepsilon})}+1)\|v_0\|_{L^2(\Omega)} \longrightarrow 0$$

(as $\varepsilon\to 0$)

Thus

$$y_\varepsilon(v_0)(0) \sim \int_0^1 v_0(\tau)d\mu_\varepsilon(\tau) \text{ where } \mu_\varepsilon = \frac{1}{\sqrt{\varepsilon}} e^{-(1/\sqrt{\varepsilon})\tau} d\tau.$$

We notice that

$$\int d\mu_\varepsilon = \frac{1}{\sqrt{\varepsilon}}\int_0^1 e^{-(1/\sqrt{\varepsilon})\tau}d\tau = \frac{1}{\sqrt{\varepsilon}}\left[\frac{e^{-(1/\sqrt{\varepsilon})\tau}}{-\frac{1}{\sqrt{\varepsilon}}}\right]_0^1$$

$$= -[e^{-(1/\sqrt{\varepsilon})}-1]$$

$$= 1 - e^{-(1/\sqrt{\varepsilon})} \longrightarrow 1 \text{ as } \varepsilon \to 0.$$

Thus, $\mu_\varepsilon \to \delta$, the Dirac measure, at zero (Fig. 3.4). If $\lim_{x\to 0} v_0(x)$ does not exist (take v_0 oscillating more and more rapidly, with $v_0 \in L^2$), then $\lim_{\varepsilon\to 0} y_\varepsilon(v_0)(0)$ does not exist!

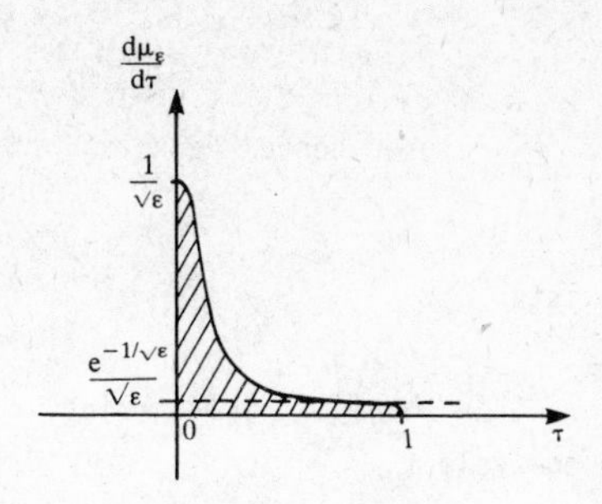

Figure 3.4

The same type of result holds for the other boundary point $\tau = 1$. If $\lim_{x\to 1} v_o(x)$ does not exist, then $\lim_{\varepsilon\to 0} y_\varepsilon(v_o)(1)$ does not exist either.

We notice, in accordance with the previous results, that if $v_o \in H^1(\Omega)$, such boundary values exist and thus such a phenomenon does not occur. Finally we have found a U_{ad} such that $\lim_{\varepsilon\to 0} \inf_{v\in U_{ad}} J^\varepsilon(v)$ does not exist!

The above result highlights a difficulty we already encountered in Theorem 1.50 and Remark 1.51. In general, when studying a convergence problem in control theory (approximation, singular perturbation, etc.) one has to do a particular study for each type of convex of constraint U_{ad}; one cannot prove a theorem explaining what happens for all U_{ad}.

3.6.2 Stability results for variational inequalities with obstacle constraints

In Theorem 2.30, we studied the general limit form of variational inequalities with varying obstacles. In some situations, like the "Fakir's bed of nails" (Example 2.28), or "blow-up of obstacles" (Example 2.29) it may happen that the limit constraint functional takes a relaxed form. Following Attouch & Picard [2], [3], we pay attention to the case where the limit constraint functional is still of obstacle type: the limit analysis problem can then be formulated in terms of Mosco-convergence of the convex sets of constraints, as shown by the following abstract result.

PROPOSITION 3.46 Let X be a reflexive C-normed space (i.e. weak convergence and convergence of the norms imply strong convergence). Let $\{F^n : X \to]-\infty,+\infty];\ n \in \mathbb{N}\}$ be a sequence of closed convex proper functions

$$F^n = \|\cdot\| + I_{K^n}$$

where for every $n \in \mathbb{N}$, K^n is a closed convex "constraint" subset of X. Let us assume that

$$\text{seq}X_w\text{-lim}_e F^n = F \text{ exists.}$$

Denoting $F = \|\cdot\| + G$, i.e. G is the limit "constraint" functional, then the following statements are equivalent

(i) G takes only the values 0 and $+\infty$ i.e. $G = I_K$ for some $K \subset X$.

(ii) The sequence $\{K^n;\ n \in \mathbb{N}\}$ Mosco-converges in X.

When (i) or (ii) is satisfied, K^n Mosco-converges to K.

Proof of Proposition 3.46

(i) ⇒ (ii) By assumption $(\|\cdot\| + I_K) = \text{seq}X_w\text{-lim}_e\ (\|\cdot\| + I_{K^n})$. Let $u \in K$. Then there exists a sequence $\{u_n;\ n \in \mathbb{N}\}$ weakly converging to u in X such that

$$\|u\|_X \geq \limsup_n \{\|u_n\|_X + I_{K^n}(u_n)\}.$$

It follows that, for every $n \in \mathbb{N}$, u_n belongs to K^n and

$$\|u\|_X \geq \limsup_n \|u_n\|_X .$$

The space X being C-normed, the sequence $\{u_n;\ n \in \mathbb{N}\}$ strongly converges to u.

On the other hand, let $\{u_n,\ n \in \mathbb{N}\}$ be a weakly converging sequence in X, such that u_n belongs to K^n for every $n \in \mathbb{N}$. By assumption,

$$\|u\| + I_K(u) \leq \liminf_n \{\|u_n\| + I_{K^n}(u_n)\} = \liminf_n \|u_n\| .$$

The second member of this inequality being bounded (since u_n weakly converges) it follows that $I_K(u) < +\infty$, that is $u \in K$.

(ii) ⇒ (i) Let us denote K to be the Mosco limit of the sequence $\{K^n;\ n \in \mathbb{N}\}$. Thus the sequence $\{I_{K^n};\ n \in \mathbb{N}\}$ Mosco-converges to I_K and $F^n = \|\cdot\| + I_{K^n}$

Mosco converges to $\|\cdot\| + I_K$. It follows that

$$\|\cdot\| + I_K = \|\cdot\| + G$$

that is, $G = I_K$.

REMARK 3.47 In the above argument, one may take instead of $\|\cdot\|_X$ any function $\rho(\|u\|)$ with $\rho:\mathbb{R}^+ \to \mathbb{R}^+$ a continuous strictly increasing function.

Let us now study the case of obstacle constraints and prove, first, that this class is closed under Mosco-convergence.

PROPOSITION 3.48 Let $g_n:\Omega \to \bar{\mathbb{R}}$ be a sequence of obstacles and $K^n \subset X = H_0^1(\Omega)$ the corresponding convex sets of constraints

$$K^n = \{u \in X/u(x) \geq g_n(x) \text{ on } \Omega\} .$$

Let us assume that

$$K^n \to K \text{ in the Mosco sense in } X, \text{ as } n \to +\infty.$$

Then, there exists a unique obstacle function $g:\Omega \to \bar{\mathbb{R}}$, which is quasi upper semicontinuous such that

$$K = Kg = \{u \in X/\tilde{u}(x) \geq g(x) \text{ quasi everywhere on } \Omega\} . \tag{3.130}$$

Proof of Proposition 3.48 Clearly K is a closed convex non empty set which satisfies:

$$\begin{cases} K + X^+ \subset K \\ K \text{ is stable for the infimum operation} \end{cases} \tag{3.131}$$

In order to verify (3.131) we notice that given $u \in K$, there exists $u_n \to u$ strongly in X, such that $u_n \in K^n$ for every $n \in \mathbb{N}$. For any $v \geq 0$

$$u_n + v \text{ still belongs to } K^n.$$

Since $u_n + v \to u+v$, it follows that $u + v \in K$.

Similarly, given u_1 and u_2 belonging to K, denoting by u_n^1 and u_n^2 two approximating sequences, $u_n^1 \wedge u_n^2$ still belongs to K^n. From $u_n^1 \wedge u_n^2 \to u_1 \wedge u_2$

it follows that $u_1 \wedge u_2 \in K$.

A closed convex set satisfying (3.131) is called a *unilateral convex set*. Following Mignot & Puel [1], Attouch & Picard [1] it can be represented as in (1.130). Uniqueness of $\tilde{g}$ (quasi upper semicontinuous) has been proved by Pierre [1].

REMARK 3.49 When the sequence of convex sets $\{K_{gn}; n \in \mathbb{N}\}$ converges in the Mosco sense to some Kg, then the corresponding solutions of variational problems

$$\min_{u \in H_0^1(\Omega)} \{\int_\Omega |\mathrm{grad}\, u|^2 dx + I_{Kg_n}(u) - \int_\Omega fu\}$$

do converge strongly in $H_0^1(\Omega)$. This has to be compared with relaxation phenomena, where there is only weak convergence of the solutions in $H_0^1(\Omega)$. Indeed, in the examples of the Fakir's bed of nails 2.28 , one can prove that $(u_n-1)^+$ converges strongly in $H_0^1(\Omega)$; it is $(u_n-1)^-$ which weakly converges!

The last natural and quite difficult question is to find minimal assumptions $g_n \xrightarrow[(n\to+\infty)]{} g$ which guarantee that $Kg_n \to K_g$ in the Mosco sense. Following Attouch & Picard [2], when denoting

$$C(A) = \inf \{\int_\Omega |\mathrm{grad}\, u|^2 dx;\ \tilde{u}(x) \geq 1 \quad \text{q.e on } A\}$$

the capacity of any subset A of , we have

THEOREM 3.50 Let $\{g_n; g:\Omega \to \bar{\mathbb{R}}\}$ be a sequence of quasi upper semicontinuous functions. Under the following hypotheses

$$\begin{cases} \int_0^\infty C(g_n-g > t)t\,dt \to 0 \text{ as } n \to +\infty \\ \forall t > 0 \quad C(g-g_n > t) \to 0 \text{ as } n \to +\infty \end{cases}$$

then, Kg_n Mosco converges to Kg in $X = H_0^1(\Omega)$.

Let us mention the following corollaries.

COROLLARY 3.51 (Boccardo & Murat [1]). Let $p > 2$. If $g_n \longrightarrow g$ weakly converges in $W_0^{1,p}(\Omega)$ then $Kg_n \xrightarrow{M} Kg$.

COROLLARY 3.52 (Attouch & Picard [2]). Let g_n, $g:\Omega \to \bar{\mathbb{R}}$ be quasi upper semi-continuous functions. Under the following hypothesis

$$\begin{cases} \forall t > 0 \quad C(|g_n - g| > t) \to 0 \text{ as } n \to +\infty \\ \exists V \in Kg \quad \exists V_n \in Kg_n \text{ s.t. } V_n \to V \text{ strongly in } H_0^1(\Omega) \\ \exists V \in L^2(C) \text{ s.t. } \tilde{V} \leq g \quad \text{q.p} \end{cases}$$

then, $Kg_n \xrightarrow{M} Kg$.

3.7 MAXIMAL MONOTONE OPERATORS. GRAPH CONVERGENCE

A main feature of epi-convergence is that it is closely related with convergence of solutions of minimization problems. Let us make this precise in the following classical situation:

Let $\{F^n : X \to]-\infty,+\infty];\ n \in \mathbb{N}\}$ be a sequence of closed convex proper functions from a reflexive Banach space X into $]-\infty,+\infty]$. Assuming these functions to be equicoercive and strictly convex, from variational properties of epi-convergence (Theorem 2.11), it follows that

(i) $F = \text{seq}X_w\text{-}\lim_e F^n$

$\Downarrow$

(ii) for every $f \in X^*$, $\inf_{x \in X} \{F^n(x) - \langle f,x \rangle\} \xrightarrow[n \to +\infty]{} \inf_{x \in X} \{F(x) - \langle f,x \rangle\}$

and the corresponding solutions weakly converge in X.

In a condensed way, for any such sequence of functionals $\{F^n;\ n \in \mathbb{N}\}$

$$F = \text{seq}X_w\text{-}\lim_e F^n \Rightarrow \forall f \in X^* \ (\partial F^n)^{-1}(f) \longrightarrow (\partial F)^{-1} f \text{ weakly in X.}$$

In the following two sections, we study relations between epi-convergence of convex functions and convergence (in a sense to be made precise) of corresponding subdifferential operators. The main result is (Theorem 3.66):

"The subdifferential operator $\partial : F \to \partial F$ is bicontinuous from the class of closed convex functions equipped with Mosco-convergence into the class of maximal monotone operators equipped with *Graph-convergence* (also called *resolvent convergence*)". So, we are naturally led to introduce the class of maximal monotone operators $A : X \to X^*$ (where X is a reflexive Banach space).

We pay particular attention to the subclass of subdifferential operators. For some approximation techniques, such as Yosida's explicit use of the norm of X, it is convenient to renorm X as a C-normed space (Definition 3.25): hence, weak convergence and convergence of the norms imply strong convergence and the duality map $H = \partial(\frac{1}{2} \|\cdot\|^2)$ is one to one bicontinuous from X onto X*.

Then we introduce the notion of graph convergence for sequences of maximal monotone operators. As already noticed by many authors, T. Kato [1] for linear maximal monotone operators, H. Brezis [1] for maximal monotone operators in Hilbert spaces, P. Benilan [1] for accretive operators in Banach spaces, F. Browder [1],..., this is the right notion for handling convergence problems for semigroups or equations associated to such operators.

In the following sections we follow the lines of H. Attouch [4] in the Hilbert case and A. Damlamian [3] for the extension to the reflexive Banach case.

3.7.1 Yosida approximation of maximal monotone operators

DEFINITION 3.53 An operator $A:X \to X^*$ (possibly multivalued, not defined everywhere) is said to be monotone if

$$\text{for every } (x_1,f_1) \in A, \quad (x_2,f_2) \in A$$

$$\langle f_2-f_1, x_2-x_1\rangle \geq 0$$

Such an operator A is said maximal monotone if it is maximal among monotone operators in the sense of inclusion. We denote

$$D(A) = \{x \in X/Ax \neq \emptyset\}$$

its domain and

$$R(A) = \{f \in X^*/A^{-1}(f) \neq \emptyset\}$$

its range.

The following theorem characterizes the maximality of a monotone operator A in terms of surjectivity of the operator H + A. It extends to the Banach reflexive case the celebrated Minty theorem:

THEOREM 3.54 (Browder [1], Rockafellar [2]) X is a C-normed space. Let $A:X \to X^*$ be a monotone operator. Then A is maximal monotone iff H + A is onto, i.e

$$R(H + A) = X^*. \tag{3.132}$$

Thanks to the above surjectivity result, the following definition of Yosida approximation of a maximal monotone operator makes sense:

DEFINITION 3.55 For every $\lambda > 0$, for every $x \in X$, there exists a unique element $J_\lambda^A x$ belonging to D(A) such that

$$H(J_\lambda^A x - x) + \lambda A(J_\lambda^A x) \ni 0. \tag{3.133}$$

We denote

$$A_\lambda x = \frac{1}{\lambda} H(x - J_\lambda^A x); \text{ thus } A_\lambda x \in A(J_\lambda^A x). \tag{3.134}$$

The operator A_λ is called the Yosida approximation of index λ of A. The operator J_λ^A is called the resolvent of index λ of A.

Among maximal monotone operators a particular subclass plays an important role: it is the subclass consisting of the subdifferentials of proper, closed convex functions F from X into $]-\infty,+\infty]$. We recall that $A = \partial F : X \to X^*$, defined by

$$f \in \partial F(x) \iff \text{for every } u \in X \quad F(u) \geq F(x) + \langle f, u-x \rangle,$$

is a maximal monotone operator (refer to Rockafellar [2]). Indeed, such an operator satisfies a stronger monotonicity property, which is called cyclic monotonicity (refer to Rockafellar [1], Brezis [1]).

Definition 3.55 is coherent with the notion of Yosida approximation introduced in Theorem 3.24. When A is the subdifferential of a closed convex function F, $A = \partial F$, then (1.133) is equivalent to saying that $J_\lambda^A x$ is the unique point minimizing on X the closed convex coercive function

$$u \to F(u) + \frac{1}{2\lambda} \|u-x\|^2 .$$

We recall that the value of this minimum is denoted

$$F_\lambda(x) = F(J_\lambda^F x) + \frac{1}{2\lambda} \|x - J_\lambda^F x\|^2 ,$$

the function F_λ being called the Moreau-Yosida approximation of index λ of F. The function F_λ is a $\mathcal{C}^1$ function whose Frechet derivative is equal to $\nabla F_\lambda = A_\lambda$, the Yosida approximation of $A = \partial F$.

More generally, let us study, in the present setting of maximal monotone operators, the properties of Yosida approximation.

PROPOSITION 3.56 Let X be a reflexive C-normed space and $A:X \to X^*$ a maximal monotone operator.

(a) The operators $J_\lambda^A:X \to X$ and $A_\lambda:X \to X^*$ *are continuous* (the spaces X and X^* being equipped with the norm topology). When X is a Hilbertian space, the resolvents J_λ are contraction mappings while A_λ, the Yosida approximation of index λ, is Lipschitzian with Lipschitz constant $1/\lambda$.

(b) For every $\lambda > 0$, A_λ is a maximal monotone operator and the family $(A_\lambda)_{\lambda>0}$ satisfies the resolvent equation

$$\forall\lambda > 0, \quad \forall\mu > 0 \quad (A_\lambda)_\mu = A_{\lambda+\mu}. \tag{3.135}$$

(c) The family $(A_\lambda)_{\lambda\to 0}$ converges in the graph sense to A as λ goes to zero: $\forall(x,f) \in A$ there exists a sequence $\{x_\lambda ; \lambda > 0\}$ strongly converging to x in X such that the sequence $\{A_\lambda x_\lambda ; \lambda > 0\}$ strongly converges to f in X^*. Moreover,

$$\text{for every } x \in D(A) \quad A_\lambda x \xrightarrow[\lambda \to 0]{} A^o x \tag{3.136}$$

(the element of minimal norm of the closed convex subset Ax of X^*).

$$\text{for every } x \in X \quad J_\lambda^A x \xrightarrow[\lambda \to 0]{} \operatorname{proj}_{\overline{D(A)}} x \tag{3.137}$$

(we recall that $\overline{D(A)}$ is a closed convex subset of X).

Proof of Proposition 3.56 We just sketch the proof of the main points in order to give an idea of the techniques used in this theory.

(a) The operator H being continuous, let us prove the continuity of J_λ^A. The continuity of A_λ will follow.

We first notice that J_λ^A and A_λ are bounded operators: Let $(x_o,y_o) \in A$. From monotonicity of A and relation (3.133),

$$\langle\frac{1}{\lambda} H(x-J_\lambda^A x) - y_o,\ J_\lambda^A x - x_o\rangle \geqslant 0.$$

Thus,

$$\langle H(x-J_\lambda^A x) - \lambda y_o,\ J_\lambda^A x-x + x-x_o\rangle \geqslant 0,$$

that is

$$\|x-J_\lambda^A x\|^2 \leqslant \|x-J_\lambda^A x\| (\| x-x_o\| + \lambda \|y_o\|) + \lambda\cdot \|y_o\| \cdot \|x-x_o\|$$

$$\|x-J_\lambda^A x\| \leqslant 2(\| x-x_o\| + \lambda \|y_o\|) \tag{3.138}$$

and J_λ^A is a bounded operator.

Let $x_n \to x$ strongly in X. The sequences $\{J_\lambda^A x_n;\ n \in \mathbb{N}\}$ and $\{A_\lambda^n x_n;\ n \in \mathbb{N}\}$ being bounded in the reflexive Banach space X, we can extract a subsequence $\{n_k;\ k \in \mathbb{N}\}$ such that

$$J_\lambda x_{n_k} \longrightarrow \alpha \text{ weakly in X, as } k \to +\infty.$$

$$A_\lambda x_{n_k} \longrightarrow \beta \text{ weakly in X*, as } k \to +\infty.$$

Let us identify α and β. For simplicity of notation, we still write $\{x_n\}$ instead of $\{x_{n_k}\}$. Let us return to the definition of J_λ, A_λ and write the relations (3.134) for x_n and x_m, m and n integers.

$$\lambda A_\lambda x_n = H(x_n-J_\lambda x_n)$$

$$\lambda A_\lambda x_m = H(x_m-J_\lambda x_m).$$

By monotonicity of A and from relation $A_\lambda u \in A(J_\lambda u)$

$$\langle H(x_n-J_\lambda x_n) - H(x_m-J_\lambda x_m),\ J_\lambda x_n-J_\lambda x_m\rangle \geqslant 0. \tag{3.139}$$

By monotonicity of H,

$$\langle H(x_n-J_\lambda x_n) - H(x_m-J_\lambda x_m),\ (x_n-J_\lambda x_n) - (x_m-J_\lambda x_m)\rangle \geqslant 0. \tag{3.140}$$

Adding (3.139) and (1.340), we obtain

$$\langle H(x_n-J_\lambda x_n) - H(x_m-J_\lambda x_m),\ x_n-x_m\rangle \geqslant 0. \tag{3.141}$$

This last quantity converges to zero as n and m go to $+\infty$ (since $x_n \to x$ strongly in X, the first member of (3.141) being weakly convergent in X*). Thus, each of the above quantities goes to zero as n and $m \to +\infty$:

$$\lim_{n,n \to +\infty} \langle H(x_n - J_\lambda x_n) - H(x_m - J_\lambda x_m), J_\lambda x_n - J_\lambda x_m \rangle = 0, \tag{3.142}$$

$$\lim_{n,m \to +\infty} \langle H(x_n - J_\lambda x_n) - H(x_m - J_\lambda x_m), (x_n - J_\lambda x_n) - (x_m - J_\lambda x_m) \rangle = 0. \tag{3.143}$$

We conclude with the help of the following lemma (Brezis, Crandall & Pazy [1]).

LEMMA 3.57 Let A be a maximal monotone operator, $A: X \to X^*$, and $(\alpha_n, \beta_n)_{n \in \mathbb{N}}$ a sequence in $X \times X^*$ which satisfies:

(i) for every $n \in \mathbb{N}$ $\beta_n \in A_{\alpha_n}$.

(ii) $\alpha_n \longrightarrow \alpha$ weakly in X, $\beta_n \longrightarrow \beta$ weakly in X*.

(iii) $\limsup_{n,m \to +\infty} \langle \beta_n - \beta_m, \alpha_n - \alpha_m \rangle \leq 0$ or $\limsup_n \langle \beta_n - \beta, \alpha_n - \alpha \rangle \leq 0$

then,

$$\langle \beta_n, \alpha_n \rangle \xrightarrow[n \to +\infty]{} \langle \beta, \alpha \rangle \text{ and } \beta \in A\alpha.$$

Moreover, if A = H is the duality map, then $\alpha_n \to \alpha$ strongly in X and $\beta_n \to \beta$ strongly in X*.

Applying the above lemma to (3.142) with $\alpha_n = J_\lambda x_n$, $\beta_n = H(x_n - J_\lambda x_n)$ noticing that from (3.134) $\beta_n \in A_{\alpha_n}$, we first obtain

$$\beta \in A\alpha.$$

Then, applying the above lemma to (3.143) with $\alpha_n = x_n - J_\lambda x_n$, we obtain

$$x_n - J_\lambda x_n \xrightarrow[(n \to +\infty)]{} x - \alpha \text{ strongly in X.} \tag{3.144}$$

From $\lambda A_\lambda x_n = H(x_n - J_\lambda x_n)$ and the two above relations, we finally obtain

$$\lambda\beta = H(x - \alpha) \text{ with } \beta \in A\alpha, \text{ i.e. } \alpha = J_\lambda x,\ \beta = A_\lambda x.$$

Thus the whole sequence $\{J_\lambda^A x_n; n \in \mathbb{N}\}$ converges to $J_\lambda x$ and from (3.144) the convergence holds for the strong topology of X.

Proof of Lemma 3.57 Each of the relations (iii) implies that

$$\limsup_{n \to +\infty} \langle \beta_n, \alpha_n \rangle \leq \langle \beta, \alpha \rangle.$$

Let $y \in Ax$ and express the monotonicity of A thus:

$$\langle y - \beta_n, x - \alpha_n \rangle \geq 0.$$

Hence

$$\langle y, x \rangle - \langle \beta_n, x \rangle - \langle y, \alpha_n \rangle + \langle \beta_n, \alpha_n \rangle \geq 0$$

Taking the limit superior as n goes to $+\infty$ and using (iii), we obtain

$$\langle y, x \rangle - \langle \beta, x \rangle - \langle y, \alpha \rangle + \langle \beta, \alpha \rangle \geq 0$$

i.e.

$$\langle y - \beta, x - \alpha \rangle \geq 0 \text{ for any } y \in Ax.$$

From the maximality of A, it follows that $\beta \in A\alpha$.

Moreover, taking $x = \alpha, y = \beta$ (which is possible since now we know $\beta \in A\alpha$) we obtain

$$\liminf_{n \to +\infty} \langle \beta_n, \alpha_n \rangle \geq -\langle \beta, \alpha \rangle + \langle \beta, \alpha \rangle + \langle \beta, \alpha \rangle = \langle \beta, \alpha \rangle$$

which, combined with the assumption (iii), implies

$$\lim_{n \to +\infty} \langle \beta_n, \alpha_n \rangle = \langle \beta, \alpha \rangle.$$

When taking A = H the duality map, we derive

$$\langle \beta_n, \alpha_n \rangle = \langle H(\alpha_n), \alpha_n \rangle = \|\alpha_n\|^2 \to \langle \beta, \alpha \rangle = \langle H(\alpha), \alpha \rangle = \|\alpha\|^2$$

i.e. the convergence of the norms. The space X being a C-normed space, the strong convergence of the sequence $\{\alpha_n; n \in \mathbb{N}\}$ follows.

(b) Let us verify the monotonicity of the operators $\{A_\lambda; \lambda > 0\}$. For every u,v belonging to X, for every $\lambda > 0$,

$$\langle A_\lambda u - A_\lambda v, u-v\rangle = \langle A_\lambda u - A_\lambda v, (u-J_\lambda^A u) - (v-J_\lambda^A v)\rangle$$
$$+ \langle A_\lambda u - A_\lambda v, J_\lambda^A u - J_\lambda^A v\rangle. \tag{3.145}$$

Since $A_\lambda x \in A(J_\lambda x)$, by monotonicity of A

$$\langle A_\lambda u - A_\lambda v, J_\lambda^A u - J_\lambda^A v\rangle \geq 0.$$

Since $A_\lambda x = \frac{1}{\lambda} H(x-J_\lambda^A x)$, by monotonicity of H

$$\langle A_\lambda u - A_\lambda v, (u-J_\lambda^A u) - (v-J_\lambda^A v)\rangle \geq 0.$$

The two last inequalities imply that the second member of (3.145) is positive i.e.

$$\langle A_\lambda u - A_\lambda v, u-v\rangle \geq 0$$

and A_λ is monotone.

The maximal monotonicity of A follows from its continuity properties: In fact, a monotone operator which is everywhere defined, hemicontinuous, and bounded from X into X*, X reflexive, is maximal monotone (refer to Browder [1]).

(c) Let us prove some of the approximating properties of the so-called Yosida approximation.

Let $y \in Ax$. By definition of $J_\lambda^A x$, $\frac{1}{\lambda} H(x-J_\lambda^A x) \in A(J_\lambda^A x)$. By monotonicity of A

$$\langle y - \frac{1}{\lambda} H(x-J_\lambda x), x-J_\lambda x\rangle \geq 0.$$

Thus

$$\frac{1}{\lambda} \langle H(x-J_\lambda x), x-J_\lambda x\rangle \leq \langle y, x-J_\lambda x\rangle$$

$$\frac{1}{\lambda} \|x-J_\lambda x\|^2 \leq \|y\| \; \|x-J_\lambda x\|.$$

This implies

$$\|A_\lambda x\|_{X^*} \leq \|y\|$$

for any $y \in Ax$, i.e.

$$\|A_\lambda x\|_{X^*} \leqslant \|A^0 x\|_{X^*} \text{ for any } x \in D(A). \tag{3.146}$$

Thus,

$$\|x - J_\lambda x\| \leqslant \lambda \, \|A^0 x\| \tag{3.147}$$

and for any $x \in D(A)$ $J_\lambda x \to x$ strongly in X as $\lambda \to 0$.

Let us now study the sequence $\{A_\lambda x; \lambda \to 0\}$, for x given in D(A). From (3.147) it is bounded in X*. Let η be a weak limit point in X* of this sequence.

$$A_{\lambda_k} x \longrightarrow \eta \quad \text{as } k \to +\infty.$$

Since

$$A_\lambda x \in A(J_\lambda x),$$

noticing that

$$J_{\lambda_k} x \longrightarrow x \text{ strongly in } X$$

$$A_{\lambda_k} x \longrightarrow \eta \text{ weakly in } X^*,$$

it follows that $\eta \in Ax$. We have used here the following important property of maximal monotone operators: if $\beta_n \in A\alpha_n$, $\beta_n \longrightarrow \beta$ weakly in X*, $\alpha_n \to \alpha$ strongly in X (or vice versa) then, $\beta \in A\alpha$.

This is a direct consequence of Lemma 3.57, since in that case $\lim_{n \to +\infty} \langle \beta_n, \alpha_n \rangle = \langle \beta, \alpha \rangle$. Using (3.146) we also have

$$\|\eta\| \leqslant \|A^0 x\|.$$

Since $\eta \in Ax$, this implies $\eta = A^0 x$, i.e. the whole sequence $\{A_\lambda x; \lambda \to 0\}$ weakly converges to $A^0 x$. Noticing that from (3.146)

$$\limsup_{\lambda \to 0} \|A_\lambda x\| \leqslant \|A^0 x\|$$

we finally conclude to the strong convergence in X* of the sequence $\{A_\lambda x; \lambda \to 0\}$ to $A^0 x$.

So pointwise convergence allows us to obtain as limit of the $\{A_\lambda; \lambda \to 0\}$

only the minimal section A^o of A. In order to obtain the whole graph of A we have to use the graph convergence concept. Indeed,

$$A_\lambda \to A \text{ in the graph sense (as } \lambda \to 0):$$

Let us prove that for any $y \in Ax$, there exists a sequence $\{x_\lambda; \lambda \to 0\}$ strongly convergent to x in X, such that $A_\lambda x_\lambda \to y$ strongly in X* as $\lambda \to 0$. In fact, let us prove one can find such a sequence $\{x_\lambda; \lambda \to 0\}$ with $A_\lambda x_\lambda \equiv y$! Take $x_\lambda = x + \lambda H^{-1}(y)$; clearly $x_\lambda \to x$ strongly as $\lambda \to 0$. On the other hand, by definition of x_λ

$$H(x-x_\lambda) + \lambda A(x) \ni 0.$$

By definition of J_λ^A, this implies $J_\lambda x_\lambda = x$ and

$$A_\lambda x_\lambda = \frac{1}{\lambda} H(x_\lambda - x) = y!$$

This naturally introduces the next section, which deals with graph convergence.

3.7.2 Graph convergence of maximal monotone operators

We introduce the graph convergence of sequences of maximal monotone operators and prove the equivalence between this notion and the pointwise convergence of the resolvents (or Yosida approximations).

DEFINITION 3.58 Let X be a reflexive Banach space and $\{A^n; n \in \mathbb{N}\}$, A, a sequence of maximal monotone operators from X into X*. The sequence $\{A^n; n \in \mathbb{N}\}$ is said to be graph-convergent to A, and we write

$$A^n \xrightarrow{G} A \text{ as } n \to +\infty,$$

if the following property holds:

$$\begin{aligned} &\text{for every } (x,y) \in A, \text{ there exists a sequence } (x_n,y_n) \in A^n \\ &\text{such that } x_n \to x \text{ strongly in X}, \; y_n \to y \text{ strongly in X*}. \end{aligned} \tag{3.148}$$

In terms of set-convergence the above property means that

$$A \subset \text{Li}A^n$$

where the operators and their graphs in $X \times X^*$ are identified. The terminology might appear a little ambiguous since the set convergence also involves the inclusion

$$\mathrm{Ls}A^n \subset A.$$

In fact, because of the maximality of A, the inclusion $\mathrm{Li}A^n \supset A$ automatically implies correspondingly $A \supset \mathrm{Ls}A^n$. We can, indeed, prove a slightly more precise result:

PROPOSITION 3.59 Let $A^n \xrightarrow{G} A$, A^n and A maximal monotone operators. Then,

$$\begin{aligned} &\text{for any sequence } (x_n,y_n) \in A^n \text{ such that } x_n \to x \text{ strongly in } X \\ &\text{and } y_n \rightharpoonup y \text{ weakly in } X^*, \text{ we have } (x,y) \in A \end{aligned} \tag{3.149}$$

(and vice versa, by exchanging strong and weak).

Proof of Proposition 3.59 Since $A^n \xrightarrow{G} A$, for every $(u,v) \in A$ there exists a sequence $(u_n,v_n) \in A^n$ such that $u_n \to u$ strongly in X and $v_n \to v$ strongly in X^*. By monotonicity of A^n

$$\langle y_n - v_n, x_n - u_n \rangle \geq 0.$$

Going to the limit as n goes to $+\infty$, since $x_n - u_n \to x - u$ strongly in X^* and $y_n - v_n \rightharpoonup y - v$ weakly in X^*, we obtain

$$\langle y-v, x-u \rangle \geq 0.$$

This being true for any $(u,v) \in A$, from the maximal monotonicity of A, it follows that

$$y \in Ax. \quad \square$$

PROPOSITION 3.60 Let $\{A^n;\ n \in \mathbb{N}\}$, A be a sequence of maximal monotone operators from X into X^*, X reflexive Banach space. The following are equivalent:

(i) $A^n \xrightarrow{G} A$, i.e. the sequence $\{A^n;\ n \in \mathbb{N}\}$ is graph-convergent to A.

(ii) $J_\lambda^n x \xrightarrow[(n \to +\infty)]{} J_\lambda x$ strongly in X, for every $x \in X$, for every $\lambda > 0$.

(iii) $A_\lambda^n x \xrightarrow[(n \to +\infty)]{} A_\lambda x$ strongly in X*, for every $x \in X$, for every $\lambda > 0$.

(iv) $A_{\lambda_o}^n x \xrightarrow[(n \to +\infty)]{} A_{\lambda_o} x$ strongly in X*, for every $x \in X$, for some $\lambda_o > 0$.

(v) $J_{\lambda_o}^n x \xrightarrow[(n \to +\infty)]{} J_{\lambda_o} x$ strongly in X, for every $x \in X$, for some $\lambda_o > 0$.

Proof of Proposition 3.60 Let us prove the sequence of implications:

(i) ⇒ (ii) ⇒ (iii) ⇒ (iv) ⇒ (v) ⇒ (i). Because of the bicontinuity of the duality map H it is enough to prove (i) ⇒ (ii) and (v) ⇒ (i).

(i) ⇒ (ii). Let us fix $\lambda > 0$ and $x \in X$. We first verify that the sequence $\{J_\lambda^n x;\ n \in \mathbb{N}\}$ remains bounded in X. Let $(x_o, y_o) \in A$ and $(x_{on}, y_{on}) \in A^n$ be an approximating sequence (since $A^n \xrightarrow{G} A$). The same estimation as (1.138) yields

$$\|x - J_\lambda^n x\| \leq 2(\|x - x_{on}\| + \lambda \|y_{on}\|)$$

Thus, the sequences $\{J_\lambda^n x;\ n \in \mathbb{N}\}$ and $\{A_\lambda^n x;\ n \in \mathbb{N}\}$ remain bounded respectively in the reflexive spaces X and X*. Let us extract a subsequence $\{n_k;\ k \in \mathbb{N}\}$ such that

$$J_\lambda^{n_k} x \longrightarrow \alpha \quad \text{weakly in } X, \text{ as } k \to +\infty$$

$$A_\lambda^{n_k} x \longrightarrow \beta \quad \text{weakly in } X^*, \text{ as } k \to +\infty.$$

We then adapt the argument developed in Proposition 3.56 and Lemma 3.57 to the present situation. For simplicity of notation we still write n instead of n_k. Let $(u,v) \in A$. Since $A^n \xrightarrow{G} A$, there exists a sequence $\{(u_n, v_n);\ n \in \mathbb{N}\}$ such that for every $n \in \mathbb{N}$, $(u_n, v_n) \in A^n$, and $u_n \to u$ strongly in X, $v_n \to v$ strongly in X*.

By definition of J_λ^n and A_λ^n, we have $A_\lambda^n x \in A^n(J_\lambda^n x)$. By monotonicity of A^n

$$\langle A_\lambda^n x - v_n,\ J_\lambda^n x - u_n \rangle \geq 0. \tag{3.150}$$

In order to go to the limit on the above inequality, we have to consider the difficulty coming from the duality bracket $\langle A_\lambda^n x, J_\lambda^n x \rangle$, the two sequences being only weakly convergent:

By monotonicity of H, for every integers n, m, since $A_\lambda x = \frac{1}{\lambda} H(x-J_\lambda x)$,

$$\langle A_\lambda^n x - A_\lambda^m x, (x-J_\lambda^n x) - (x-J_\lambda^m x)\rangle \geq 0,$$

i.e.

$$\langle A_\lambda^n x - A_\lambda^m x, J_\lambda^m x - J_\lambda^n x\rangle \geq 0$$

$$\langle A_\lambda^m x, J_\lambda^m x\rangle + \langle A_\lambda^n x, J_\lambda^n x\rangle \leq \langle A_\lambda^n x, J_\lambda^m x\rangle + \langle A_\lambda^m x, J_\lambda^n x\rangle.$$

Let us first fix m and let n go to + ∞

$$\langle A_\lambda^m x, J_\lambda^m x\rangle + \limsup_n \langle A_\lambda^n x, J_\lambda^n x\rangle \leq \langle \beta, J_\lambda^m x\rangle + \langle A_\lambda^m x, \alpha\rangle.$$

Then, let m go to + ∞,

$$\limsup_{n\to+\infty} \langle A_\lambda^n x, J_\lambda^n x\rangle \leq \langle \beta, \alpha\rangle. \qquad (3.151)$$

Returning to (3.150), we obtain

$$\langle \beta - v, \alpha - u\rangle \geq 0 \text{ for every } (u,v) \in A.$$

From maximal monotonicity of A, it follows that

$$\beta \in A\alpha. \qquad (3.152)$$

Then, taking precisely $(u,v) = (\alpha,\beta)$ which is known now to belong to A, and returning to (3.150), we obtain

$$\liminf_k \langle A_\lambda^{n_k} x, J_\lambda^{n_k} x\rangle \geq \langle \beta, \alpha\rangle.$$

Combining this inequality with (3.151),

$$\lim_{k\to+\infty} \langle A_\lambda^{n_k} x, J_\lambda^{n_k} x\rangle = \langle \beta, \alpha\rangle.$$

Equivalently

$$\lim_{k\to+\infty} \langle A_\lambda^{n_k} x, \frac{1}{\lambda}(x - J_\lambda^{n_k} x)\rangle = \langle \beta, \frac{1}{\lambda}(x-\alpha)\rangle.$$

Using

$$A_\lambda^{n_k} x = \frac{1}{\lambda} H(x - J_\lambda^{n_k} x)$$

and

$$A_\lambda^{n_k} x \longrightarrow \beta \text{ weakly in } X^*, \ \frac{1}{\lambda}(x - J_\lambda^{n_k} x) \longrightarrow \frac{1}{\lambda}(x-\alpha) \text{ weakly in } X,$$

from the maximal monotonicity of H and Lemma 3.57 it follows that

$$A_\lambda^{n_k} x \to \beta \text{ strongly in } X^*$$

$$J_\lambda^{n_k} x \to \alpha \text{ strongly in } X.$$

The duality map H being continuous,

$$\beta = \frac{1}{\lambda} H(x-\alpha),$$

which, combined with (3.152),

$$\beta \in A\alpha,$$

implies that $\alpha = J_\lambda x$, $\beta = A_\lambda x$. By uniqueness of the limit point, the whole sequences $\{J_\lambda^n x;\ n \in \mathbb{N}\}$ and $\{A_\lambda^n x;\ n \in \mathbb{N}\}$ strongly converge respectively to $J_\lambda x$ and $A_\lambda x$.

(v) ⇒ (i). Let $(u,v) \in A$. Then, let us define $x = u + \lambda_0 H^{-1}(v)$. From $v = \frac{1}{\lambda_0} H(x-u)$ and $(u,v) \in A$, by definition of $J_{\lambda_0}^A$, it follows that

$$J_{\lambda_0}^A x = u \text{ and } A_{\lambda_0} x = v.$$

By assumption, the sequence $\{J_{\lambda_0}^{A^n} x;\ n \in \mathbb{N}\}$ strongly converges to $J_{\lambda_0}^A x$ in X. Take $u_n = J_{\lambda_0}^n x$ and $v_n = A_{\lambda_0}^n x$; clearly $(u_n,v_n) \in A^n$ and $(u_n,v_n) \to (u,v)$ strongly in $X \times X^*$.

3.7.3 Topology of the resolvent (or graph) convergence on the class of maximal monotone operators

DEFINITION 3.61 Let X be a reflexive Banach space and M the class of maximal monotone operators from X into X*.

By definition, the topology of the resolvent convergence on M is the weakest topology on M which makes continuous all the applications

$\{\Gamma_{\lambda,x};\ \lambda > 0,\ x \in X\}$ from M into X:

$$\Gamma_{\lambda,x}(A) = J^A_\lambda x.$$

We denote by M_R the class M equipped with the topology of the resolvent convergence.

Let us summarize in the following theorem the properties of this topology.

THEOREM 3.62 Let X be a reflexive separable Banach space. Then M_R, the class of maximal monotone operators from X into X* equipped with the resolvent convergence, is a *metrizable, separable, complete space.* An example of a metric which induces the topology of the resolvent convergence and which is complete is

$$d(A,B) = \sum_{k \in \mathbb{N}} \frac{1}{2^k} \inf \{1, \| J^A_{\lambda_o} x_k - J^B_{\lambda_o} x_k \| \}, \tag{3.153}$$

where λ_o is taken strictly positive and $\{x_k;\ k \in \mathbb{N}\}$ is a dense subset of X. For any sequence $\{A^n;\ n \in \mathbb{N}\}$ of maximal monotone operators, the following equivalences hold (i) $\Leftrightarrow$ (ii) $\Leftrightarrow$ (iii) $\Leftrightarrow$ (iv):

(i) $A^n \xrightarrow[(n \to +\infty)]{} A$ in M_R

(ii) $A^n \xrightarrow[(n \to +\infty)]{G} A$, i.e. A^n is graph-convergent to A.

(iii) $J^{A^n}_\lambda x \xrightarrow[(n \to +\infty)]{} J^A_\lambda x$ for every $\lambda > 0$, for every $x \in X$

(iv) $J^{A^n}_{\lambda_o} x \xrightarrow[(n \to +\infty)]{} J^A_{\lambda_o} x$ for some $\lambda_o > 0$ and x belonging to a dense subset of X.

Proof of Theorem 3.62 The implications (i) $\Rightarrow$ (ii) $\Rightarrow$ (iii) $\Rightarrow$ (iv) are easy consequences of the definition of the topology of M_R and of Proposition 3.60. The new point is that (iv) $\Rightarrow$ (i), i.e. it is enough to ask *one* resolvent to converge on a dense subset of X in order to obtain graph convergence.

So let us assume that for every $u \in E$ a dense subset of X

$$J^n_{\lambda_o} u \to J_{\lambda_o} u,$$

and prove that this property can be extended to all of $x \in X$.

Let us fix a point $x \in X$. In order to get the boundedness of the sequence

$\{J^n_{\lambda_o} x; n \in \mathbb{N}\}$, following the same argument as in Proposition 3.60, it is enough to exhibit a sequence $(u_n, v_n) \in A^n$ which remains bounded: take $u_n = J^n_{\lambda_o} u_o$, $v_n = A^n_{\lambda_o} u_o$ for some $u_o \in E$.

Let us extract a subsequence

$$J^{n_k}_{\lambda_o} x \longrightarrow \alpha \text{ weakly in } X,$$

$$A^{n_k}_{\lambda_o} x \longrightarrow \beta \quad \text{weakly in } X^*, \text{ as } k \to +\infty.$$

By monotonicity of A^n, for every $u \in E$,

$$\langle A^n_{\lambda_o} x - A^n_{\lambda_o} u, J^n_{\lambda_o} x - J^n_{\lambda_o} u\rangle \geqslant 0. \tag{3.154}$$

As in Proposition 3.60, we easily derive

$$\langle \beta, \alpha\rangle \geqslant \limsup_n \langle A^n_{\lambda_o} x, J^n_{\lambda_o} x\rangle.$$

Thus, going to the limit on (3.154), as $n \to +\infty$,

$$\langle \beta - A_{\lambda_o} u, \quad \alpha - J_{\lambda_o} u\rangle \geqslant 0 \text{ for every } u \in E. \tag{3.155}$$

The operators A_{λ_o} and J_{λ_o} being continuous the above inequality can be extended to all of X. We then notice that any couple $(x,y) \in A$ can be written as

$$x = J_{\lambda_o} u, \; y = A_{\lambda_o} u \text{ by taking } u = x + \lambda_o H^{-1}(y).$$

Thus

$$\langle \beta - y, \alpha - x\rangle \geqslant 0 \text{ for every } (x,y) \in A,$$

and by maximal monotonicity of A, $\beta \in A\alpha$.

Returning to (3.154), for every $u \in E$,

$$\liminf_k \langle A^{n_k}_{\lambda_o} x, J^{n_k}_{\lambda_o} x\rangle \geqslant \langle A_{\lambda_o} u, \alpha\rangle + \langle \beta, J_{\lambda_o} u\rangle - \langle A_{\lambda_o} u, J_{\lambda_o} u\rangle.$$

We first extend by continuity this inequality to all of X, then take precisely

$u = \alpha + \lambda_0 H^{-1}(\beta)$. Using that $\beta \in A\alpha$, it follows that

$$J_{\lambda_0} u = \alpha \text{ and } A_{\lambda_0} u = \beta.$$

Thus

$$\liminf_k \langle A_{\lambda_0}^{n_k} x, J_{\lambda_0}^{n_k} x\rangle \geq \langle \beta, \alpha\rangle$$

which, combined with (3.155), yields

$$\lim_{k\to+\infty} \langle A_{\lambda_0}^{n_k} x, J_{\lambda_0}^{n_k} x\rangle = \langle \beta, \alpha\rangle.$$

We then complete the argument as in Proposition 3.60, (i) ⇒ (ii): by application of Lemma 3.57 it follows that

$$A_{\lambda_0}^{n_k} x \to \beta \text{ strongly in } X^*$$

$$J_{\lambda_0}^{n_k} x \to \alpha \text{ strongly in } X$$

and

$$\beta = \frac{1}{\lambda} H(x-\alpha).$$

Combined with $\beta \in A\alpha$ it follows that

$$A_{\lambda_0}^{n} x \to A_{\lambda_0} x \text{ and } J_{\lambda_0}^{n} x \to J_{\lambda_0} x. \qquad \square$$

So the metric d defined by (3.153) induces the topology M_R of the resolvent convergence. Let us now verify that M_R *is a complete metric space* when equipped with this metric d.

Let $\{A^n;\, n \in \mathbb{N}\}$ be a Cauchy sequence in (M_R, d): for some $\lambda_0 > 0$, for every $x \in E$, a dense subset of X.

$$\lim_{n,m\to+\infty} \| J_{\lambda_0}^{n} x - J_{\lambda_0}^{m} x \| = 0.$$

The space X being complete, for every $x \in E$ the following limits exist:

$$Tx \doteq \lim_{n\to+\infty} {}_{\lambda_0} x$$

$$Ux \doteq \lim_{n\to+\infty} \frac{1}{\lambda_0} H(x - J_{\lambda_0}^{n} x),$$

these two quantities being linked by the relation

$$Ux = \frac{1}{\lambda_o} H(x-Tx) \text{ for every } x \in E. \tag{3.156}$$

Let us introduce the operator $B:X \to X^*$, where graph is equal to

$$B = \{(Tx,Ux) \in X \times X^*/x \in E\}, \tag{3.157}$$

and

$$A = \bar{B}$$

which, as we shall verify, is the G-limit of the sequence $\{A^n;\ n \in \mathbb{N}\}$. Noticing that

$$\frac{1}{\lambda_o} H(x-J^n_{\lambda_o} x) \in A^n(J^n_{\lambda_o} x)$$

(refer to (3.133)) it follows from monotonicity of the $\{A^n;\ n \in \mathbb{N}\}$ that B is monotone. The operator A, whose graph is the closure of the graph of B in $X \times X^*$, is still monotone. Let us verify that it is maximal monotone.

From the surjectivity criteria (Theorem 3.54) this is equivalent to the solvability for every $x \in X$ of the following equation: we look for $\alpha \in D(A)$ satisfying

$$H(\alpha-x) + \lambda_o A\alpha \ni 0. \tag{3.158}$$

If x belongs to E, by definition of B

$$U(x) = B(Tx)$$

which, combined with (3.156) yields

$$H(Tx-x) + \lambda_o B(Tx) = 0. \tag{3.159}$$

Noticing that $A \supset B$, equation (3.158) is satisfied by taking $\alpha = Tx$. For a general $x \in X$, the proof is completed by a density argument: From the density of E in X, there exists a sequence $\{x_p;\ p \in \mathbb{N}\}$ converging to x. Let us prove that the sequence $\{Tx_p;\ p \in \mathbb{N}\}$ actually converges strongly in X and that its limit

$$\alpha = \lim_{p\to+\infty} Tx_p$$

is a solution of (3.158).

Picking some $x_0 \in E$ and noticing that the sequence $(J^n_{\lambda_0} x_0, A^n_{\lambda_0} x_0) \in A^n$ is bounded in $X \times X^*$, it follows from estimation (3.138) that the operators $\{J^n_{\lambda_0}; n \in \mathbb{N}\}$ are uniformly bounded on bounded subsets of X.

Thus, the sequence $\{Tx_p; p \in \mathbb{N}\}$ is bounded in X and from (3.156) so is the sequence $\{Ux_p; p \in \mathbb{N}\}$ in X^*. After extracting some subsequence, we can assume that

$$\begin{aligned} Tx_p &\longrightarrow \alpha \text{ weakly in } X \\ Ux_p &\longrightarrow \beta \text{ weakly in } X^*. \end{aligned} \tag{3.160}$$

Let us verify that these convergences are strong convergences. From (3.159)

$$H(Tx_p - x_p) + \lambda_0 B(Tx_p) = 0$$

and definition of A, it will follow that

$$H(\alpha - x) + \lambda_0 A(\alpha) \ni 0,$$

which is (3.158).

The strong convergence property results again from application of Lemma 3.57. From

$$\begin{aligned} 0 &\leq \langle x_p - Tx_p - x_q + Tx_q, Ux_p - Ux_q\rangle \quad \text{(from (3.156) and monotonicity of H)} \\ &= \langle x_p - x_q, Ux_p - Ux_q\rangle - \langle Tx_p - Tx_q, Ux_p - Ux_q\rangle \\ &\leq \langle x_p - x_q, Ux_p - Ux_q\rangle \quad \text{(from monotonicity of B and (3.157))} \end{aligned}$$

and the strong convergence of $x_p \xrightarrow[(p \to +\infty)]{} x$, it follows that

$$\lim_{p,q \to +\infty} \langle (x_p - Tx_p) - (x_q - Tx_q), Ux_p - Ux_q\rangle = 0.$$

Then apply Lemma 3.57.

So A is a maximal monotone operator. Returning to (3.159) we obtain that for every $x \in E$,

$$H(Tx - x) + \lambda_0 A(Tx) \ni 0$$

i.e $\quad Tx = J^A_{\lambda_o} x.$

Thus, for every $x \in E$

$$J^A_{\lambda_o} x = \lim_{n\to+\infty} J^{A^n}_{\lambda_o} x$$

which from Theorem 3.62, (iv) ⇒ (i), implies that $A^n \xrightarrow{G} A$.

The *separability of* M_R is an easy consequence of the implication (iv)⇒(i): Given some $\lambda_o > 0$ and E a dense denumerable subset of X, let us consider the application

$$J: M_R \to X^{\mathbb{N}}$$

$$A \to \{J^A_{\lambda_o} x;\ x \in E\}.$$

This application is one to one: if

$$J^A_{\lambda_o} x = J^B_{\lambda_o} x \text{ for every } x \in E,$$

by continuity of the resolvents $J^A_{\lambda_o}$ and $J^B_{\lambda_o}$ on X (Proposition 3.56), it can be extended to all of X:

$$J^A_{\lambda_o} = J^B_{\lambda_o} \text{ and } A_{\lambda_o} = B_{\lambda_o}.$$

Noticing that any element $(x,y) \in A$ can be written $x = J^A_{\lambda_o} u$, $y = A_{\lambda_o} u$, with $u = x + \lambda_o H^{-1}(y)$, it follows that $A = B$.

From Theorem 3.62, (iv) ⇒ (i), the application J is homeomorphism of M_R onto $J(M_R)$, its image (equipped with the topology induced by the product topology of $X^{\mathbb{N}}$).

The space X being metric separable is a second countable space and so is $X^{\mathbb{N}}$, and M_R as a subset is still second countable and hence separable. □

<u>COROLLARY 3.63</u> The subset of M_R of the maximal monotone continuous operators from X into X* is dense in M_R.

<u>Proof of Corollary 3.63</u> From Proposition 3.56, for any maximal monotone operator A, the sequence $\{A_\lambda;\ \lambda \to 0\}$ of its Yosida approximates G-converges

to A as λ goes to zero. From the continuity properties of the A_λ (Lipschitz property in the Hilbertian setting) and the topological interpretation of the G-convergence (Theorem 3.62) the above conclusion follows. □

3.8 CONVERGENCE OF SUBDIFFERENTIAL OPERATORS

In this section, we study the graph convergence on the subclass of subdifferential operators ∂F and pay particular attention to the relationship with the convergence - in fact, epi-convergence - of the corresponding functions F.

3.8.1 The class of subdifferential operators is closed for G-convergence

Let us first recall the following characterization of subdifferential operators among maximal monotone operators.

Let $A = \partial F$ be the subdifferential of a closed convex proper function F, $F:X \to]-\infty, +\infty]$, where X is a reflexive Banach space. Then, for every closed chain in $D(\partial F)$, $x_o, x_1, \ldots, x_{\ell-1}$, $x_\ell = x_o$ and any $y_i \in \partial F(x_i)$, the following inequalities hold:

$$F(x_1) \geqslant F(x_o) + \langle y_o, x_1 - x_o \rangle$$

$$F(x_2) \geqslant F(x_1) + \langle y_1, x_2 - x_1 \rangle$$

$$\vdots$$

$$F(x\;) \geqslant F(x_{\ell-1}) + \langle y_{\ell-1}, x_\ell - x_{\ell-1} \rangle.$$

Adding all these inequalities, and using that $x_\ell = x_o$, we obtain:

$$\begin{aligned} &\text{for any closed chain } x_o, x_1, \ldots, x_{\ell-1}, x_\ell = x_o \text{ in } D(A) \text{ and} \\ &\text{any } y_i \in Ax_i \ (i = 1, \ldots, \ell) \\ &\sum_{i=0}^{\ell-1} \langle y_i, x_i - x_{i+1} \rangle \geqslant 0. \end{aligned} \tag{3.161}$$

An operator A satisfying (3.161) is said to be *cyclically monotone*. When taking $\ell = 2$ in (3.161) we obtain

$$\langle y_o - y_1, x_o - x_1 \rangle \geqslant 0$$

for every (x_o,y_o), $(x_1,y_1) \in A$, which is the definition of a monotone operator.

So cyclic monotonicity is a stronger proper than monotonicity, and indeed it characterizes among maximal monotone operators the subdifferential operators. More precisely,

PROPOSITION 3.64 (Rockafellar [2]) Let $A: X \to X^*$ be a maximal monotone operator. Then A is the subdifferential of a closed convex proper function $F: X \to]-\infty,+\infty]$ iff A is cyclically monotone.

In that case, the following *"integration" formula* holds: $A = \partial F$ with

$$\text{for every } x \in X,\ F(x) = \sup_{x_o \sim x} \{F(x_o) + \sum_{i=1}^{\ell} \langle Ax_{i-1}, x_i - x_{i-1}\rangle\}, \tag{3.162}$$

the supremum being taken over all finite chain $x_o \sim x$ (that is $x_o, x_1, \ldots, x_{\ell-1}$, $x_\ell = x$ such that $x_o, x_1, \ldots, x_{\ell-1}$ belong to $D(A)$).

The primitive F is defined up to an additive constant: in the above formula, one can start from any point $x_o \in D(A)$ and choose the value $F(x_o)$.

As a consequence of the above characterization of subdifferential operators, noticing that the cyclic monotonicity property (3.161) is preserved by G-convergence, we obtain

COROLLARY 3.65 The subclass of subdifferential operators is closed in M_R. In other words, if

$$\partial F^n \xrightarrow{G} A,$$

then $A = \partial F$ for some closed convex proper function F.

3.8.2 Equivalence between G-convergence of $\{\partial F^n;\ n \in \mathbb{N}\}$ and Mosco convergence of $\{F^n;\ n \in \mathbb{N}\}$

In Theorem 3.26, we proved, via Moreau-Yosida approximation, the equivalence between Mosco convergence of a sequence of closed convex proper functions, $F^n \to F$, and the pointwise convergence of the resolvents of the subdifferential operators $A^n = \partial F^n$, $A = \partial F$, $J_\lambda^{A^n} x \to J_\lambda^{A} x$ strongly in X, for every $\lambda > 0$, for every $x \in X$. From Theorem 3.62, this can be reformulated in terms of G-convergence as follows:

THEOREM 3.66 Let X be a reflexive Banach space. For any sequence $\{F^n; n \in \mathbb{N}\}$, $F : X \to]-\infty,+\infty]$ of closed convex proper functions, the following properties are equivalent:

(i) $F^n \xrightarrow{M} F$ (Mosco-convergence of F^n to F)

(ii) $\begin{cases} \partial F^n \xrightarrow{G} \partial F & \text{(G-convergence of } \partial F^n \text{ to } \partial F) \\ \exists (u,f) \in \partial F \ \ \exists (u_n,f_n) \in \partial F^n \text{ such that } u_n \xrightarrow{X_s} u,\ f_n \xrightarrow{X^*_s} f,\ F^n(u_n) \to F(u). \end{cases}$

The second sentence of (ii) is just a normalization assumption, since F^n is determined by ∂F^n up to an additive constant.

In the proof of Theorem 3.66 via Moreau-Yosida approximation, we used in an essential way the renorming of X as a C-normed space and obtained in addition the equivalence of (i) and (ii) with

(iii) $\forall \lambda > 0 \ \ \forall x \in X \ \ F^n_\lambda(x) \to F_\lambda(x)$.

Since the equivalence (i) $\Leftrightarrow$ (ii) is only of a topological nature and has nothing to do with the choice of a precise norm of X, we like to give an alternative proof of it. Implication (ii) $\Rightarrow$ (i) relies on integration formula (3.162), while implication (i) $\Rightarrow$ (ii) relies on a result of Brondsted and Rockafellar (cf. Moreau [2] Proposition 10i), about ε-subdifferentials. These arguments are of independent interest and can be extended as we shall see to more general situations.

Proof of Theorem 3.66 Let us prove implication (ii) $\Rightarrow$ (i) via integration formula (3.162): We first verify that

$$F \leqslant \text{seq}X_w\text{-li}_e F^n. \tag{3.163}$$

Let $x_n \longrightarrow x$ be a weakly converging sequence in X. For any finite chain $(u_1,f_1),\ldots,(u_\ell,f_\ell)$ of elements of ∂F, from $\partial F^n \xrightarrow{G} \partial F$, there exist approximating chains $(u^n_1,f^n_1),\ldots,(u^n_\ell,f^n_\ell)$ belonging to ∂F^n such that

$$\text{for every } i = 1,\ldots,\ell \qquad u^n_i \to u_i \text{ strongly in X}$$
$$f^n_i \to f_i \text{ strongly in } X^*.$$

From

$$F^n(x_n) \geq F^n(u_n) + \langle u_1^n - u_n, f_n\rangle + \langle u_2^n - u_1^n, f_1^n\rangle + \ldots + \langle x_n - u_\ell^n, f_\ell^n\rangle,$$

where (u_n, f_n) are given by (ii), by going to the limit as n goes to $+\infty$, we obtain

$$\liminf_{n\to+\infty} F^n(x_n) \geq F(u) + \langle u_1 - u, f\rangle + \langle u_2 - u_1, f_1\rangle + \ldots + \langle u_\ell - u_{\ell-1}, f_{\ell-1}\rangle + \langle x - u_\ell, f_\ell\rangle.$$

This inequality being satisfied by any chain $u \sim x$, taking the supremum and using formula (3.162) it follows that

$$\liminf_n F^n(x_n) \geq F(x),$$

which is (3.163).

Let us now verify that

$$F \geq X_s\text{-}ls_e F^n. \tag{3.164}$$

Noticing that, for every $x \in D(F)$, $J_\lambda^F x \to x$ strongly in X as $\lambda \to 0$ and $F(J_\lambda x) \xrightarrow[(\lambda\to 0)]{} F(x)$ (refer to Theorem 3.24), from diagonalization Lemma 1.18 it is enough to prove that

$$\begin{aligned}&\text{for every } x \in D(\partial F)\text{, there exists a strongly convergent}\\ &\text{sequence } x_n \to x \text{ such that } F(x) \geq \limsup_n F^n(x_n).\end{aligned} \tag{3.165}$$

Let us fix some $y \in \partial F(x)$. By assumption (ii), there exists an approximating sequence $\{(x_n, y_n);\ n \in \mathbb{N}\}$ such that

$$\text{for every } n \in \mathbb{N},\quad y_n \in \partial F^n(x_n),$$

$$x_n \to x \text{ strongly in X, as } n \to +\infty$$

$$y_n \to y \text{ strongly in } X^*\text{, as } n \to +\infty.$$

Let us prove that such a sequence $\{x_n;\ n \in \mathbb{N}\}$ satisfies (3.165): For any chain $(u_\ell, f_\ell), \ldots, (u_1, f_1)$ of elements of ∂F, let us introduce approximating sequences $(u_\ell^n, f_\ell^n), \ldots, (u_1^n, f_1^n)$ of elements of ∂F^n, given by assumption (ii). Then, let us write

$$F^n(u_\ell^n) \geq F^n(x_n)+\langle y_n,u_\ell^n-x_n\rangle$$

$$F^n(u_{\ell-1}^n) \geq F^n(u_\ell^n) + \langle f_\ell^n,u_{\ell-1}^n-u_\ell^n\rangle$$

$$\vdots$$

$$F^n(u_n) \geq F^n(u_1^n) + \langle f_1^n,u_n-u_1^n\rangle$$

and add

$$F^n(u_n) \geq F^n(x_n) + \langle y_n,u_\ell^n-x_n\rangle + \langle f_\ell^n,u_{\ell-1}^n-u_\ell^n\rangle +\ldots+ \langle f_1^n,u_n-u_1^n\rangle.$$

Then, take the limit superior as n goes to $+\infty$,

$$F(u) \geq \limsup_n F^n(x_n) + \langle y,u_\ell-x\rangle + \langle f_\ell,u_{\ell-1}-u_\ell\rangle +\ldots+ \langle f_1,u-u_1\rangle,$$

and the supremum with respect to all finite chains of elements of ∂F. By integration formula (3.162),

$$F(u) \geq \limsup_n F^n(x_n) + F(u) - F(x),$$

which is (3.165).

Let us now prove implication (i) $\Rightarrow$ (ii).

Let $(u,f) \in \partial F$. Equivalently

$$F(u) + F^*(f) - \langle f,u\rangle = 0. \tag{3.166}$$

Since $F^n \xrightarrow{M} F$, from Proposition 3.19,

there exists a sequence $u_n \to u$ strongly in X such that $F^n(u_n) \to F(u)$

there exists a sequence $f_n \to f$ strongly in X* such that $F^{n*}(f_n) \to F^*(f)$.

From (3.166) it follows that, for every $n \in \mathbb{N}$,

$$F^n(u_n) + F^{n*}(f_n) - \langle f_n,u_n\rangle = \varepsilon_n$$

where $\{\varepsilon_n; n \in \mathbb{N}\}$ is a sequence of positive numbers converging to zero. From the Bronsted & Rockafellar theorem, for every $n \in \mathbb{N}$, one can find an $(x_n,y_n) \in \partial F^n$ such that

$$\|x_n - u_n\| < \sqrt{\varepsilon_n}$$

$$\|y_n - f_n\| < \sqrt{\varepsilon_n}.$$

It follows that $x_n \to u$ in X_s, $y_n \to f$ in X^*_s, i.e. $\partial F^n \xrightarrow{G} \partial F$. □

3.8.3 Epi-convergence for the weak topology of functions and convergence of subdifferentials

It is a natural question to ask: given a sequence of closed convex functions $\{F^n; n \in \mathbb{N}\}$, F from a reflexive Banach space X into $]-\infty, +\infty]$ and a convergence result

$$F = \tau\text{-}\lim_e F^n,$$

what is the corresponding convergence result for the subdifferential operators, that is in what sense does $\partial F^n \to \partial F$?

In Theorem 3.66, we gave the answer when considering Mosco-convergence (that is, epi-convergence for both strong and weak topologies) of $F^n \xrightarrow{M} F$.

We now examine this question when taking τ = weak topology of X.

THEOREM 3.67 Let $F^n, F: X \to]-\infty, +\infty]$ be a sequence of closed convex proper functions. The following implication holds:

(i) $F = \text{seq}X_w\text{-}\lim_e F^n$

⇓

(ii) $\begin{cases} \partial F^n \xrightarrow{G(w,s)} \partial F \\ \exists (u,f) \in \partial F \quad \exists (u_n, f_n) \in \partial F^n \text{ such that } u_n \xrightarrow{X_w} u,\ f_n \xrightarrow{X^*_s} f,\ F^n(u_n) \to F(u) \end{cases}$

where G(w,s) convergence for a sequence of operators $A^n \xrightarrow{G(w,s)} A$ from X into X* means:

$$\begin{cases} \forall (x,y) \in A \quad \exists (x_n, y_n) \in A^n \text{ such that } x_n \xrightarrow{X_w} x,\ y_n \xrightarrow{X^*_s} y \\ \forall \{n_k; k \in \mathbb{N}\},\quad \forall (x_k, y_k) \in A^{n_k},\ x_k \xrightarrow{X_w} x,\ y_k \xrightarrow{X^*_s} y \text{ imply } (x,y) \in A. \end{cases} \tag{3.167}$$

When X is separable and the $\{F^n; n \in \mathbb{N}\}$ satisfy a uniform coerciveness property (refer to Theorem 3.11, assumption (3.32)) then, the equivalence

(i) $\Leftrightarrow$ (ii) holds. Indeed

$$F = \text{seq}X_w\text{-lim}_e F^n \Leftrightarrow F = X_w\text{-lim}_e F^n \Leftrightarrow \partial F^n \xrightarrow{G(w,s)} \partial F + \text{Normalization.}$$

(since in that case $\text{seq}X_w\text{-lim}_e$ and $X_w\text{-lim}_e$ are equivalent notions).

Proof of Theorem 3.67 The proof is quite similar to the one developed in Theorem 3.66. One can go to the limit on all duality brackets since sequences are weakly convergent in X and strongly convergent in X^* (we use Theorem 3.11). The only difficulty comes from (ii) $\Rightarrow$ (i): similarly, for every $(x,y) \in \partial F$ one can find approximating sequences

$$x_n \xrightarrow{X_w} x, \quad y_n \xrightarrow{X_s^*} y$$

such that $F^n(x_n) \to F(x)$.

Then for x belonging to dom (F), we follow an approximation argument: taking advantage of the convergence

$$J_\lambda x \xrightarrow{X_s} x \text{ as } \lambda \to 0,$$

$$F(J_\lambda x) \to F(x) \text{ as } \lambda \to 0,$$

we apply the preceding argument to each $J_\lambda x \in D(\partial F)$. We introduce, as in Lemma 3.12, $\{x_k^*;\ k \in \mathbb{N}\}$ a dense denumerable subset of the unit sphere of X^* and the metric d on X

$$d(x,y) = \sum_{k \in \mathbb{N}} 2^{-k} \, |\langle x_k^*, x-y \rangle|.$$

We have the following scheme

$$\begin{array}{ccc} (x_n^\lambda, F^n(x_n^\lambda)) & \xrightarrow[n \to +\infty]{X_d \times \mathbb{R}} & (J_\lambda x,\ F(J_\lambda x)) \\ & & \Big\downarrow X_d \times \mathbb{R} \\ & & (x,\ F(x)) \end{array}$$

By diagonalization Lemma 1.18 , there exists a map $n \to \lambda(n)$, with $\lambda(n) \xrightarrow[n \to +\infty]{} 0$, such that

$$x_n^{\lambda(n)} \xrightarrow{X_d} x \text{ and } F^n(x_n^{\lambda(n)}) \to F(x).$$

From the equi-coerciveness property (3.32), the sequence $x_n = x_n^{\lambda(n)}$, is bounded. Since on bounded sequences, X_d metric induces weak convergence, the conclusion follows. □

The same type of arguments and Theorem 3.11 yield the following result:

PROPOSITION 3.68 Let X be a reflexive Banach space and $\{F^n; n \in \mathbb{N}\}$, F a sequence of closed convex proper functions from X into $]-\infty, +\infty]$. The following implication holds:

(ii) $\partial F^n \xrightarrow{G(s,w)} \partial F$ + Normalization condition

⇓

(i) $F = X_s\text{-}\lim_e F^n$.

When the $\{F^n;\ n \in \mathbb{N}\}$ satisfy the equicoerciveness assumption (3.32) and X is separable, then the equivalence (i) ⟺ (ii) holds, i.e.

$$F = X_s\text{-}\lim_e F^n \Longleftrightarrow \partial F^n \xrightarrow{G(s,w)} \partial F + \text{Normalization condition.}$$

3.9 CONVERGENCE OF OPERATORS. EXAMPLES

From Theorems 3.66 and 3.67, all epi-convergence results proved in this book can be translated in terms of graph-convergence of corresponding subdifferential operators. In this section, we study for converging sequences of maximal monotone operators the convergence of mathematical objects which are naturally linked, such as the spectrum, semi-groups, etc.

3.9.1 G-convergence of elliptic operators. Convergence of the spectrum

Let us consider for any $\lambda > 0$ the class of linear elliptic operators

$$\mathbb{A}_\lambda = \{A: H_o^1(\Omega) \to H^{-1}(\Omega);\ \text{for every } u \in H_o^1(\Omega)\quad Au = -\Sigma \frac{\partial}{\partial x_i}(a_{ij}(x)\frac{\partial u}{\partial x_j})$$

where the a_{ij}, $i,j = 1,2,\dots,N$ belong to $L^\infty(\Omega)$ and satisfy (3.168)

$a_{ij} = a_{ji}$ for every $i,j = 1,2,\dots,N$

$\lambda|z|^2 \leq \Sigma\, a_{ij}(x)z_iz_j$ for every $z \in \mathbb{R}^N$, a.e. $x \in \Omega\}$.

For any such operator A, we denote

$$F_A(u) = \int_\Omega \Sigma\, a_{ij}(x) \frac{\partial u}{\partial x_i} \frac{\partial u}{\partial x_j}\, dx,$$

that is, $A = \partial F_A$ and A_H denotes the (not everywhere defined) operator from $H = L^2(\Omega)$ into $L^2(\Omega)$ equal to

$$D(A_H) = \{u \in H_o^1(\Omega)/Au \in L^2(\Omega)\}, \quad A_H u = Au.$$

From Theorems 3.66, 3.67 (assuming Ω to be bounded) we derive the following equivalent formulations of G-convergence:

PROPOSITION 3.69 For any sequence $\{A^n; n \in \mathbb{N}\}$ of operators of $\mathbb{A}_\lambda$ the following statements are equivalent:

(i) $F = H_o^1(\Omega)_w\text{-}\lim_e F^n$ (where F_A^n stands for F^n, F_A for F).

(ii) $F^n \to F$ in the Mosco sense in $L^2(\Omega)$

(iii) $A_H^n \xrightarrow{G} A_H$ in $L^2(\Omega) \times L^2(\Omega)$

(iii) $A^n \xrightarrow{G(w,s)} A$ in $H_o^1(\Omega) \times H^{-1}(\Omega)$

(iv) for every $f \in H^{-1}(\Omega)$, $(A^n)^{-1}f \longrightarrow A^{-1}f$ weakly in $H_o^1(\Omega)$.

(v) $\lim\limits_{n\to+\infty} \|(A^n)^{-1} - A^{-1}\|_{\mathcal{L}(L^2(\Omega),L^2(\Omega))} = 0.$

Following the terminology of Spagnolo [4] the sequence of operators $\{A^n; n \in \mathbb{N}\}$ is then said to be G-convergent.

Proof of Proposition 3.69 Let us just explain how to derive (v) from (iv). Let us consider the operators $\{A_n^{-1}, n \in \mathbb{N}\}$ from $H^{-1}(\Omega)$ into $H_o^1(\Omega)$. They satisfy a uniform contraction property:

$$A^n u = f \Rightarrow \langle A^n u, u\rangle = \langle f, u\rangle$$

$$\Rightarrow \lambda\, |\text{grad}\, u|^2_{L^2(\Omega)} \leq \|f\|_{H^{-1}} \cdot \|u\|_{H_o^1}.$$

From the Poincare inequality, $\|u\|_{H_o^1} \leq C|\text{grad}\, u|_{L^2}$, it follows that

$$\|u\|_{H_0^1(\Omega)} \leq \frac{C^2}{\lambda} \|f\|_{H^{-1}(\Omega)}.$$

From compact injections of $H_0^1(\Omega)$ into $L^2(\Omega)$ and of $L^2(\Omega)$ into $H^{-1}(\Omega)$ and by the Ascoli theorem it follows that

$$\sup_{|f|_{L^2(\Omega)} \leq 1} |(A^n)^{-1}f - (A)^{-1}f|_{L^2(\Omega)} \to 0 \text{ as } n \to +\infty .$$

Comments This last formulation (v) of G-convergence is interesting in different types of applications:

(a) It suggests a natural way of defining a rate of convergence for such converging sequences $\{F^n; n \in \mathbb{N}\}$ (refer to Attouch & Wets [5] for a general approach and to Babuska [1] Biroli [2] for the homogenization problem).

(b) It allows us to embed the class

$$\mathcal{F}_{\lambda,\Lambda}\{F:H_0^1(\Omega) \to \mathbb{R}^+/F(u) = \int_\Omega \Sigma\, a_{ij}(x) \frac{\partial u}{\partial x_i} \frac{\partial u}{\partial x_j}\, dx$$

with

$$\lambda|z|^2 \leq \Sigma\, a_{ij} z_i z_j \leq \Lambda|z|^2$$

as a subclass of

$$E = \mathcal{L}(L^2(\Omega); L^2(\Omega)),$$

the space of linear continuous maps from $L^2(\Omega)$ into $L^2(\Omega)$. More precisely

COROLLARY 3.70 The mapping

$$i:\mathcal{F}_{\lambda,\Lambda} \longrightarrow E = \mathcal{L}(L^2(\Omega); L^2(\Omega))$$

$$F \longrightarrow A = (\partial F)^{-1}$$

is a one to one bicontinuous map from $\mathcal{F}_{\lambda,\Lambda}$ equipped with the Mosco topology of $L^2(\Omega)$ onto $i(\mathcal{F}_{\lambda,\Lambda})$, which is a compact subset of the Banach space E equipped with the norm convergence.

Let us now examine for such converging sequences of linear elliptic operators the convergence of the corresponding spectrum. For a general approach to this problem, refer to Kato [1], Dunford & Schwartz [1], Boccardo & Marcellini [1] and in the particular case of homogenization of elliptic eigenvalue problems refer to S. Kesavan [3] and M. Vanninathan [1], (problems with holes).

In order to obtain uniform estimates on the eigenvalues of the operators we consider a "bounded subclass" $\mathbb{A}_{\lambda,\Lambda}$ of $\mathbb{A}_{\lambda}$:

$$\mathbb{A}_{\lambda,\Lambda} = \{A \in \mathbb{A}_{\lambda};\ Au = -\sum_{i,j=1}^{N} \frac{\partial}{\partial x_i}\left(a_{ij}(x)\frac{\partial u}{\partial x_j}\right) \text{ with}$$
$$\lambda|z|^2 \leqslant \sum a_{ij}(x)z_i z_j \leqslant \Lambda|z|^2\}. \tag{3.169}$$

For any such operator A, A^{-1} is a linear continuous compact self-adjoint positive definite operator from $L^2(\Omega)$ into $L^2(\Omega)$. From classical spectral theory (refer to Courant & Hilbert [1], for example) there exists a sequence $\{\lambda^\ell;\ \ell \in \mathbb{N}\}$ of positive real numbers and a sequence $\{u^\ell;\ \ell \in \mathbb{N}\}$ in $H_o^1(\Omega)$ such that:

(i) for any $\ell \in \mathbb{N}$ $Au^\ell = \lambda^\ell u^\ell$

(ii) $0 < \lambda^1 \leqslant \lambda^2 \leqslant \ldots \leqslant \lambda^\ell \leqslant \ldots \to +\infty.$ (3.170)

(iii) the multiplicity of each λ^ℓ is finite

(iv) the sequence $\{u^\ell;\ \ell \in \mathbb{N}\}$ is an orthonormal basis for $L^2(\Omega)$.

Moreover, introducing the Rayleigh quotient

$$\forall v \in H_o^1(\Omega) \quad v \neq 0 \quad R(v) = \frac{\langle Av,v\rangle}{|v|^2_{L^2}}$$

the minimax principle (cf. Courant & Hilbert [1]) states that

$$\lambda^\ell = R(u^\ell) = \max_{v\in E^\ell} R(v) = \min_{\substack{v\in H_o^1 \\ v\perp E^{\ell-1}}} R(v) = \min_{W\in D^\ell}\ \max_{v\in W} R(v) \tag{3.171}$$

where E^ℓ is the subspace of $H_o^1(\Omega)$ spanned by $\{u^1,u^2,\ldots,u^\ell\}$ and D^ℓ is the class of subspace of $H_o^1(\Omega)$ of dimension ℓ.

We can now state the convergence theorem for a spectrum, where given a sequence $A^n \xrightarrow{G} A$ of such operators is denoted by $\{\lambda_n^\ell;\ \ell \in \mathbb{N}\}$ and $\{u_n^\ell; \ell \in \mathbb{N}\}$ (resp. λ^ℓ and u^ℓ) the corresponding eigenvalues and eigenvectors.

THEOREM 3.71 Let $\{A^n;\ n \in \mathbb{N}\}$ be a sequence of linear elliptic operators belonging to the same class $\mathbb{A}_{\lambda,\Lambda}$, $0 < \lambda \leqslant \Lambda < +\infty$. The following are equivalent:

(i) $A^n \xrightarrow{G} A$ as n goes to $+\infty$.

(ii) for every $\ell \in \mathbb{N}$,

(a) $\lim\limits_{n\to+\infty} \lambda_n^\ell = \lambda^\ell$

(b) denoting m_ℓ the multiplicity of λ^ℓ, i.e.

$$\lambda^{\ell-1} < \lambda^\ell = \lambda^{\ell+1} = \ldots = \lambda^{\ell+m_\ell-1} < \lambda^{\ell+m_\ell}$$

the sequence of subspaces of dimension m_ℓ generated by $\{u_n^\ell,\ldots,u_n^{\ell+m_\ell-1}\}$ Mosco-converges in $L^2(\Omega)$ to the eigenspace of A relative to $\lambda^\ell = \{u^\ell,\ldots,u^{\ell+m_\ell-1}\}$.

Moreover, when λ^ℓ is a simple eigenvalue of A, there exists $N(\ell)$ such that λ_n^ℓ is a simple eigenvalue of A^n for $n \geqslant N(\ell)$. Consequently $u_n^\ell \to u^\ell$ strongly in $L^2(\Omega)$ (and weakly in $H_o^1(\Omega)$) as $n \to +\infty$, in that case.

Proof of Theorem 3.71 (i) $\Rightarrow$ (ii) From (3.169), for every $v \in H_o^1(\Omega)$, for every $n \in \mathbb{N}$

$$\lambda\ |\text{grad} v|_{L^2}^2 \leqslant \langle A^n v,v\rangle \leqslant \Lambda |\text{grad} v|_{L^2}^2.$$

Hence

$$\lambda \frac{|\text{grad} v|_{L^2}^2}{|v|_{L^2}^2} \leqslant R_n(v) \leqslant \Lambda \frac{|\text{grad} v|_{L^2}^2}{|v|_{L^2}^2}$$

and from the minimax principle (3.171), for every $\ell \in \mathbb{N}$ and $n \in \mathbb{N}$,

$$\lambda\ \mu^\ell \leqslant \lambda_n^\ell \leqslant \Lambda\mu^\ell, \tag{3.172}$$

where μ^ℓ is the ℓth eigenvalue of the Laplacian for the Dirichlet boundary data. Thus, for each $\ell \in \mathbb{N}$, the sequences $\{\lambda_n^\ell;\ n \in \mathbb{N}\}$ are bounded and so are the sequences $\{u_n^\ell;\ n \in \mathbb{N}\}$ in $H_0^1(\Omega)$.

Extracting converging subsequences respectively in $\mathbb{R}$ and in weak-$H_0^1(\Omega)$, and applying a diagonalization argument, we can extract a subsequence (still denoted $(\lambda_n^\ell, u_n^\ell)$) such that:

$$\text{for every } \ell \in \mathbb{N},\quad \lambda_n^\ell \longrightarrow \bar{\lambda}^\ell,\ u_n^\ell \longrightarrow \bar{u}^\ell \text{ weakly in } H_0^1(\Omega). \tag{3.173}$$

The problem is to prove that $\bar{\lambda}^\ell = \lambda^\ell$. Since the sequence $\{A^n;\ n \in \mathbb{N}\}$ is G-convergent to A in $L^2(\Omega)$, from Proposition 3.59, noticing that

$$A^n u_n^\ell = \lambda_n^\ell u_n^\ell$$

and

$$u_n^\ell \to \bar{u}^\ell,\ \lambda_n^\ell u_n^\ell \to \bar{\lambda}^\ell \bar{u}^\ell \text{in } L^2(\Omega),$$

it follows that

$$A\bar{u}^\ell = \bar{\lambda}^\ell \bar{u}^\ell. \tag{3.174}$$

Thus the pair $(\bar{u}^\ell, \bar{\lambda}^\ell)$ satisfies the limit eigenvalue problem and $\bar{\lambda}^\ell \in \{\lambda^k;\ k \in \mathbb{N}\}$. Since $0 < \lambda_n^1 \leq \lambda_n^2 \leq \ldots$ it follows that

$$0 \leq \bar{\lambda}^1 \leq \bar{\lambda}^2 \leq \ldots .$$

Moreover from (3.172), for every $\ell \in \mathbb{N}$,

$$\lambda \cdot \mu^\ell \leq \bar{\lambda}^\ell \leq \Lambda \cdot \mu^\ell.$$

Thus, the sequence $\{\bar{\lambda}^\ell;\ \ell \in \mathbb{N}\}$ tends to $+\infty$ as ℓ goes to $+\infty$:

$$0 < \bar{\lambda}^1 \leq \bar{\lambda}^2 \leq \ldots \to +\infty. \tag{3.175}$$

What might happen is that the sequence $\{\bar{\lambda}^\ell;\ \ell \in \mathbb{N}\}$ is strictly included in the sequence $\{\lambda^\ell;\ \ell \in \mathbb{N}\}$. So, in order to complete the proof of (i) $\Rightarrow$ (ii)a, we need to prove that there is no eigenvalue of the limit operator A other than those in the sequence $\{\bar{\lambda}^\ell;\ \ell \in \mathbb{N}\}$.

Let us argue by contradiction and assume that $\lambda \neq \bar{\lambda}^{\ell}$ for every $\ell \in \mathbb{N}$ is an eigenvalue of A. Let u be a corresponding normalized ($|u|_{L^2(\Omega)} = 1$) eigenvector

$$Au = \lambda u. \tag{3.176}$$

From (3.175), there exists some integer $m \in \mathbb{N}$ such that

$$\lambda < \bar{\lambda}^{m+1}. \tag{3.177}$$

Let us introduce for every $n \in \mathbb{N}$, w_n the solution of

$$A^n w_n = \lambda u \tag{3.178}$$

By G-convergence of A^n to A (Proposition 3.69, (iv)) and definition of u (3.176) it follows that

$$w_n \rightharpoonup u \text{ weakly in } H_o^1(\Omega) \text{ and strongly in } L^2(\Omega). \tag{3.179}$$

Since

$$\langle A^n w_n, w_n \rangle = \lambda \langle u, w_n \rangle \to \lambda |u|_{L^2}^2 = \lambda$$

and

$$|w_n|_{L^2} \to |u|_{L^2} = 1,$$

it follows that

$$R_n(w_n) \to \lambda \quad \text{as } n \to +\infty. \tag{3.180}$$

Let us introduce the following sequence $\{v_n;\ n \in \mathbb{N}\}$

$$v_n = w_n - \sum_{k=1}^{m} (w_n, u_n^k) u_n^k.$$

Noticing that

$$(w_n, u_n^k) \to (u, \bar{u}^k) \text{ as } n \to +\infty$$

and that by assumption $\lambda \notin \{\bar{\lambda}^k\}$, it follows from orthogonality of the eigenvectors of A that

$(w_n, u_n^k) \to 0$ for every $k \in \mathbb{N}$.

Thus

$$v_n \longrightarrow u \text{ weakly in } H_o^1(\Omega), \text{ and } R_n(v_n) \to \lambda . \tag{3.181}$$

On the other hand, by definition of v_n,

$$\langle v_n, u_n^k \rangle = 0 \text{ for every } k = 1,\dots,m.$$

Hence by the minimax principle,

$$\lambda_n^{m+1} \leqslant R_n(v_n). \tag{3.182}$$

Going to the limit on (3.182), using (3.181), we obtain

$$\bar{\lambda}^{m+1} \leqslant \lambda,$$

which is a clear contraction with $\lambda < \bar{\lambda}^{m+1}$.

Let us now prove point (b) of (ii): fixing some $\ell \in \mathbb{N}$ we denote

$$E = [u^\ell, u^{\ell+1},\dots,u^{\ell+m_\ell-1}]$$

the eigensubspace of A relative to λ^ℓ, and

$$E_n = [u_n^\ell, u_n^{\ell+1},\dots,u_n^{\ell+ m_\ell-1}].$$

We stress the fact that the multiplicity of λ_n^ℓ may be strictly less than that of λ^ℓ; consequently E_n may be strictly larger than the eigensubspace of λ_n^ℓ! We first observe that because of the uniform coerciveness of the operators $\{A^n;\ n \in \mathbb{N}\}$ for every $k \in \mathbb{N}$, the sequences$\{u_n^k;\ n \in \mathbb{N}\}$ are bounded in $H_o^1(\Omega)$ and hence relatively compact in $L^2(\Omega)$. By a diagonalization argument we can extract a subsequence $\{n_h;\ h \in \mathbb{N}\}$ such that for every k, the sequence $\{u_{n_h}^k\ ;\ h \in \mathbb{N}\}$ strongly converges in $L^2(\Omega)$ to some u^k.

From G-convergence of A^n to A, $\lambda_n^k \longrightarrow \lambda^k$, and

$$A^{n_h}(u_{n_h}^k) = \lambda_{n_h}^k\ u_{n_h}^k$$

it follows that

$$A(u^k) = \lambda^k u^k.$$

From linearity of A^n it follows that $\operatorname{Lim\,sup}_n E_n \subset E$. The eigenvectors $\{u^k;\ k = \ell, \ell+1, \ldots, \ell+m_\ell-1\}$ being still orthogonal and E being of dimension m_ℓ

$$\operatorname{Lim\,sup}_n E_n = E. \tag{3.183}$$

On the other hand, let $v \notin \operatorname{Lim\,inf}_n E_n$. Then, there exists some subsequence $\{n_h;\ h \in \mathbb{N}\}$ such that $v \notin \operatorname{Lim\,sup} E_{n_h}$. From the preceding argument it would follow

$$v \notin \operatorname{Lim\,sup} E_{n_h} = E.$$

Thus, $(v \notin \operatorname{Lim\,inf} E_n) \Rightarrow (v \notin E)$, that is

$$E \subset \operatorname{Lim\,inf} E_n. \tag{3.184}$$

So, conclusion $E = \operatorname{Lim} E_n$ holds (for the weak topology of $H_0^1(\Omega)$, or equivalently for Mosco-convergence of $L^2(\Omega)$). For implication (ii) $\Rightarrow$ (i) refer to Boccardo & Marcellini. □

3.9.2 Convergence of semi-groups

For simplicity in this section, we work in a single space H, which is assumed to be a real Hilbert space. Then the theory of the generation of contraction semigroups is well posed: Given A a maximal monotone operator $A: D(A) \subset H \to H$, we can associate the semigroup $\{S^A(t);\ t \geq 0\}$ via the exponential formula

$$\forall x \in \overline{D(A)} \quad S^A(t)x = \lim_{k\to+\infty} \left(I + \frac{t}{k}A\right)^{-k} x. \tag{3.185}$$

For every x belonging to D(A), $u(t) = S^A(t)x$ satisfies the evolution equation

$$\frac{du}{dt} + A\,u(t) \ni 0,\ u(0) = u_o.$$

Conversely, given any such continuous contraction semigroup, there is a unique maximal monotone operator A which generates it via (3.185); refer to H. Brezis [1].

So the correspondence $A \to \{S^A(t);\ t \geq 0\}$ is a one to one correspondence between maximal monotone operators and continuous semigroups of contractions.

The following theorem from Brezis [2] explains in what sense this correspondence is continuous. Given A a maximal monotone operator we denote A^o its minimal section, $A^o x = \text{proj}_{Ax} 0$ and $J_\lambda^A x = (I + \lambda A)^{-1} x$ its resolvents.

THEOREM 3.72 Let $\{A^n;\ n \in \mathbb{N}\}$, A be a sequence of maximal monotone operators from H into itself. The following are equivalent:

(i) for every $x \in D(A)$, there exists a sequence $x_n \to x$ strongly in H such that

$$(A^n)^o x_n \to A^o x \text{ strongly in H.}$$

(ii) for every $x \in \overline{D(A)}$, for every $\lambda > 0$, $J_\lambda^{A^n} x \xrightarrow[(n \to +\infty)]{} J_\lambda^A x$ strongly in H.

(iii) for every $x \in \overline{D(A)}$, there exists a sequence $\{x_n;\ x_n \in \overline{D(A^n)}$ for every $n \in \mathbb{N}\}$ such that

$$x_n \to x \text{ strongly in H, as } n \to +\infty$$

$$S^{A^n}(t)x_n \to S^A(t)x \text{ strongly in H, as } n \to +\infty.$$

Comments Proof of (ii) ⇒ (iii) relies on exponential formula (3.185) and an estimation of the difference $|S^A(t)x - (I + \frac{t}{k} A)^{-k} x|$ for k large.

We notice that in (ii) one asks the resolvents to converge only for x belonging to $\overline{D(A)}$! So from Proposition 3.60, G-convergence of A^n to A in H, $A^n \xrightarrow{G} A$ does imply convergence of the corresponding semigroups as in (iii). For further results concerning the relation between G-convergence and convergence of semigroups refer to Attouch [1].

When considering $A^n = \partial F^n$ as subdifferential operators, one can prove the following equivalence.

THEOREM 3.73 Let F^n, $F: H \to]-\infty,+\infty]$ be closed convex proper functions. The following are equivalent:

(i) for every $\lambda > 0$, for every $x \in \overline{D(F)}$ $F_\lambda^n(x) \to F_\lambda(x)$ as $n \to +\infty$.

(ii) $\begin{cases} S^{A^n}(t) \to S^A(t) \text{ as } n \to +\infty \text{ in the sense of Theorem 3.72.} \\ \text{Normalization condition: } \exists x_n \to x \text{ such that } (\partial F^n)^o x_n \to (\partial F)^o x, \\ F^n(x_n) \to F(x). \end{cases}$

Let us end this section with the following striking result (refer to Attouch [2]).

THEOREM 3.74 Let $A^n = \partial F^n \xrightarrow{G} A = \partial F$ be a G-convergent sequence of subdifferential operators in an Hilbert space H. Let us denote u_n (resp. u) the solution of the evolution equation

$$\frac{du_n}{dt} + \partial F^n(u_n) \ni f_n, \quad u_n(0) = x_n \quad (x_n \in \overline{D(F^n)}) \tag{3.186}$$

resp.

$$\frac{du}{dt} + \partial F(u) \ni f, \qquad u(0) = x \qquad (x \in \overline{D(F)}).$$

(1) Assuming that $x_n \to x$ strongly in H and $f_n \to f$ in $L^2(0,T;H)$, then

$$u_n \to u \text{ uniformly on } [0,T]$$

and

$$\int_0^T t\left|\frac{du_n}{dt} - \frac{du}{dt}\right|^2 dt \to 0 \text{ as } n \to +\infty.$$

(2) Assuming moreover $x_n \in D(F^n)$, $x \in D(F)$ and $F^n(x_n) \to F(x)$ as $n \to +\infty$, then

$$\frac{du_n}{dt} \to \frac{du}{dt} \quad \text{strongly in } L^2(0,T;H)$$

and

$$F^n(u_n) \to F(u) \text{ uniformly on } [0,T].$$

Comments We first notice that because of the smoothing effect on the initial condition of equations governed by subdifferentials (refer to H. Brezis [1]), (3.186) makes sense. The surprising fact is that, although one cannot expect a better estimate on $\{\frac{du_n}{dt} ; n \in \mathbb{N}\}$ than to be bounded in

$L^2(0,T;H)$, there is indeed strong convergence of du_n/dt to du/dt in $L^2_{loc}(0,T;H)$!

The idea is to multiply (3.186) by $\frac{du_n}{dt}$ in $L^2(0,T)$; for simplicity let us consider case (2):

$$\left|\frac{du_n}{dt}\right|^2_{L^2(0,T;H)} + F^n(u_n(T)) - F^n(x_n) - \int_0^T \langle f_n(t),u_n(t)\rangle dt = 0.$$

Since

$$\left|\frac{du}{dt}\right|_{L^2} \leq \liminf_n \left|\frac{du_n}{dt}\right|_{L^2}$$

$$F(u(T)) \leq \liminf_n F^n(u_n(T)) \quad \text{and} \quad F^n(x_n) \to F(x)$$

$$\int_0^T \langle f,u\rangle dt \leq \liminf_n \int_0^T \langle f_n,u_n\rangle dt$$

using the equality

$$\left|\frac{du}{dt}\right|^2 + F(u(T)) - F(x) - \int_0^T \langle f(t), u(t)\rangle dt = 0$$

it follows from an elementary lemma that

$$\int_0^T \left|\frac{du_n}{dt}\right|^2 \to \int_0^T \left|\frac{du}{dt}\right|^2,$$

that is

$$\frac{du_n}{dt} \to \frac{du}{dt} \text{ in } L^2(0,T;H).$$

Refer to Baillon & Haraux [1], Veron [1] for similar results. □

Bibliography

E. ACERBI

[1] Convergence of second order elliptic operators in complete form; Boll. Un. Mat. Ital., (5) 18-B, 359-556 (1981).

E. ACERBI, N. FUSCO

[1] Semicontinuity problems in calculus of variations; Archive for Rat. Mech. and Anal. (to appear).

F. ACKER, M.A. PRESTEL

[1] Convergence d'un schema de minimisation alternée; Ann. Fac. Sc. Toulouse, 2, 1-9 (1980).

A. AMBROSETTI, C. SBORDONE

[1] Γ-convergenza e G-convergenza per problemi non lineari di tipo ellittico; Boll. Un. Mat. Ital. (5), 13-A, 352-362 (1976).

A. ANCONA

[1] Théorie du potentiel dans les espaces fonctionnels à forme coercive. Cours 3eme cycle. Paris VI, (1973).

H. AROSIO

[1] Asymptotic behaviour as $t \to +\infty$ of solutions of linear parabolic equations; Comm. Part. Diff. eq., 4, 769-794 (1979).

M. ARTOLA, G. DUVAUT

[1] Homogénéisation d'une plaque renforcée C.R.A.S. Paris, 284, 707-710 (1977).

[2] Homogénéisation d'une classe de problèmes non linéaires. C.R.A.S. Paris, 288, 775-778 (1979).

Z. ARTSTEIN

[1] Continuous dependence on parameters; J. Diff. Equations, 19, 214-225 (1975).

Z. ARTSTEIN, S. HART

[1] Law of large number for random sets and allocation processes. Mathematics of Operations Research 6, 482-492 (1981).

H. ATTOUCH

[1] Convergence de fonctions convexes, des sous-differentiels et semi-groupes

associés; C.R.A.S. Paris 284, 539-542 (1977).

[2] Convergence de fonctionnelles convexes; Proc. Journées d'Anal. non linéaire, Besançon (1977) Lecture Notes in Math. Springer 665, 1-40 (1978).

[3] Convergence des solutions d'inéquations variationnelles avec obstacles; Proc. Int. Meeting on"Recent Methods in Nonlinear Analysis", Rome, May 8-12 (1978), ed. De Giorgi, Magenes, Mosco, Pitagora ed., Bologna, 101-114 (1979).

[4] Famille d'opérateurs maximaux monotones et mésurabilité; Ann. Mat. Pura Appl. (4) 120, 35-111 (1979).

[5] Sur la Γ-convergence. Collège de France. Seminar on Nonlinear partial differential equations (1978-79) Vol. 1, ed. H. Brezis, J.-L. Lions, Research Notes in Math., Pitman, London.

[6] Introduction a l'homogènéisation d'inéquations variationnelles. Rend. Sem. Mat. Univers. Politecn. Torino (Italy) 40, 2 (1981).

[7] An energetic approach to homogenization problems with rapidly oscillating potentials, Tech. Report M.R.C, University of Madison-Wisconsin, #1989 (1979).

[8] A Galerkin approximation for the minimization of a convex lower semi-continuous function on a reflexive Banach space, SIAM Journal on control and optimization (to appear).

[9] Théorie de la Γ-convergence. Applications ā des inéquations variationnelles de la mécanique. Séminaire Gaulaouic-Meyer-Schwarz. Ecole Polytechnique. Palaiseau (1982-83).

[10] Variational properties of epi-convergence. Applications to limit analysis problems in mechanics and duality theory. International Conference on multifunctions and integrands, Catane (Sicilia) 1983. Editors R.T. Rockefellar, G. Salinetti, M. Valadier. Lecture Notes in Math (Springer), to appear.

H. ATTOUCH, H. BREZIS

[1] Duality for the sum of convex functions in general Banach spaces. "Aspects of mathematics and its applications" North-Holland (to appear).

H. ATTOUCH, B. CORNET

[1] Solutions lentes d'une classe d'équations d'évolution multivogues en économie. Cahiers de Math. de la décision. Cérémade. Paris-Dauphine (to appear).

H. ATTOUCH, A. DAMLAMIAN

[1] Homogenization for a Volterra equation. Math. Research Center, Univ. Madison-Wisconsin M.R.C. Technical Report No. 2566 (1983).

[2] Application des méthodes de convexité et monotonie à l' étude de certaines équations quasi-linèaires. Proc. Royal Society of Edinburgh, 79A, 197-129 (1977).

H. ATTOUCH, A. DAMLAMIAN, F. MURAT, C. PICARD

[1] Le problème de la passoire de Neumann. Rend. Sem. Mat. Univers. Polite n. Torino (Italy) (1983).

H. ATTOUCH, Y. KONISHI

[1] Convergence d'operateurs maximaux monotones et inéquations variationnelles; C.R.A.S. Paris, 282, 467-469 (1976).

H. ATTOUCH, F. MURAT

[1] Potentiels fortement oscillant (in preparation)

[2] Homogénéisation de milieux fissurés (in preparation)

H. ATTOUCH, C. PICARD

[1] Problèmes variationnels et théorie du potentiel non linéaire; Ann. Fac. Sci. Toulouse, 1, 89-136 (1979).

[2] Inéquations variationnelles avec obstacles et espaces fonctionnels en théorie du potentiel; Applicable Analysis, 12, 287-306 (1981).

[3] Asymptotic Analysis of variational problems with constraints of obstacle type; Publications Mathematiques d'Orsay (1982).

[4] Variational inequalities with varying obstacles. The general form of the limit problem; Journal of functional analysis, 50 (3) 1255 (1983).

H. ATTOUCH, C. SBORDONE

[1] Asymptotic limits for perturbed functionals of calculus of variations; Ricerche di Matematica, 29, 85-124 (1980).

[2] A general homogenization formula for functionals in calculus of variations (to appear).

H. ATTOUCH, R. WETS

[1] Approximation and convergence in nonlinear optimization; Nonlinear programming 4. Ed. by Magasarian, Meyer, Robinson. Academic Press (1981) Computer Science Department, University of Wisconsin-Madison 367-394.

[2] A convergence for bivariate functions aimed at the convergence of saddle values; Mathematical theory of optimization, Cecconi, Zolezzi, ed. Springer Verlag, Lecture Notes in Mathematics, 979 Berlin (1983).

[3] A convergence theory for saddle functions; Trans. Amer. Math. Soc. (1983).
[4] Convergence de points min/sup et de points fixes; CRAS Paris, Vol. 296 (1983).
[5] A rate of convergence for sequences of maximal monotone operators and convex functions; (to appear).

J.P. AUBIN
[1] Mathematical methods of game and economic theory. North-Holland (1979).
[2] Applied functional analysis, Wiley (1979).

D. AZE
[1] Homogénéisation primale et duale par epi-convergence. Applications à l'élasticité. Publication AVAMAC (Perpignan) 1984.

I. BABUSKA
[1] Solution of interface problems by homogenization; Siam Journ. Math. Anal., 7, 603-645 (1976); 8, 923-937 (1977).
[2] Homogenization approach in engineering; Lecture Notes in Economics and Math systems, Springer, 134, 137-153 (1976).

K. BACK
[1] Continuity of the Fenchel transform of convex functions, preprint. Department of Economics, Northwestern University, Evanston, Illinois (1983).

N.S. BAHVALOV
[1] Averaging of non linear partial differential equations with rapidly oscillating coefficients: Soviet Math Dokl; 16, 1469-1473 (1975); Translated from Dokl Akad Nauk 225 (2), 249-252 (1975).
[2] Homogenization and perturbation problems; Proc. Meeting on "Computing Methods in applied sciences and engineering" Versailles, (1979) ed. by Glowinski, J.-L. Lions, North-Holland, 645-658 (1980).

J.B. BAILLON, A. HARAUX
[1] Comportement à l'infini pour les équations d'évolution avec forcing periodique, Archive Rat. Mech. Anal. 67, 101-109 (1977).

C. BAIOCCHI
[1] Free boundary problems in the theory of fluid flow through porous media. Proc. Int. Congress on Math. 1974, Vancouver, vol 2 (1975).

C. BAIOCCHI, A. CAPELO
[1] Disequazioni variazionali e quasi variazionali; Applicazioni a problemi di frontiera libera. Pitagora Editrice, Bologna (1978) (English translation in preparation).

A. BAMBERGER

[1] Approximation des coefficients d'opérateurs elliptiques, stable pour la G-convergence, Rapport interne, Ecole Polytechnique, 15 (1977).

V. BARBU, G. DA PRATO

[1] Global existence for Hamilton-Jacobi equations. Ann. Sc. N. Sup. Pisa, VIII 2, 251-284 (1981).

V. BARBU, Th. PRECUPANU

[1] Convexity and Optimization in Banach spaces; Sijthoff and Noordhoff (1978).

G. BEER

[1] A natural topology for upper semicontinuous functions and a Baire category dual for convergence in measure; Pacific J. Math. 96, 251-263 (1981).

[2] Upper semicontinuous functions and the Stone approximation theorem; J. Approximation theory, 34, 1-11 (1982).

[3] On uniform convergence ... and top convergence of sets; to appear in Canadian Math. Bull.

Ph. BENILAN

[1] Equations d'évolution dans un Banach quelconque et applications. Thèse, Univ. Orsay (1972).

Ph. BENILAN, M. CRANDALL

[1] Continuity in L^1 of $\beta \to -\Delta(\beta)$

M.L. BENNATI

[1] Perturbazioni variazionali di problemi di controllo; Boll. Un. Mat. Ital (5) 16-B, 910-922 (1979).

A. BENSOUSSAN

[1] Un résultat de perturbation singulière pour systèmes distribués instables; CRAS. Paris, I. 296, 469-472 (1983).

A. BENSOUSSAN, J.-L. LIONS, G. PAPANICOLAOU

[1] Asymptotic analysis for periodic structures; North-Holland, Amsterdam (1978).

V.L. BERDICHEVSKII

[1] On the averaging of periodic structures; transl. in J. Appl. Math. Mech. 41 (1977), 1010-1023.

A.L. BERDICHEVSKII, V.L. BERDICHEVSKII

[1] Flow of an ideal fluid around a periodic system of obstacles; Izv. Akad.

Nauk, 6, 3-18 (1978).
C. BERGE
[1] Espaces topologiques. Fonctions multivoques. Dunod Paris (1966).
R. BERGSTROM, L. MACLINDEN
[1] Preservation by convergence of convex sets and functions..., Trans. Am. Math. Soc. 268, 127-142 (1981).
M.L. BERNARD-MAZURE
[1] Equi-sci, Γ-convergence et convergence simple; Sèm. d'analyse convexe Montpellier-Perpignan, No. 7 (1981).
M. BIROLI
[1] Sur la G-convergence pour des inéquations quasi-variationnelles; Boll. Un. Mat. Ital (5) 14-A, 540-550 (1977).
[2] An estimate on convergence for an homogenization problem for variational inequalities; Rend. Sem. Mat. Univ. Padova, 58, 9-22 (1977).
[3] G-convergence for elliptic equations..., Rend Sem. Mat Univ. Milano, 47, 269-328 (1977).
[4] An estimate on convergence of approximations by iterations; Ann. Mat. Pure Appl. (4) 112, 83-101 (1979).
[5] Homogenization for variational and quasi variational inequalities, Proc. Meeting on "Free boundary problems", Pavia, Sett. Ott. 1979, Istituto Nazionale di Alta Matematica, Roma, Vol 2, 45-61 (1980).
[6] Homogenization for variational and quasi variational inequalities, Rend. Sem. Mat. Univ. Milano (to appear).
[7] G-convergence for elliptic variational and quasi variational inequalities, Proc. Meeting on Recent Methods in non linear analysis, ed. De Giorgi, Magenes, Mosco, ed. Pitagora, Bologne, 361-388, (1979).
M. BIROLI, U. MOSCO
[1] Stability and homogenization for nonlinear variational inequalities with irregular obstacles and quadratic growth. Ceremade 8110. Cahiers de Paris-Dauphine.
L. BOCCARDO
[1] Aleuni problemi al contorno con vincoli unilaterali dependenti da un parametro; Boll. Un. Mat. Ital. (4) 8, 97-109 (1973).
[2] Perturbationi di funzionali del calcolo delle variazioni; Ricerche Mat, 29, 213-242 (1980).

L. BOCCARDO, I. CAPUZZO DOLCETTA

[1] G-convergenza e problema di Dirichlet unilaterale; Boll. Un. Mat. Ital (4) 12, 115-123 (1975).

[2] Stabilita delle soluzioni di disequationi variazionali...; Ann. Un. Ferrara, Sez VII, 24, 99-111 (1978).

[3] Disequazioni quasi variazionali con funzioni d'ostacolo...; Boll. Un. Mat. Ital (5) 15-B, 370-385 (1978).

L. BOCCARDO, I. CAPUZZO DOLCETTA, M. MATZEU

[1] Nouveau résultats d'existence, perturbation et homogènéisation...; Proc. Int. meeting on "Recent methods in nonlinear analysis", Rome, May 1978, ed. De Giorgi, Magenes, Mosco, Pitagora ed. Bologna, 115-130 (1979).

L. BOCCARDO, F. DONATI

[1] Existence and stability results for solutions of some strongly nonlinear constrained problems; Nonlinear Analysis, Th. Meth. Appl, 5, 975-988 (1981).

L. BOCCARDO, P. MARCELLINI

[1] Sulla convergenza delle soluzioni di disequazionali variazionali; Ann. Mat. Pura Appl. (4) 110, 137-159 (1976).

L. BOCCARDO, F. MURAT

[1] Nouveaux résultats de convergence dans des problèmes unilatéraux. Nonlinear partial differential equations and their applications, Collège de France seminar, Vol II, ed. H. Brezis & J.L. Lions. Research Notes in Math. No. 60, Pitman (1982), pp. 64-85.

[2] Homogénéisation de problèmes quasi linéaires. Atti del Convegno Studio di problemi-limite dell' analisi funzionale, Bressonone, sett. 81. Pitagora ditrice, 13-51 (1982).

[3] Remarques sur l'homogénéisation de certaines problèmes quasi-linéaires. Portugaliae Math., 41 (1982) (Volume dédié ā la mémoire de J. Sebastião E. Silva), ā paraître.

A. BOSSAVIT, A. DAMLAMIAN

[1] Homogenization of the Stefan problem and application to composite magnetic media, I.M.A. Journal of Applied Math., 27, 319-334 (1981).

J.F. BOURGAT

[1] Numerical experiments of the homogenization..., Proc. Int. Colloq. on "Computing methods in applied sciences and engineering" Versailles, Dec. 1977, Lecture Notes Math. 704 Springer, 330-356 (1979).

H. BREZIS

[1] Opérateurs maximaux monotones et semi groupes de contractions dans les espaces Hilbert, North-Holland (1973).

[2] New results ... monotone operators and nonlinear semigroups. Kyoto University, November (1975).

[3] Analyse fonctionnelle. Théorie et applications. Coll. Math. appliquées Masson (1983).

H. BREZIS, L.A. CAFFARELLI, A. FRIEDMAN

[1] Reinforcement problems for elliptic equations and variational inequalities; Ann. Mat. Pura Appl.

A. BRILLARD

[1] Quelques questions de convergence... en calcul des variations: Thèse Orsay, Paris Sud (February 1983).

[2] Ecoulements lents de fluides incompressibles dans des milieux poreux Publications AVAMAC (Perpignan, France) (1984).

R. BRIZZI, J.P. CHALOT

[1] Homogénéisation de frontière. Thèse Université de Nice (1978).

F. BROWDER

[1] Nonlinear operators and nonlinear equations of evolution; Proc. Symp. Pure Math. 18 (2). Amer. Math. Soc. (1976).

[2] Topological degree for pseudo-monotone operators. Conférence Collège de France. Séminaire H. Brezis, J.-L. Lions (1983).

F. BROWDER, P. HESS

[1] Nonlinear mappings of monotone type in Banach spaces; J. Funct. Analysis, 11 251-294 (1974).

G. BUTTAZZO

[1] Su una definizione generale dei Γ-limiti; Boll. Un. Mat. Ital. (5) 14-B, 722-744 (1977).

G. BUTTAZZO, G. DAL MASO

[1] Γ-limit of a sequence of nonconvex and non equilipschitz integral functionals; Ricerche Mat., 27, 235-251 (1978).

[2] Γ-limit of integral functionals; J. Analyse Math., 37, 145-185 (1980).

[3] Integral representation on $W^{1,\alpha}(\Omega)$ and $BV(\Omega)$ of limits of variational integrals; Atti. Acad. Naz. Lincei, Cl. Sci. Mat. Fis. Nat. (8) 66, 338-344 (1979).

[4] Γ-convergence and optimal control problems. J.Optim.Theory Appl. 38, 385-407 (1982).

[5] Γ-convergence et perturbation singulière; CRAS Paris, 296, 649-651 (1983).

[6] On Nemyckii operators and integral representation of local functionals to appear in Rend. Mat. (preprint Scuola Normale Superiore, Pisa, (1982).

G. BUTTAZZO, D. PERCIVALE

[1] On the approximation of the elastic bounce problem on Riemannian manifolds; Scuola Normale Pisa, (1981); to appear in J. Differential equations.

G. BUTTAZZO, M. TOSQUES

[1] Γ-convergenza per aleune classi di funzionali; Ann. Univ. Ferrara, Sez. VII, 23, 257-267 (1977).

V.F. BUTUZOV

[1] The asymptotics ... in chemical kinetics, taking diffusion into account; Soviet. Math. Dokl. 19, 1979-1083 (1978).

L.A. CAFFARELLI, A. FRIEDMAN

[1] Asymptotics estimates for the dam problem with several layers; Indiana Univ. Math. J., 27, 551-580 (1978).

[2] Reinforcement problems in elasto-plasticity; Rocky Mountain J. Math. 10, 155-184 (1980).

D. CAILLERIE

[1] Comportement limite d'une inclusion mince de grande rigidité dans un corps élastique CRAS Paris, série A, Vol. 287, 675-678 (1978).

[2] Equations aux dérivées partielles à coeff. périod. dans des domaines cylindriques aplatis. CRAS Paris, série A, Vol. 290, 143-146 (1980).

[3] The effect of a thin inclusion of high rigidity in an elastic body; Math. Meth in the Appl. Sci. 2, 251-270 (1980).

[4] Homogénéisation des equations de la diffusion stationnaire dans des domaines cylindriques aplatis. R.A.I.R.O. Analyse numérique Vol. 15 (4), 295-319 (1981).

[5] Equations de diffusion ... inclusions aplaties de grande conductivité, CRAS Paris, série I, 292, 115-118 (1981).

[6] Homogénéisation d'un corps élastique renforcé par des fibres de grande rigidité..., CRAS Paris, série II, 292, 477-480 (1981).

[7] Etude de quelques problèmes de perturbation en théorie de l'élasticité et conduction thermique. Thèse, Paris VI (1982).

[8] Etude de la conductivité stationnaire dans un domaine comportant une répartition périodique d'inclusions minces de grande conductivite RAIRO Analyse Numérique Vol. 17, No. 2, 137-159 (1983).

L. CARBONE

[1] Sur la Γ-convergence des intégrales du type de l'énergie à gradient borné; J. Math. Pures Appl. (9), 56, 79-84 (1977).

[2] Γ-convergence d'integrales sur des fonctions avec des contraintes sur le gradient; Comm. Part. Diff. Equat., 2, 627-651 (1977).

[3] Sull'omogeneizzazione ... con vincoli sul gradiente; AHi. Acad. Naz. Lincei, Rend. Cl. Sci. Mat. Fis. Natur. (8) 63, 10-14 (1978).

[4] Sur un probleme d'homogénéisation avec des contraintes sur le gradient; J. Math. Pures. Appl., 58, 275-297 (1979).

L. CARBONE, F. COLOMBINI

[1] On convergence of functionals with unilateral constraints; J. Math. Pures Appl., 59, 465-500 (1980).

L. CARBONE, S. SALERNO

[1] On a problem of homogenization with quickly oscillating constraints on the gradient; J. Math. Anal. and Appl., 90, 219-250 (1982).

[2] Further results on a problem of homogenization with constraints on the gradient; to appear in J. d'Analyse Math.

L. CARBONE, C. SBORDONE

[1] Un teorema di compattezza per la Γ-convergenza di funzionali non coercitivi; Atti. Accad. Naz. Lincei, Rend. Sci. Mat. Fis. Nat. (8) 62, 744-748 (1977).

[2] Some properties of Γ-limits of integral functionals; Ann. Mat. Pura Appl (4) 122, 1-60 (1979).

M. CARRIERO, E. PASCALI

[1] Γ-convergenza di integrali non negativi maggiorati da funzionali del tipo dellavea; Ann. Univ. Ferrara., 24, 51-64 (1978).

[2] Il problema del rimbalzo unidimensionale... approssimazioni con penalizzazioni; Rendiconti di Matematica (4) 13, 541-554 (1980).

[3] Uniqueness of the one-dimensional bounce problem as a generic property in $L^1(0,\Gamma;\mathbb{R})$; Boll. Un. Mat. Ital. (6) 1-A, 87-91 (1982).

C. CASTAING, M. VALADIER

[1] Convex Analysis and Measurable Multifunctions; Lect. Notes in Math., Springer No. 580, (1977).

E. CAVAZZUTI

[1] Γ-convergenza multipla, convergenza di punti di sella e di max-min. Boll. Un. Mat. Ital (6) 1-B, 251-274 (1982).

J. CEA

[1] Approximation variationnelle des problèmes aux limites. Ann. Inst. Fourier, 14, 345-444 (1964).

G. CHOQUET

[1] Convergences; Ann. Univ. de Grenoble, 23, 55-112 (1947-48).

P. CIARLET, S. KESAVAN

[1] Les équations des plaques élastiques comme limite des équations de l'élasticité tridimensionnelle. Problèmes aux valeurs propres. Proc. Fourth Int. Symp. Compt. Methods..., Versailles, France (1979).

G. CICOGNA

[1] Un 'analisi categoriale delle nozioni di convergenza; Boll. Un. Mat. Ital. (5) 17-A, 531-536 (1980).

D. CIORANESCU

[1] Calcul des variations sur des sous-espaces variables. CRAS Paris, 291, 19-22, 87-90 (1980).

D. CIORANESCU, L. HEDBERG, F. MURAT

[1] Le bilaplacien et le tapis du fakir. (to appear).

D. CIORANESU, F. MURAT

[1] Un terme étrange venu d'ailleurs, I & II. Nonlinear partial differential equations and their applications, Collège de France Séminar, Vol. II & III, ed. H. Brezis & J.L. Lions. Research Notes in Math. Nos. 60 & 70, Pitman, 98-138, 154-178 (1982).

D. CIORANESCU, J. SAINT JEAN PAULIN

[1] Homogènéisation dans des ouverts á cavités; CRAS Paris, 284 857-860, (1977).

[2] Homogénéisation de problèmes d'évolution dans des ouverts à cavités; CRAS Paris (286), 889-902 (1978).

[3] Homogénéisation in open sets with holes; J. Math. Pures Appl. 71, 590-607 (1979).

M. CODEGONE

[1] Problèmes d'homogénéisation en théorie de la diffraction; CRAS Paris, Serie A, 288, 387-389 (1979).

[2] Scattering of elastic waves through a heterogenous medium; Math. Met. in Appl. Sci., 2, 271-287 (1980).

[3] On the acoustic impedence condition for ondulated boundary; Ann. Inst. H. Poincaré Sect A, 36, 1, 1-18 (1982).

M. CODEGONE, A. NEGRO

[1] Homogenization of the nonlinear quasistationary Maxwell equations with degenerated coefficients (1983). To appear in Applicable Analysis.

M. CODEGONE, J.R. RODRIGUES

[1] On the homogenization of the rectangular dam problem; Rendiconti Sem. Mat. Univ. Pol. Torino 39 (2), 125-136 (1981).

[2] Convergence of the coincidence set in the homogenization of the obstacle problem; Ann. Fac. Sc. Toulouse, 3, 275-285 (1981).

J.D. COLE

[1] Perturbation methods in applied mathematics. Blaisdell, Toronto (1968).

F. COLOMBINI, S. SPAGNOLO

[1] Sur la convergence des solutions d'équations paraboliques. J. Math. Pures Appl., 56, 263-306 (1977).

[2] On the convergence of solutions of hyperbolic equations; Comm. Partial Diff. Equations, 3, 77-103 (1978).

C. CONCA

[1] On the application of the homogenization theory to a class of problems arising in fluid mechanics. Publications Lab. d'Analyse Numérique Paris VI (1983).

R. COURANT, H. HILBERT

[1] Methods of Mathematical Physics I and II. Interscience, New York (1953).

N. CRESSIE

[1] A strong limit theorem for random sets. Advances in Applied Probability, 10, 36-46 (1978).

[2] Random set theory and modelling. Statistics/Probability Research Report. The Flinders University of South Australia (1983).

G. DAL MASO

[1] Aleuni teoremi sui Γ-limiti di misure; Boll. Un. Mat. Ital (5) 15-B, 182-192 (1978).

[2] Integral representation on $BV(\Omega)$ of Γ-limits of variational integrals; Manuscripta Math. 30, 387-416 (1980).

[3] Asymptotic behaviour of minimum problems with bilaterial obstacles. Ann. Mat. Pura Appl. (4) 129, 327-366 (1981).

[4] Limiti di problemi di minimo con ostacoli. Atti del Convegno "Studio di problemi-limite della omalisi funzionale", Biessanone sett 81, Pitagora editrice, Bologna, 79-100 (1982).

[5] On the integral representation of certain local functionals. Universita Udine. To appear in Ricerche Mat. (1983).

[6] Limits of minimum problems for general integral functionals with unilateral obstacles. Atti Accad. Naz. Lincei, Rend Cl. Sci. Fis. Mat. Nat. (8) 74 (1983) (Preprint University of Udine (1983)).

G. DAL MASO, E. DE GIORGI

[1] Γ-convergence and calculus of variations. "Mathematical Theories of Optimization", Proceedings S. Margherita Ligure 1981, Lecture Notes in Mathematics 979, Springer, Berlin, 121-143 (1983).

G. DAL MASO, P. LONGO

[1] Γ-limits of obstacles; Ann. Mat. Pura Appl. (4) 128, 1-50 (1980).

G. DAL MASO, L. MODICA

[1] Un criterio generale per la convergenza dei minimi locali; Rendiconti di Matematica (7) 1, 81-94 (1981).

[2] Sulla convergenza dei minimi localli; Boll. Un. Mat. Ital. (6) 1-A, 55-61 (1982).

[3] A general theory of variational functionals. "Topics in functional analysis 1980-81", by F. Strocchi, E. Zarantonello, E. De Giorgi, G. Dal Maso, L. Modica, Scuola Normale Superiore, Pisa, 149-221 (1981).

[4] Nonlinear Stochastic Homogenization. Universita degli Studi di Pisa Dipartimento di Matematica. No. 49, February (1984).

A. DAMLAMIAN

[1] Homogènéisation du problème de Stefan; C.R.A.S. Paris, Vol. 289, 9-11 (1979).

[2] How to homogenize a non linear diffusion equation: Stefan's problem. SIAM J. Math. Anal., 12 (3) 306-313 (1981).

[3] Convergence et dualité des fonctions convexes dans les espaces de Banach reflexifs (to appear).

[4] Homogenization for Eddy currents 6 (1981) Delft progress report.

E. DE GIORGI

[1] Sulla convergenza di alcune successioni di integrali del tipo dell'area; Rendiconti di Matematica (4) 8, 277-294 (1975).

[2] Γ-convergenza e G-convergenza; Boll. Un. Mat. Ital. (5) 14-A, 213-220 (1977).

[3] Convergence problems for functionals and operators; Proc. Int. Meeting on "Recent Methods in Nonlinear Analysis", Rene 1978, ed. E. De Giorgi, Magenes, Mosco Pitagora, Bologna, 131-188 (1979).

[4] Γ-limiti di ostacoli; Atti S.A.F.A.N, Napoli, 1980, ed. Confora, Rionero, Sbordone, Trombetti, Liguori, Napoli, 51-84 (1981).

[5] New problems in Γ-convergence and G-convergence; Proc. Meeting on "Free Boundary problems", Pavia 1979, Istituto Nazionale di Atta Matematica, Roma, Vol. II 183-194 (1980).

[6] Quelques problèmes de Γ-convergence; Proc. Meeting on "Computing Methods in Applied Sciences and Engineering", Versailles, 979, ed. Glowinski, J.-L. Lions, North-Holland, 637-643 (1980).

[7] Operatori elementari di limite ed applicazioni al Calcolo delle Variazioni; Atti. del Convegno su "Studio di problem-limite in Analisi funzionale" Bressanone, 1981, ed. Putuzzo e. Steffe C.L.E.U.P., Padova (1982).

[8] Atti. S.A.F.A. V, Catania (to appear) (1981).

E. DE GIORGI, G. DAL MASO, P. LONGO

[1] Γ-limiti di ostacoli; Atti Accad. Naz. Lincei, Rend. Cl. Sc. Mat. (8) 68, 481-487 (1980).

E. DE GIORGI, T. FRANZONI

[1] Su un tipo di convergenza variazionale; Atti. Accad. Naz. Lincei, Rend. Cl. Sc. Mat. (8) 58, 842-850 (1975).

[2] Su un tipo di convergenza variazionale; Rend. Sem. Mat. Brescia, 3, 63-101 (1979).

E. DE GIORGI, G. LETTA

[1] Une notion générale de convergence faible pour des fonctions croissante d'ensemble; Ann. Sc. Norm. Sup. Pisa Cl. Sci (4) 4, 61-99 (1977).

E. DE GIORGI, A. MARINO, M. TOSQUES

[1] Problemi di evoluzione in spazi metrici e curve di massima pendenza; Atti Accad. Naz. Lincei, Rend Cl. Sci. Fis. Mat. (8), 68, 180-187 (1980).

E. DE GIORGI, L. MODICA

[1] Γ-convergenza e superfici minime; Pubblicazione interna Scuola Normale Pisa (1979).

E. DE GIORGI, S. SPAGNOLO

[1] Sulla convergenza degli integrali dell'energia per operatori ellitici del II ordine; Boll. Un. Mat. Ital. (4) 8, 391-411 (1973).

I. DEL PRETE, M. LIGNOLA

[1] On the variational properties of $\Gamma^-(d)$ convergence; Ricerche di Matematica (to appear).

Z. DENKOWSKI

[1] The convergence of generalized sequences of sets and functions in locally convex spaces; I. Zeszyty Naukowe UJ, 22, 37-58 (1980).

J. DENY

[1] 6 Exposés C.I.M.E. Stresa (1969).

A. DERIEUX

[1] A perturbation study of the obstacle problem by means of a generalized implicit function theorem; Annali di Mat. pura Appl. IV, 127, 321-364 (1981).

M. DOLCHER

[1] Topologie a strutture di convergenza, Ann. Sc. Norm. Sup. Pisa Sci (3) 14, 63-92 (1960).

S. DOLECKI

[1] Tangency and differentiation: some applications of convergence theory; Annali Mat. Pura ed. Appl., 30, 223-255 (1982).

[2] Convergence of global minima and infima. Universita degli Studi di Trento (1983).

S. DOLECKI, G. SALINETTI, R. WETS

[1] Convergence of functions: Equi-semicontinuity; Trans. Amer. Math. Soc., 276, 409-429 (1983).

P. DONATO

[1] Una stima per la differenza di H-limiti ... Univ. Salerno, preprint.

N. DUNFORD, J.T. SCHWARZ

[1] Linear Operators. Interscience, New York (1958).

G. DUVAUT

[1] Etude de matériaux composites élastiques à structure périodique. Homogénéisation. Proc. Congress of theoretical and applied mathematics. Delft (1976). Ed. Koiter, North-Holland.

[2] Comportement macroscopique d'une plaque perforée périodiquement in "Singular perturbations and boundary layer theory", Lecture Notes in Mathematics No. 594, Springer, 131-145 (1977).

[3] Homogénéisation des plaques ... J. Analyse non linéaire Besancon. Lecture Notes in Math. Springer No. 665, 56-69 (1977).

G. DUVAUT, J.-L. LIONS

[1] Les inéquations en mécanique et en physique. Dunod, Paris (1972).

G. DUVAUT, A.M. METELLUS

[1] Homogénéisation d'une plaque mince. CRAS Paris, 283, 947-950 (1976).

I. EKELAND, R. TEMAM

[1] Convex analysis and variational problems; North-Holland (1978).

H.W. ENGL, R. KRESS

[1] A singular perturbation problem for linear operators with an application to electrostatic and magnetostatic boundary and transmission problems. Math. Meth. in the Appl. Sci. 3, 249-274 (1981).

H.W. ENGL, W. ROMISCH

[1] Convergence of approximate solutions of nonlinear random operator equations with non-unique solutions. Institutsbericht No. 230 J. Kepler Universitat Linz. Institut für Mathematik, February (1983).

[2] Approximate solutions of nonlinear random operator equations: convergence in distribution. Institutsbericht No. 249. J. Kepler Universitat Linz. Institut für Mathematik, October (1983).

I.C. EVANS

[1] A convergence theorem for a chemical diffusion-reaction system (Preprint). University of Lexington-Kentucky.

F. FLEURY, G. PASA, D. POLYSENSHI

[1] Homogénéisation de corps composites; CRAS Paris, Serie B, 289, 241-244 (1979).

C. FOIAS, L. TARTAR

[1] Opérateurs de Tocplitz généralisés et homogènéisation; CRAS Paris 291, 15-18 (1980).

A. FOUGERES

[1] Cours sur l'epi-convergence. Perpignan (1981-82).

A. FOUGERES, J.C. PERALBA

[1] Application au calcul des variations de l'optimisation intégrale convexe; Journées d'Analyse non linéaire Besancon. Lecture Notes in Math.

Springer 665 (1977).
S. FRANCAVIGLIA
[1] Generalized Kuratowski limits for multifunctions (to appear).
N. FUSCO
[1] Dualita et semicontinuita per integrali del tipo dell'area; Rend. Acad. Sci. Fis. Mat. Napoli (4) 46, 81-90 (1979).
[2] Γ-convergenza unidimensionale; Boll. Un. Mat. Ital. (5) 16-B, 74-86 (1979).
E. GINER
[1] Epi-limites de fonctionnelles intégrales (to appear).
E. GIUSTI
[1] Minimal surfaces and functions of bounded variation; Notes on Pure Mathematics Vol. 10, Australian Nat. Univ. Canberra (1977).
C. GOFFMAN, J. SERRIN
[1] Sublinear functions of measures and variational integrals; Duke Math. J. 31 159-178 (1964).
R.B. GONZALEZ DE LA PAZ
[1] Sur un problème d'optimisation de domaine. Séminaire d'Analyse convexe Montpellier. France, Vol. 10, Exposes 4, 11 (1980).
G.H. GRECO
[1] Limites et fonctions d'ensembles. Rend. Sem. Mat. Padova 72 (1984).
[2] Decomposizioni di semifiltri e Γ-limiti sequenziali... Univ. Trento, preprint (1983).
A. HARAUX, F. MURAT
[1] Perturbations singulières et problèmes de contrôle optimal: deux cas bien posés, CRAS Paris, 297, 21-24 (1983).
[2] Perturbations singulières et problèmes de contrôle optimal: un cas mal posé; CRAS Paris 297, 93-96 (1983).
[3] Influence of a singular perturbation on the infimum of some functionals (to appear).
Z. HASHIN, S. SHTRIKMAN
[1] A variational approach to the theory of the elastic behaviour of multi-phase materials; J. Mech. Phys. Solids 11, 127-140 (1963).
C. HESS
[1] Theorème ergodique et loi forte des grands nombres pour des ensembles aléatoires; CRAS. Paris, 288, 519-522 (1979).

E. Ya. HRUSLOV

[1] The method of orthogonal projections and the Dirichlet problem in domains with a fine grained boundary; Math USSR Sb., 17, 37-59 (1972).

[2] The asymptotic behaviour... fragmentation of the boundary; Math. USSR Sb. 35, 266-282 (1979).

D. HUET

[1] Approximation d'un espace de Banach et perturbations singulières; CRAS Paris, 289, 69-70 (1979).

J.L. JOLY

[1] Une famille de topologies sur l'ensemble des fonctions convexes pour lesquelles la polarité est bicontinue; J. Math. Pures Appl., 52, 421-441 (1973).

J.L. JOLY, T. DE THELIN

[1] Convergence of convex integrals in L^p space; J. Math. Anal. Appl. 54, 230-244 (1976).

T. KATO

[1] Perturbation theory for linear operators. Springer, Berlin (1966).

S. KESAVAN

[1] Homogénéisation et valeurs propres; CRAS Paris, 285, 229-232 (1977).

[2] Homogénéisation d'un problème de contrôle optimal; CRAS Paris, 285, 441-444 (1977).

[3] Homogénéisation of elliptic eigenvalues problem; Appl. Math. Optim, 5 163-167; 197-216 (1979).

KHA T'EN NGOAN

[1] On the convergence of solutions of boundary value problems for sequences of elliptic equations; Uspekhi Mat. Nauk., 32, 183-184 (1977).

[2] On the convergence ... sequences of elliptic systems; Moskow Univ. Math. Bull, 32, 66-74 (1977).

KHA T'EN NGOAN, S.M. KOZLOV, O.A. OLEINIK, V.V. ZHIKOV

[1] Averaging and G-convergence of differential operators; Math surveys, 34, 69-147 (1979).

S.M. KOZLOV

[1] The averaging of differential operators with almost periodic rapidly oscillating coefficients; Math. USSR Sb., 35, 481-498 (1978).

[2] Averaging random structures, Sov. Math. Dokl., 19, 950-954 (1978).

[3] Averaging of random operators; Math. USSR Sb., 37, 167-180 (1980).

E. KRAUSS

[1] On convergence properties of families of maximal monotone operators; Nonlinear Analysis, Theory and Applications; Proc. Seventh Int. Summer School, Berlin (1979).

[2] A degree for operators of monotone type (to appear in Math. Nachr.).

K. KURATOWSKI

[1] Topology; Academic Press, New York (1966).

H. LANCHON

[1] Torsion élastoplastique d'une barre ...; Journal de mécanique, 12, 151-171 (1973).

F. LENE

[1] Comportement macroscopique de matériaux élastiques comportant des inclusions rigides ou des trous répartis périodiquement; CRAS Paris, 286, 75-78 (1978).

Th. LEVY

[1] Equations et conditions d'interface pour des phénomènes acoustiques dans des milieux poreux; in Singular perturbations and boundary layer theory, Lecture Notes in Math., Vol. 594, Springer, Berlin, 301-311 (1977).

[2] Acoustic phenomena in elastic porous media; Mech. Res. Comm. 4, 253-257 (1977).

Th. LEVY, E. SANCHEZ PALENCIA

[1] Equations and interface conditions for acoustic phenomena in porous media. Jour. Math. Anal. Appl., 61,813-834 (1977).

J.-L. LIONS

[1] Cours Collège de France. Homogénéisation. Perturbations singulières en contrôle (1975-1983).

[2] Perturbations singulières dans les problèmes aux limites et en contrôle optimal; Lecture Notes in mathematics, Springer, 323 (1973).

[3] Asymptotic behaviour of variational inequalities with highly oscillating coefficients; Lecture Notes in Mathematics, Springer 503, 30-55 (1976)

[4] Homogénéisation non locale; Proc. Int. Meeting on "Recent Methods in Nonlinear Analysis", Rome 1978, ed. E. De Giorgi, E. Magenes, U. Mosco, Pitagora ed., Bologna 189-204 (1979).

[5] Some methods in the Mathematical Analysis of systems and their control, Science Press, Beijing, China; Gordon and Breach Inc., New York (1981).

P.L. LIONS

[1] Two remarks on convergence of convex functions...; Nonlinear Analysis, 2, 553-562 (1978).

M. LOBO HIDALGO, E. SANCHEZ-PALENCIA

[1] Perturbation of spectral properties for a class of stiff problems; Proc. Meeting on "Computing methods in Applied Sciences and Engineering" Versailles, 1979, North-Holland, 683-697 (1980).

R. LUCCHETI, F. PATRONE

[1] Metodo epsilon e convergenza di Mosco; Rend. Sem. Mat. Univ. Padova 57, 1-15 (1977).

R. LUCCHETI, F. MIGNANEGO

[1] Variational perturbations of the minimum effort problem; Publ. Math. Genova, sec. series, 260, 0-13 (1978).

W.H. McCONNELL

[1] On the approximation of elliptic operators with discontinuous coefficients; Ann. Sc. Norm. Sup. Pisa, Cl. Sci. 3, 139-156 (1976).

L. MacLINDEN

[1] An application of Ekeland's theorem to minimax problems; Nonlinear Analysis, T.M.A, 6 (2), 189-196 (1982).

P. MARCELLINI

[1] Su una convergenza di funzioni convesse; Boll. Un. Mat. Ital. (4), 8, 137-158 (1973).

[2] Un teorema di passagio al limite per la somma di funzioni convesse; Boll. Un. Mat. Ital. (4) 11, 107-124 (1975).

[3] Periodic solutions and homogenization of nonlinear variational problems; Ann. Mat. Pura. Appl. (4), 117, 139-152 (1978).

[4] Some problems of semicontinuity; Proc. Int. Meeting on "Recent Methods in Nonlinear Analysis", Rome 1978, ed. E. De Giorgi, E. Magenes, U. Mosco, Pitagora ed., Bologna, 205-222 (1979).

[5] Convergence of second order linear elliptic operators; Boll. Un. Mat. Ital. (5) 16-B, 278-290 (1979).

P. MARCELLINI, C. SBORDONE

[1] Relaxation of non convex variational problems, Atti. Accad. Naz. Lincei, Rend. Cl. Sci. Fis. Mat. Nat. (8) 63, 341-344 (1977).

[2] Sur quelques questions de G-convergence et d'homogénéisation non linéaire; CRAS. Paris, 284, 535-537 (1977).

[3] An approach to the asymptotic behaviour of elliptic-parabolic operators, J. Math. Pures Appl. (9), 56, 157-182 (1977).

[4] Dualita e perturbazioni di funzionali integrali; Ricerche Mat., 26, 383-421 (1977).

[5] Homogenization of non uniformly elliptic operators; Applicable analysis, 68, 101-114 (1978).

[6] Semicontinuity problems in the calculus of variations; Nonlinear Anal. 4, 241-257 (1980).

A.V. MARCHENKO, E. Ya. HRUSLOV

[1] Boundary value problems in domains with close-grained boundaries (Russian). Naukova Dumka, Kiev (1974).

[2] New results in the theory of boundary value problems for regions with close-grained boundaries, Uspekhi Mat. Nauk., 33, 127 (1978).

S. MARCHI

[1] Stime di omogeneizzazione per equazioni ellittiche; SAFA IV, Napoli, Liguori ed., Napoli, 357-361 (1981).

A. MARINO, S. SPAGNOLO

[1] Un tipo di approssimazione dell' operatore $\Sigma D_i(a_{ij}D_j)$ con operatori $\Sigma D_i(\beta D_i)$; Ann. Sc. Norm. Sup. Pisa, Cl. Sci. (3) 23, 657-673 (1969).

A. Marino, M. Tosques

[1] Curves of maximal slope for a certain class of non regular functions; Boll. Un. Mat. Ital. (to appear).

[2] Existence and properties of the curves of maximal slope (to appear).

V.G. MARKOV, O.A. OLEINIK

[1] On propagation of heat in one-dim dispense media; J. Appl. Math. Mech. 39, 1028-1037 (1975).

M. MATZEU

[1] Su un tipo di continuita dell' operatore subdifferenziale; Boll Un. Mat. Ital. (5), 14-B, 480-490 (1977).

N.G. MEYERS, J. SERRIN

[1] H=W; Proc. Nat. Acad. Sci. USA., 51, 1055-1056 (1964).

E. MICHAEL

[1] Topologies on spaces of subsets; Trans. Amer. Math. Soc. 71, 151-182 (1951).

F. MIGNOT, J-P. PUEL

[1] Homogénéisation d'un problème de bifurcation; Proc. Rome 1978, "Recent

Methods in Nonlinear analysis", Pitagora ed., Bologna, 281-310 (1979).

F. MIGNOT, J.P. PUEL, P.M. SUQUET

[1] Flambage de plaques élastiques perforées: bifurcation et homogénéisation; Ann. F. sc. Toulouse, 3, 1-58 (1981).

[2] Homogenization and bifurcation; Int. J. Engineering Sci., 18, 409-414 (1980).

A. MILANI

[1] On a singular perturbation problem for Maxwell equations; Rend. Sem. Mat. Univ. Pol. Torino, 38 (1980).

M. MIRANDA

[1] Distribuzioni aventi derivate misure. Insemio di perimetro Pinito; Ann. Sc. Normale Sup. Pisa, Cl. Sci. Fis. Mat. (3), 18, 27-56 (1964).

[2] Un teorema di esistenza ed unicita per il problema dell' area in n variabili; Ann. Sc. Norm. Sup. Pisa, Cl. Sci. Fis. Mat. (3), 19, 233-249 (1965).

[3] Sul minimo dell' integrale del gradiente di una funzione; Ann. Sc. Norm. Sup. Pisa Cl. Sci. Fis. Mat. (3) 19, 627-665 (1965).

[4] Comportamento dell successioni convergenti di frontiere minimali; Rend. Sem. Mat. Univ. Padova, 38, 238-257 (1967).

[5] Dirichlet problem with L^1 data for the non-homogeneous minimal surfaces equation; Indiana Univ. Math. J., 24, 227-241 (1974).

L. MODICA

[1] Γ-convergence to minimal surfaces problem and global solutions to $\Delta u = 2(u^3-u)$; Proc. Int. Meeting on "Recent Methods in Nonlinear Analysis" Rome,1978, Pitagora ed. Bologna, 223-244 (1979).

[2] Omogeneizzazione con coefficients casuali; Atti del Convegno su "Studio di problemi-limite in Analisi Funzionale", Bressanone, 1981, ed. da P. Patuzzo, S. Stelle C.L.E.U.P., Padova (1982).

L. MODICA, S. MORTOLA

[1] Un esempio di Γ^- convergenza; Boll Un. Mat. Ital. (5), 14-B, 285-299 (1977).

[2] Il limite nella Γ-convergenza di una famiglia di funzionali ellitici; Boll. Un. Mat. Ital. (5) 14-A, 526-529 (1977).

J.J. MOREAU

[1] Proximité et dualité dans un espace hilbertien; Bull. Soc. Mat. France, 93, 273-299 (1965).

[2] Fonctionnelles convexes, Séminaire Collège de France (1966-67).

[3] Inf-convolution ... convexité des fonctions numériques; J. Math. Pures et Appl., 49, 109-154 (1970).

[4] On unilateral constraints, friction and plasticity C.I.M.E. New Variational technics in Math. Physics, Bressanone. Ed. Cremonese, Rome (1974).

C.B. MORREY

[1] Multiple integrals in calculus of variations; Springer Verlag, Berlin (1966).

S. MORTOLA, A. PROFETI

[1] On the convergence of minimum points for non equicoercive functionals; Comm. Partial Diff. Equations (to appear).

G. MOSCARIELLO

[1] Su una convergenza di successioni di integrali del Calcolodelle variazioni; Atti. Accad. Naz. Lincei, Rend. Cl. Sci. Fis. Mat. Nat. (8) 61, 368-375 (1976).

[2] Γ-convergenza negli spazi sequenziali; Rend. Accad. Sci. Fis. Mat. Napoli (4) 43, 333-350 (1976).

U. MOSCO

[1] Approximation of the solutions of some variational inequalities, Ann. Scuola Normale Sup. Pisa, 21, 373-394 (1967).

[2] Convergence of convex sets and of solutions of variational inequalities; Advances in Math., 3, 510-585 (1969).

[3] On the continuity of the Young-Fenchel transformation. J. Math. Anal. Appl. 35, 518-535 (1971).

[4] An introduction to the approximate solution of variational inequalities. C.I.M.E. II Circlo, Erice 1971, Ed. Cremonese, Rome (1973).

F. MURAT

[1] Contre-exemple pour divers problèmes où le contrôle intervient dans les coefficients; Ann. Mat. Pura Appl. (4) 112, 49-68 (1977).

[2] Compacité par compensation; Ann. Sc. Norm. Sup. Pisa, Cl. Sci (4) 5, 489-507 (1978).

[3] Compacité par compensation II; Proc. Int. Meeting on "Recent Methods in nonlinear analysis", Rome 1978, ed. E. De Giorgi, E. Magenes, U. Mosco, Pitagora ed. Bologna, 245-256 (1979).

[4] Compacité par compensation: condition nécessaire et suffisante de continuité faible sous une hypothèse de rang constant; Ann Sc. Norm. Sup. Pisa. Cl. Sci. (4) 8, 69-102 (1981).

[5] Sur l'homogénéisation d'inéquations elliptiques du 2ème ordre, relatives au convexe $K(\psi_1,\psi_2) = \{v \in H_0^1(\Omega);\ \psi_1 < v < \psi_2$ p.p. dans $\Omega\}$; Publications Université Paris VI, Thèse d'état (1976).

[6] L'injection du cone positif de H^{-1} dans $W^{-1,q}$ est compacte pour $q < 2$. J. Math. Pures Appl. 60, 309-322. (1981).

[7] H-convergence. Rapport du séminaire d'analyse fonctionnelle et numérique de l'Université d'Alger (1978).

[8] Control in coefficients. Encyclopedia of systems and control. Pergamon Press (1983).

F. MURAT, J. SIMON

[1] Sur le contrôle par un domaine géometrique. Publication No. 76015 du Laboratoire d'Analyse Numérique du l'Université Paris VI, (1976).

A. NEGRO

[1] Remarks on stability and rate of convergence for approximation of linear semigroups in Banach space. Università di Torino (to appear).

T. NORANDO

[1] Stime di omogeneizzazione per disequazioni quasi-variazionali; Proc. "Recent methods in nonlinear analysis and applications" S.A.F.A. IV Napoli, Liguori ed. Napoli, 367-370 (1981).

O.A. OLEINIK

[1] On the convergence of solutions of elliptic and parabolic equations under weak convergence of coefficients; Uspekhi Mat. Nauk., 30, 257-258 (1975).

G.C. PAPANICOLAOU

[1] Introduction to the asymptotic analysis of stochastic equations; Mod. Model. Cont. Phenom., Lect. Appl. Math., 16, 109-147 (1977).

[2] Heat conduction in a one-dimensional random medium; Comm. Pure Appl. Math., 31, 583-592 (1978).

[3] A limit theorem for turbulent diffusion; Comm. Math. Phys., 65, 97-128 (1979).

G.C. PAPANICOLAOU, S.R.S. VARADHAN

[1] Boundary value problems with rapidly oscillating random coefficients; Coll. Math. Soc. Janos Bolyai (1979).

[2] Diffusion in regions with many small holes. Springer Lecture Notes in Information and control, No. 25, B. Griglionis ed.190-206 (1980).

P. PATUZZO GREGO

[1] Sulla G-convergenza delle equazioni differenziale ordinarie nel caso di domini illimitati; Boll. Un. Mat. Ital (5) 16-B, 466-479 (1979).

R. PEIRONE

[1] Γ-convergenza per funzionali a valori vettoriali; Tesi di laurea, Genova (1981).

J.P. PENOT

[1] Calcul sous differentiel et optimisation; J. Funct. Anal., 27, 248-276 (1978).

[2] Compact filters, nets and relations; in J. Math. Annal. Appl. (to appear).

H. PHAM HUY, E. SANCHEZ-PALENCIA

[1] Phénomènes de transmission à travers des couches minces de conductivité élevée; J. Math. Anal. Appl., 47, 284-309 (1974).

C. PICARD

[1] Comportement limite de suites d'inéquations variationnelles avec obstacles. Thèse Orsay, Université Paris Sud (1984).

C. PICARD, QUENTIN de GROMARD

[1] Tapis du fakir pour les surfaces minima (to appear).

L.C. PICCININI

[1] G-convergence for ordinary differential equations with Peano phenomenon; Rend. Sem. Mat. Univ. Padova, 58, 65-86 (1977).

[2] Homogenization for ordinary differential equations; Rend. Cir. Mat. Palermo (2) 27, 95-112 (1978).

[3] G-convergence for ordinary differential equations; Proc. Int. Meeting on "Recent methods in non linear analysis", Rome 1978, Pitagora ed., Bologna, 257-280 (1979).

[4] Ergodic properties in the theory of homogenization; Atti del convegno su "Studio di problemi-limite in Analisi Funzionale", Bressanone, 1981, ed. da P. Patuzzo et S. Steffe, C.L.E.U.P, Padova (1982).

G. PIERI

[1] Variational perturbations of the linear-quadratic problem; J. Optimization theory Appl., 22, 63-77 (1977).

M. PIERRE

[1] "Equations d'évolution non linéaires, inéquations variationnelles et et potentiels paraboliques Thèse d'état, Paris VI (1979).

O. PIRONNEAU, C. SAGUEZ

[1] Asymptotic behaviour with respect to the domain of solutions of P.D.E. Rapport No. 218, I.R.I.A. (1977).

G. PORRU

[1] Su un teorema di compattezza rispetto alla G-convergenza per operatori ellittici; Rend. Sem. Fac. Sci. Univ. Cagliari, 48, 67-80, (1978).

V.I. PRJAZINSKII, V.G. SUSKO

[1] Small parameter asymptotics ... for a quasi-linear parabolic equation; Soviet Math. Dokl, 20, 698-700 (1979).

A. PROFETI, B. TERRENI

[1] Su alcuni metodi per lo studio della convergenza di equazioni paraboliche; Boll. Un. Mat. Ital. (5) 17-B, 675-691 (1980).

[2] Uniformita per una convergenza di operatori parabolici nel caso dell' omogeneizzazione; Boll. Un. Mat. Ital (5) 16-B, 826-841 (1979).

J. RAUCH, M. TAYLOR

[1] Potential and scattering theory on widely perturbed domains; J. Funct. Analysis, 18, 27-59 (1975).

[2] Electrostatic screening. J. Math. Phys., 16, 284-288 (1975).

P.A. RAVIART, J.M. THOMAS

[1] Introduction a l'analyse numérique des equations aux dérivées partielles; Collection Math. appliquées à la Maîtrise, Masson, (1983).

V.D. REPNIKOV, S.D. EIDELMAN

[1] A new proof of the stabilization theorem for solutions of the Cauchy problem for the heat-conduction equation; Mat. Sb., 73, 155-159 (1967); transl. in Math. USSR. Sb, 2, 135-139 (1967).

Y. RESHETNIAK

[1] General theorems on semicontinuity and on convergence with a functional; Siberian Math. J., 8, 801-816 (1968).

B. RICCERI

[1] Applications de théorèmes de semi-continuité inférieure; CRAS Paris, 295, 75-78 (1982).

R. ROBERT

[1] Convergence de fonctionnelles convexes; J. Math. Anal. Appl., 45, 533-555 (1974).

R.T. ROCKAFELLAR

[1] Characterization of the subdifferentials of convex functions; Pacific J.

of Math., 17, 497-510 (1966).

[2] Convex functions, monotone operators and variational inequalities; Theory and applications of monotone operators, ed. A. Ghizzetti, Tipografia Oderisi Editrice, Gubbio (Italy) 34-65 (1969).

[3] Measurable dependance of convex sets and functions on parameters; J. Math. Anal. Appl., 28, 4-25 (1969).

[4] Integral functionals, normal integrands and measurable selections; Non-linear Operators and Calculus of variations, Lecture Notes in Math. 543, Springer Verlag, Berlin (1977).

[5] Monotone operators and the proximal point algorithm; SIAM J. Control, 14, 877-898 (1976).

[6] Generalized directional derivatives and subgradients of non convex functions; Can. J. Math., 32, 257-280 (1980).

J.F. RODRIGUES

[1] Stabilité de la frontière libre dans le problème de la digue; CRAS. Paris, 294-I, 565-568 (1982).

[2] Sur le comportement asymptotique de la solution et de la frontière libre d'une inéquation variationnelle parabolique; Ann. Fac. Sc. Toulouse, 4 (1982).

[3] Free boundary convergence in the homogenization of the one phase Stefan problem, Trans. Amer. Math. Soc. (1982).

[4] Stability of the free boundary in the obstacle problem for a minimal surface, Portugaliae Math. (to appear).

C. SAGUEZ

[1] Contrôle optimal d'un système gouverné par une inéquation variationnelle parabolique, Observation du domaine de contact; CRAS. Paris, 287-A, 957-959 (1978).

J. SAINT JEAN PAULIN

[1] Etude de quelques problèmes de mécanique et d'électrotechnique liés aux méthodes d'homogénéisation. Thèse d'état, Université Paris VI (1981).

[2] Homogénéisation et perturbations; C.R.A.S. Paris, 291, 23-26 (1980).

G. SALINETTI

[1] Convergence for measurable multifunctions... . "Math. Programming and its economical applications". ed. G. Castellani and P. Mazzoleni, Angeli, Milano (1981).

G. SALINETTI, R. WETS

[1] On the relation between two types of convergence for convex functions; J. Math. Anal. Appl., 60, 211-226 (1977).

[2] On the convergence of sequences of convex sets in finite dimension; SIAM Review, 21, 18-33 (1979).

[3] On the convergence of closed-valued measurable multifunctions; Trans. Amer. Math. Soc., 266, 275-289 (1981).

[4] On the convergence in distribution of measurable multifunctions, normal integrands, stochastic processes and stochastic infima (to appear).

[5] Convergences of sequences of closed sets (to appear).

J. SANCHEZ-HUBERT

[1] Etude de certaines équations intégrodifférentielles issues de la théorie de l'homogénéisation. Boll. Un. Mat. Ital. (5) 16-B, 857-875 (1979).

J. SANCHEZ-HUBERT, E. SANCHEZ-PALENCIA

[1] Sur certains problèmes physiques d'homogénéisation donnant lieu à des phénomènes de relaxation; CRAS. Paris, 286, 903-906 (1978).

E. SANCHEZ-PALENCIA

[1] Comportement limite d'un problème de transmission à travers une plaque mince et faiblement conductrice; CRAS. Paris, 270, 1026-1028 (1970).

[2] Solutions périodiques par rapport aux variables d'espace et applications CRAS Paris, 271, 1129-1132 (1970).

[3] Equations aux dérivées partielles dans un milieu hétérogène; CRAS, Paris 272, 1410-1413 (1970).

[4] Problèmes de perturbations liées aux phénomènes de conduction à travers des couches minces de grande resistivité; J. Math. Pures Appl., 53, 251-270 (1974).

[5] Comportement local et macroscopique d'un type de milieux physiques hétérogènes; International J. Engineering Sci., 12, 331-351 (1974).

[6] Perturbations spectrales; Lecture Notes in Mathematics, Springer, 294, 437-455 (1977).

[7] Méthode d'homogénéisation pour l'étude de matériaux hétérogènes. Phénomènes de mémoire; Rend. Sem. Mat. Univ. e Politecnic Torino, 36, 15-26 (1977-78).

[8] Phénomènes de relaxation dans les écoulements des fluides visqueux dans les milieux poreux; CRAS. Paris, 286, 185-188 (1978).

[9] Nonhomogenous media and vibration theory; Lecture Notes in Physics, Springer, 127 (1980).

[10] Problèmes aux limites dans des domaines contenant des parois perforées. Collège de France. Seminar Lions-Brezis. Research Notes in Math. Pitman (1980-81).

C. SBORDONE

[1] Alcune questioni di convergenza per operatori differenziali del 2° ordine; Boll. Un. Mat. Ital. (4) 10, 672-682 (1974).

[2] Sulla G-convergenza di equazioni ellittiche e paraboliche; Ricerche Mat. 24, 76-136 (1975).

[3] Su alcune applicazioni di un tipo di convergenza variazionale; Ann. Sc. Norm. Sup. Pisa Cl. Sci., (4) 2, 617-638 (1975).

[4] Sur une limite d'intégrales polynomiales du calcul des variations; J. Math. Pures. Appl. (9) 56, 67-77 (1977).

[5] Closure and duality of saddle functions (to appear).

M. SCHATZMAN

[1] A hyperbolic problem of second order with unilateral constraints, J. Math. Pures Appl., 73, 138-191 (1980).

[2] Un problème hyperbolique...: la corbe vibrante avec obstacle ponctuel; J. Diff. Equat., 36, 295-334 (1980).

E. SCHECHTER

[1] One sided continuous dependance of maximal solutions; J. of Differential Equations, 39, 413-425 (1981).

[2] Existence and limits of Caratheodory-Martin evolutions; Nonlinear Analysis, T.M.A, 5(8), 897-930 (1981).

P.K. SENATOROV

[1] The stability of Dirichlet's problem for elliptic equations with respect to perturbation in measure of its coefficients; Diff. Equations, 6, 1312-1313 (1970).

[2] The stability of the eigenvalues and eigenfunctions of a Sturm-Liouville problem; Diff. Equations, 7, 1266-1269 (1971).

R. SENTIS

[1] Analyse Asymptotique d'équations de transport; thèse d'état Paris-Dauphine (1981).

[2] Approximation et homogénéisation; CRAS. Paris, 289, 567 (1979).

J. SERRIN

[1] A new definition of the integral for non-parametric problems in Calculus of variations; Acta Math., 102, 23-32 (1959).

[2] On the definition and properties of certain variational integrals; Trans. Amer. Math. Soc., 101, 137-167, (1961).

SHIH SHU CHUNG

[1] Sur les problèmes de contrôle bon marché. CRAS Paris, 291, 635-638 (1980).

L. SIMON

[1] On G-convergence of elliptic operators; Indiana Math. Univ. J., 28, 587-594 (1979).

Y. SONNTAG

[1] Convergence au sens de Mosco; théorie et applications à l'approximation des solutions d'inéquations. Thèse d'Etat. Université de Provence. Marseille (1982).

S. SPAGNOLO

[1] Sul limite delle soluzioni di problemi di Cauchy relativi all'equazione del calore; Ann. Sc. Norm. Sup. Pisa. Cl. Sci. (3), 21, 657-699 (1967).

[2] Sulla convergenza delle soluzioni di equazioni paraboliche ed ellittiche; Ann. Sc. Norm. Sup. Pisa. Cl. Sci. Fis. Mat. (3), 22, 575-597 (1968).

[3] Some convergence problems; Symp. Math., 18, 391-398 (1976).

[4] Convergence in energy for elliptic operators, Proc. III Symp. Numer. Solut. Partial Diff. Eq. College Park, 1975, ed. Hubbard, Academic Press, 469-498 (1976).

[5] Convergence of parabolic equations; Boll. Un. Mat. Ital. (5) 14-B, 547-568 (1977).

[6] Convergence de solutions d'équations d'évolution; Proc. Int. Meeting on "Recent Methods in Nonlinear Analysis" Rome 1978, Pitagora ed. Bologna, 311-328 (1979).

[7] Some type of perturbation for evolution equations; Proc. Meeting on "Computing Methods in Applied Sciences and Engineering" ed. R. Glowinski, J.-L. Lions, North-Holland, Amsterdam, 675-682 (1980).

J.E. SPINGARN

[1] Partial inverse of a monotone operator. Appl. Math. Optim. 10, 247-265 (1983).

S. STEFFE

[1] Un teorema di omogeneizzazione per equazioni differenziali ordinarie a coefficienti quasi-periodici; Boll. Un. Mat. Ital. (5) 17-B, 757-764 (1980).

[2] Atti del Convegno su "Studio di problemi-limite in Analisi Funzionale" Bressanone, Sett. 1981, ed. da Patuzzo e. S. Steffe, C.L.E.U.P, Padova (1982).

P. SUQUET

[1] Plasticité et homogenéisation. Thèse d'état en Mathématiques (Mécanique théorique) Université Pierre et Marie Curie, Paris VI (1982).

[2] Une méthode duale en homogénéisation: application aux milieux élastiques; Journal de Mécanique théorique et appliquée (to appear).

L. TARTAR

[1] Convergence d'opérateurs différentiels; Quaderno di ricerca del C.N.R. "Analisi convessa e applicazioni", Roma, 101-104 (1974).

[2] Cours Peccot au Collège de France, Paris (1977).

[3] Homogénéisation en hydrodynamique; Lecture Notes in Math., Springer (594), 474-481 (1977).

[4] "Non linear constitutive relations and homogenization" in Contemporary Development", ed. La Penha, Medeiros; North-Holland, Amsterdam (1978).

[5] Compensated compactness and applications to P.D.E; "Non-linear Analysis and Mechanics. Heriot Watt symposium, "Vol. 4, Ed. Pitman, London, 136-212.

[6] Homogénéisation et compacité par compensation; Séminaire Goulaouic-Schwartz Ecole Polytechnique, Exposé No. 9 (1978).

R. TEMAM

[1] An existence theorem for a variational problem of plasticity; "Non linear problems of analysis in geometry and analysis". Pitman, 57-70 (1981).

R. TEMAM, G. STRANG

[1] Duality and relaxation in plasticity; Journal de Mécanique, 19, 493-527 (1980).

A. TRUFFERT

[1] Propriété de Fatou-Vitali: application à l'epi-convergence des fonctionnelles intégrales Thèse, Université Perpignan (1983).

M. TSUKADA

[1] Convergence of closed convex sets and σ-fields; Z. Wahrscheinlichtkeitstheorie verw. Gebiete, 62, 137-146 (1983).

M.M. VAINBERG

[1] Variational methods for the study of non-linear operators; Holden-Day, San Francisco (1964).

M. VALADIER

[1] On conditional expectation of random sets; Annali di Mat. Pura ed Appl., 126, 81-91 (1980).

B. VAN CUTSEM

[1] Martingales de convexes fermés aléatoires en dimension finie, Ann. Inst. H. Poincare Sec. B,8, 365-385 (1972).

[2] Eléments aléatoires à valeurs convexes compactes, Thèse, Grenoble, France (1971).

M. VANNINATHAN

[1] Homogénéisation des valeurs propres dans les milieux perforés; CRAS Paris, 287, 403-406 (1978).

L. VERON

[1] Some remarks on the convergence of approximate solutions of non linear evolution equations in Hilbert space, Mathematics of computation, Vol. 39, 160, 325-337 (1982).

W. VERVAAT

[1] Une compactification des espaces fonctionnels C et D; une alternative pour la démonstration des théorèmes limites fonctionnels; CRAS Paris, 292, 441-444 (1981).

D. WALKUP, R. WETS

[1] Continuity of some convex cone valued mappings; Proc. A.M.S, 18, 229-253 (1967).

R. WETS

[1] Duality relations in stochastic programming, in Symposia Mathematica XIX, Academic Press, New York, 341-355 (1976).

[2] On the convergence of random convex sets, in "Convex Analysis and its Applications" Lecture Notes in Economics and Mathematical systems, Springer Verlag, 144, 191-206 (1977).

[3] Convergence of convex functions, variational inequalities and convex optimization problems, in Variational Inequalities and Complementary problems, eds. P. Cottle, F. Giannessi, J.-L. Lions, Wiley, Chichester (UK), 375-403 (1980).

[4] A formula for the level sets of epi-limits and some applications, in Mathematical Theory of optimization, eds. P. Cecconi and T. Zolezzi, Springer Verlag, Lecture Notes in Mathematics (1983).

[5] On a compactness theorem for epi-convergent sequences of functions, in Proc. Rio de Janeiro Congress on Mathematical Programming, ed. M. Kelmanson, North-Holland (1983).

R.A. WIJSMAN

[1] Convergence of Sequences of convex sets, cones and functions; Bull. A.M.S. (70), 186-188 (1964).

[2] Convergence of Sequences of convex sets,... II; Trans. Amer. Math. Soc., 123, 32-45 (1966).

K. YOSIDA

[1] Functional analysis; Springer Verlag (3rd edition) (1971).

V.V. YURINSKIJ

[1] Averaging elliptic equations with random coefficients; Sib. Math. J, 20, 611-623 (1980).

E. ZARANTONELLO

[1] Projections on convex sets, in Contributions to non linear Funct. Anal. Symp. Madison; Academic Press, NoY (1971).

V.V. ZHIKOV

[1] On the stabilization of solutions of parabolic equations; Mat. Sb., 104, 597-616 (1977). Transl. in Math. USSR-Sb., 33, 519-537 (1977).

[2] A point stabilization criterion for second order parabolic equations with almost periodic coefficients; Mat. Sb., 110 (1979); transl. in Math USSR-Sb., 38, 279-292 (1981).

T. ZOLEZZI

[1] On the convergence of minima; Boll. Un. Mat. Ital., 8, 246-257 (1973).

[2] Characterization of some variational perturbations of the abstract linear-quadratic problem; SIAM. J. Control Optim., 16, 106-121 (1978).

[3] Variational perturbations of linear optimal control problems; Proc. Int. Meeting on "Recent Methods in Nonlinear Analysis", Rome 1978, Pitagora ed. Bologna, 329-338 (1979).

[4] Approssimazioni e perturbazioni di problemi di ottimizzazione (to appear).